ADVANCES IN CHEMICAL ENGINEERING
Volume 23

Process Synthesis

ADVANCES IN
CHEMICAL ENGINEERING

Editor-in-Chief
JAMES WEI
School of Engineering and Applied Science
Princeton University
Princeton, New Jersey

Editors

JOHN L. ANDERSON
Department of Chemical Engineering
Carnegie Mellon University
Pittsburgh, Pennsylvania

KENNETH B. BISCHOFF
Department of Chemical Engineering
University of Delaware
Newark, Delaware

MORTON M. DENN
College of Chemistry
University of California at Berkeley
Berkeley, California

JOHN H. SEINFELD
Department of Chemical Engineering
California Institute of Technology
Pasadena, California

GEORGE STEPHANOPOULOS
Department of Chemical Engineering
Massachusetts Institute of Technology
Cambridge, Massachusetts

ADVANCES IN CHEMICAL ENGINEERING
Volume 23

Process Synthesis

Edited by
JOHN L. ANDERSON
Department of Chemical Engineering
Carnegie Mellon University
Pittsburgh, Pennsylvania

ACADEMIC PRESS
San Diego   New York   Boston   London   Sydney   Tokyo   Toronto

TP
145
.D7
v.23
1996

This book is printed on acid-free paper. ∞

Copyright © 1996 by ACADEMIC PRESS, INC.

All Rights Reserved.
No part of this publication may be reproduced or transmitted in any form or by any means, electronic or mechanical, including photocopy, recording, or any information storage and retrieval system, without permission in writing from the publisher.

Academic Press, Inc.
A Division of Harcourt Brace & Company
525 B Street, Suite 1900, San Diego, California 92101-4495

United Kingdom Edition published by
Academic Press Limited
24-28 Oval Road, London NW1 7DX

International Standard Serial Number: 0065-2377

International Standard Book Number: 0-12-008523-2

PRINTED IN THE UNITED STATES OF AMERICA
96 97 98 99 00 01 QW 9 8 7 6 5 4 3 2 1

CONTENTS

CONTRIBUTORS vii
PREFACE ix

Industrial Applications of Chemical Process Synthesis
JEFFREY J. SIIROLA

I.	Introduction	2
II.	Industrial Chemical Process Innovation	3
III.	Synthesis in Conceptual Process Engineering	10
IV.	A Total Flowsheet Example	20
V.	Energy Efficient Distillation Synthesis	28
VI.	Residue Curve Maps and the Separation of Azeotropic Mixtures . . .	39
VII.	Challenges for the Future	58
VIII.	Conclusions	61
	Acknowledgments	61
	References	61

The Synthesis of Distillation-Based Separation Systems
ARTHUR W. WESTERBERG AND OLIVER WAHNSCHAFFT

I.	Introduction	64
II.	The Richness of the Solution Space	66
III.	Assessing the Behavior of a Mixture	69
IV.	Separating Nearly Ideal Systems	75
V.	Separating Highly Nonideal Mixtures	90
VI.	Synthesis Discussion	94
VII.	Pre-analysis Methods	98
VIII.	Synthesis Method for Nonideal Mixtures	107
IX.	More Advanced Pre-analysis Methods	131
X.	Post-analysis Methods: Column Design Calculations	166
	Acknowledgments	167
	References	167

Mixed-Integer Optimization Techniques for Algorithmic Process Synthesis
IGNACIO E. GROSSMANN

I.	Introduction	172
II.	Overview of Previous Work	172
III.	Mathematical Programming Approach	178
IV.	Representation of Alternatives	180

V.	MINLP Modeling	187
VI.	MINLP Algorithms	197
VII.	Solution Strategies for MINLP Synthesis Problems	213
VIII.	Applications	224
IX.	Concluding Remarks	237
	Acknowledgments	239
	References	239

Chemical Reactor Network Targeting and Integration: An Optimization Approach

SUBASH BALAKRISHNA AND LORENZ T. BIEGLER

I.	Introduction	248
II.	Geometric Concepts for Attainable Regions	250
III.	Reactor Network Synthesis: Isothermal Systems	254
IV.	Reactor Network Synthesis: Nonisothermal Systems	265
V.	Energy Integration of Reactor Networks	274
VI.	Simultaneous Reaction, Separation, and Energy System Synthesis	283
VII.	Summary and Conclusions	295
	Acknowledgments	298
	References	298

Operability and Control in Process Synthesis and Design

STEVE WALSH AND JOHN PERKINS

I.	Introduction	302
II.	General Techniques	306
III.	Neutralization of Waste Water	342
IV.	Modeling of Waste Water Neutralization Systems	353
V.	Design Examples	360
VI.	Conclusions	391
	Notation	393
	Appendix A: Worst-Case Design Algorithm	396
	Appendix B: Example of Progress of Algorithm	398
	References	401

INDEX	403
CONTENTS OF VOLUMES IN THIS SERIES	411

CONTRIBUTORS

Numbers in parentheses indicate the pages on which the authors' contributions begin.

SUBASH BALAKRISHNA, *Department of Chemical Engineering, Carnegie Mellon University, Pittsburgh, Pennsylvania 15213* (247)

LORENZ T. BIEGLER, *Department of Chemical Engineering, Carnegie Mellon University, Pittsburgh, Pennsylvania 15213* (247)

IGNACIO E. GROSSMANN, *Department of Chemical Engineering, Carnegie Mellon University, Pittsburgh, Pennsylvania 15213* (171)

JOHN PERKINS, *Centre for Process Systems Engineering, Imperial College of Science, Technology and Medicine, London SW7 2BY, England* (301)

JEFFREY J. SIIROLA, *Eastman Chemical Company, Kingsport, Tennessee 37660* (1)

OLIVER M. WAHNSCHAFFT, *Department of Chemical Engineering and the Engineering Design Research Center, Carnegie Mellon University, Pittsburgh, Pennsylvania 15213* (63)

STEVE WALSH, *Centre for Process Systems Engineering, Imperial College of Science, Technology and Medicine, London SW7 2BY, England* (301)

ARTHUR W. WESTERBERG, *Department of Chemical Engineering and the Engineering Design Research Center, Carnegie Mellon University, Pittsburgh, Pennsylvania 15213* (63)

PREFACE

Over the past thirty years, *process synthesis* has had a major impact on the development, design, and operation of chemical processes. This field exploits key physical and chemical phenomena in the process, as well as their interactions, and it requires a systematic approach to address these phenomena. Process synthesis strategies have been developed for the design of heat-exchanger networks, utility systems, separation sequences, reactor networks, and control systems. While many strategies consider the design of these homogeneous systems separately, a key research question is the interaction of these subsystems and exploitation of this synergy for the overall synthesis of process flowsheets.

Synthesis methods began with the application of *heuristics* gathered by specific process knowledge and experience. This led naturally to early application of artificial intelligence tools and expert systems. Rigorous and elegant approaches then evolved through *problem representations* (i.e., conceptual/graphical representations), which are generally geometric in nature and are based on physical insights in the process. Unlike heuristics, these representations allow the development of provable ways to synthesize a process and demonstrate its superiority over an *ad hoc* procedure. Finally, quantitative approaches are also needed, especially in assessing trade-offs among design criteria and interactions between subsystems that could not be addressed directly with simple rules or simple representations. As optimization strategies were developed and refined to handle larger and more difficult problems, *optimization-based* formulations of these problems led to powerful strategies for process synthesis.

Today it is recognized that all three approaches (heuristics-based selection, geometric representation, and optimization methods) are useful, and indeed required, for complex process synthesis strategies. This follows because different applications lend themselves to quite different representations. This volume addresses a variety of these synthesis strategies for process subsystems, but represents only a sampling of the state-of-the-art of process synthesis research. The five chapters in this volume address quite different process subsystems and application areas but still combine basic concepts related to a systematic approach.

The first chapter, by Siirola, reviews the impact of process synthesis in industry and shows how process synthesis fits into the innovation process within industrial manufacturing and research. It also highlights a number of industrial successes leading to substantial energy savings and overall cost reductions. Most of these savings are in the areas of distillation sequences, and examples include heat-integrated separation sequences and separation of azeotropic systems.

The second chapter, by Westerberg and Wahnschafft, further develops the synthesis of nonideal separation sequences through the use of physical insights, artificial intelligence, shortcut models, and geometric constructions. Using a combination of these approaches, as illustrated with a number of examples, the strategy in this chapter yields complex separation sequences that guarantee the separation of nonideal mixtures into desired products.

The third chapter, by Grossmann, develops an overall framework for algorithmic process synthesis. This framework is applied to heat-exchanger network synthesis, separation sequences, and superstuctures for total flowsheets. These examples are formulated and solved as mixed integer nonlinear programs (MINLP) which deal with the optimization of discrete (structural) and continuous decisions. Illustrated with numerous process synthesis formulations, the chapter reviews MINLP algorithms and also discusses the incorporation of logic constraints and heuristics in developing a qualitative/quantitative framework for process synthesis.

The fourth chapter, by Balakrishna and Biegler, deals with the difficult problem of reactor network synthesis. These systems are generally very nonlinear and nonconvex, and both heuristic- and optimization-based approaches can lead to nonunique and only locally optimal solutions—an undesirable situation. The paper combines geometric concepts from attainable regions (AR) of the reactor network with an optimization-based approach. The AR concepts, recently developed by Glasser and co-workers, lead to insights that offer smaller, simpler, and superior NLP (nonlinear programming) and MINLP formulations for this system. This approach is demonstrated on numerous examples, including some that interact with other flowsheet subsystems.

The last chapter, by Walsh and Perkins, deals with an optimization-based approach for operability and control in process synthesis and design. Process control is often performed after the design is completed, without considering control and operability at the design stage. This chapter shows strong interactions of design and control and develops a comprehensive strategy for these systems. Centered around the optimization of dynamic systems with uncertainty, a strategy is developed to guarantee good control structures over a large variety of disturbances. This approach is applied to an industrial wastewater treatment process with impressive results.

All of the chapters develop highly successful synthesis methods for their respective cutting-edge applications. Nevertheless, they also highlight many unresolved issues in process synthesis and give guidelines for future research. As a result, there are still many challenging research issues in this active field. It is our hope that this volume points these out and spurs future research in this area.

Lorenz T. Biegler

INDUSTRIAL APPLICATIONS OF CHEMICAL PROCESS SYNTHESIS

Jeffrey J. Siirola

Eastman Chemical Company
Kingsport, Tennessee

I. Introduction	2
II. Industrial Chemical Process Innovation	3
A. Multiple Levels of Detail	5
B. A Journey through the Innovation Process	8
III. Synthesis in Conceptual Process Engineering	10
A. Systematic Generation: Means–Ends Analysis	13
B. Hierarchical Approaches	16
C. The Adaptive Initial DEsign Synthesizer	17
IV. A Total Flowsheet Example	20
A. Conventional Conceptual Process Design	20
B. Evolutionary Modification	23
C. The Hierarchical Process Synthesis Procedure	26
V. Energy Efficient Distillation Synthesis	28
A. Heat-Integrated Distillation Trains	28
B. Complex Columns for Binary Separations	36
VI. Residue Curve Maps and the Separation of Azeotropic Mixtures	39
A. Dehydration of Ethanol	41
B. Separations System Synthesis Method for Nonideal Mixtures	50
C. Production of Diethoxymethane	54
VII. Challenges for the Future	58
VIII. Conclusions	61
Acknowledgments	61
References	61

Systematic approaches for the invention of conceptual chemical process designs have been proposed and discussed for more than twenty-five years. During that same time, the importance of front-end engineering, especially conceptual design, to product quality, health and safety, environmental impact, energy consumption, operability, capital and operating costs, and overall competitiveness has become ever more apparent. A number of process synthesis frameworks, approaches, methods, and tools have now been developed to the point of industrial application. This chapter describes a framework for the industrial chemical plant innovation process, showing how process synthesis fits into that structure and how that framework has in turn influenced the development of systematic process synthesis methods. It also describes a number of industrial case studies in which process synthesis techniques have been successfully applied to the conceptual design of total process flowsheets, as well as to specific design subproblems including heat-integrated distillation trains, multiple-effect distillation, and the separation of azeotropic systems. Typical energy savings of 50% and net present cost reductions of 35% have been achieved in industrial practice using systematic process synthesis methodologies. Even greater benefits are expected to be realized as the next generation of approaches currently being developed is transferred to industry.

I. Introduction

The manufacturing sector of the chemical processing industry is generally in the business of making materials rather than making artifacts. This is done in response to perceived needs and the belief that the materials offered will satisfy these needs in a valuable manner.

In contrast with artifact-making, material-making tends to involve more conversion and transformation than assembly, and is generally more capital- and energy-intensive than labor-intensive. Material-making sometimes involves substances that are toxic or otherwise hazardous to the environment. Furthermore, since material-making equipment costs are often sublinear functions of capacity, material-making facilities tend to be large and aggregated to take advantage of economies of scale, integrated material and energy flows, and centralized environmental mitigation. Many material-making facilities have very long operating lives, much longer than the life of many artifacts and most artifact-making machinery.

Chemical manufacturing plants come into existence through a series of actions sometimes called the *innovation process*. This process leads from the identification of a need to the operation of a material-making facility. The characteristics of material-making—relatively few but fairly large manufacturing facilities, very long operating lives, high initial capital costs, high continuing operating costs, and potential environmental impacts—place special importance on sound implementation of the innovation process and, in particular, on making good engineering design decisions within that process.

All existing chemical process designs were somehow invented. Some, especially lower volume products involving complex chemistry, have been implemented by a rather straightforward extrapolation of the laboratory procedure used to experimentally demonstrate the transformation of available raw materials into the desired product. Larger volume products, on the other hand, are more often implemented as continuous processes that bear little resemblance to either the procedure or the equipment used in the laboratory. Generally there is a combinatorially very large number of alternative pieces of equipment and interconnections among these pieces of equipment that will feasibly implement the desired chemistry. Identifying better process alternatives is a key activity within the innovation process.

Successful design engineers seem to build and evolve conceptual process flowsheets from a rich repertoire of past experiences and design heuristics. Such an experience base generally includes an extensive knowledge of available equipment, simple and complex unit operations, standard tricks or patterns (for example, strategies for breaking azeotropes), encyclopedias of complete flowsheets for existing chemicals, and some sense of hierarchy of which process design problems to tackle in what order and at what level of detail. This wealth of background information is copied directly or modified as necessary to fit the situation at hand. Generally time and resource constraints limit the number of conceptual process alternatives that may be generated and evaluated by the designer to a tiny fraction of the total number feasible. The key to discovering alternatives with superior economics is the judicious use of modern methods and tools together with a little good luck.

Occasionally, process designs are produced that are conceded by those skilled in the art as being *clever*. Perhaps most, if not all, world-beating designs exhibiting superior economics exploit something clever. What is technically feasible, what is competent, and what is clever process design? Is there a *best* design that cannot be beaten? Can the invention of chemical process designs be organized, systematized, or even automated? How can more or better alternatives be generated? These questions have been the focus of process synthesis research over the last twenty-five years. It is not the intention of this chapter to discuss the latest advances in that research. Rather, it is to illustrate that some of these results are beginning to have a real impact on industrial practice.

II. Industrial Chemical Process Innovation

To understand chemical process synthesis in an industrial environment, it may be useful to first discuss the chemical innovation process. *Invention* is discovery. A new material composition may be invented. The chemistry to transform raw materials into this new composition may be invented. The process flowsheet to implement this chemistry on an industrial scale may also be invented. But the invention of a flowsheet does not guarantee that the chemistry will be reduced to practice in an economical manner. That is accomplished by *innovation,* an organized multistage goal-directed process, which leads from the identification of a customer's need to the operation of a facility to produce a material believed to address that need.

The innovation process may be implemented in a number of different ways. The specific details and emphasis within the innovation process may differ, depending on whether the objective is to build a pioneering facility for a new chemical or an improved facility for an existing chemical. But in general, a product that addresses the need must be (or has already been) identified; a chemical route must be found from available raw materials that produces this desired product; and a facility that implements this chemistry must be conceived, designed, constructed, started, operated, and maintained (Fig. 1). Of special interest in the context of conceptual process design are the stages *basic chemistry* (in which the fundamental reaction chemistry is selected), *detailed chemistry* (in which supporting chemical details of catalysis, solvents, and reaction conditions

- **Need Identification**
- **Manufacturing Decision**
- **Basic Chemistry**
- **Detailed Chemistry**
- **Task Identification**
- **Unit Operations**
- **Basic Plant Engineering**
- **Detailed Engineering**
- **Vendor Specifications**
- **Component Acquisition**
- **Construction Plan and Schedule**
- **Plant Construction**
- **Operating Procedures**
- **Commissioning and Start-up**
- **Production Plan and Schedule**
- **Plant Operation and Maintenance**

FIG. 1. Innovation process sequence.

are defined), *task identification* (in which the physical operations to prepare raw materials for reaction and isolate reaction products for sale are identified), *unit operations* (in which the chemical and physical operations previously identified are associated with actual pieces of equipment), and *basic plant engineering* (where the supporting utilities and other facilities infrastructure are defined).

Each stage of the innovation process is in a sense implemented by all of the stages that follow. Consequently, there is a great deal of interaction among the stages. For example, the choices that can be made at one stage are clearly limited by selections made during previous stages. At the same time, the optimal choice to be made among alternatives identified at any stage may well depend on costs associated with subsequent stages that implement that choice but have yet to be addressed. Earlier stages are both less well-defined and less constrained than later stages. However, decisions made in these earlier stages typically prove to have a greater impact on the overall economic outcome of the entire venture. Because of the interacting nature of the stages, it is often necessary to revisit stages or iterate among the stages in order to converge to an acceptable solution.

A. MULTIPLE LEVELS OF DETAIL

Because of the interacting nature of the innovation process stages, it is possible that each visit to a stage may be approached with a slightly different objective or even conducted at different levels of detail. For example, each stage of the innovation process might be visited four times (Fig. 2).

1. The first pass through any stage of the innovation process is at the lowest level of detail, which might be called the *targeting* level. Its purpose might be to get a rough indication of what is to be accomplished at that stage of the innovation process as well as to see what is likely to be feasible given the choices that have been made previously in the process. Targeting is already familiar in the context of heat-integration networks where fairly simple analytical procedures can give much information about what is technically and perhaps economically feasible by bounding the expected performance even before the generation of any heat-exchanger network alternatives.

2. The second pass through each innovation stage might be called the *preliminary* or *conceptual* level. Here, a tentative, not too detailed solution to the given innovation problem is conceived. One general design paradigm to accomplish this essentially consists of a four-block procedure, as shown in Fig. 3. In the first *formulation* block, the goals for the particular stage of the innovation process are specified. This is followed by an iteration of three blocks consisting of *synthesis* (generation of a solution

	Target	Conceptual	Refined	Final
Need Identification	●	●	●	●
Manufacturing Decision	●	●	●	●
Basic Chemistry	●	●	●	●
Detailed Chemistry	●	●	●	●
Task Identification	●	●	●	●
Unit Operations	●	●	●	●
Basic Plant Engineering	●	●	●	●
Detailed Engineering	●	●	●	●
Vendor Specifications	●	●	●	●
Component Acquisition	●	●	●	●
Construction Plan and Schedule	●	●	●	●
Plant Construction	●	●	●	
Operating Procedures	●	●	●	
Commissioning and Start-up	●	●	●	●
Production Plan and Schedule	●	●		
Operation and Maintenance	●	●	●	●

FIG. 2. Innovation process levels of detail.

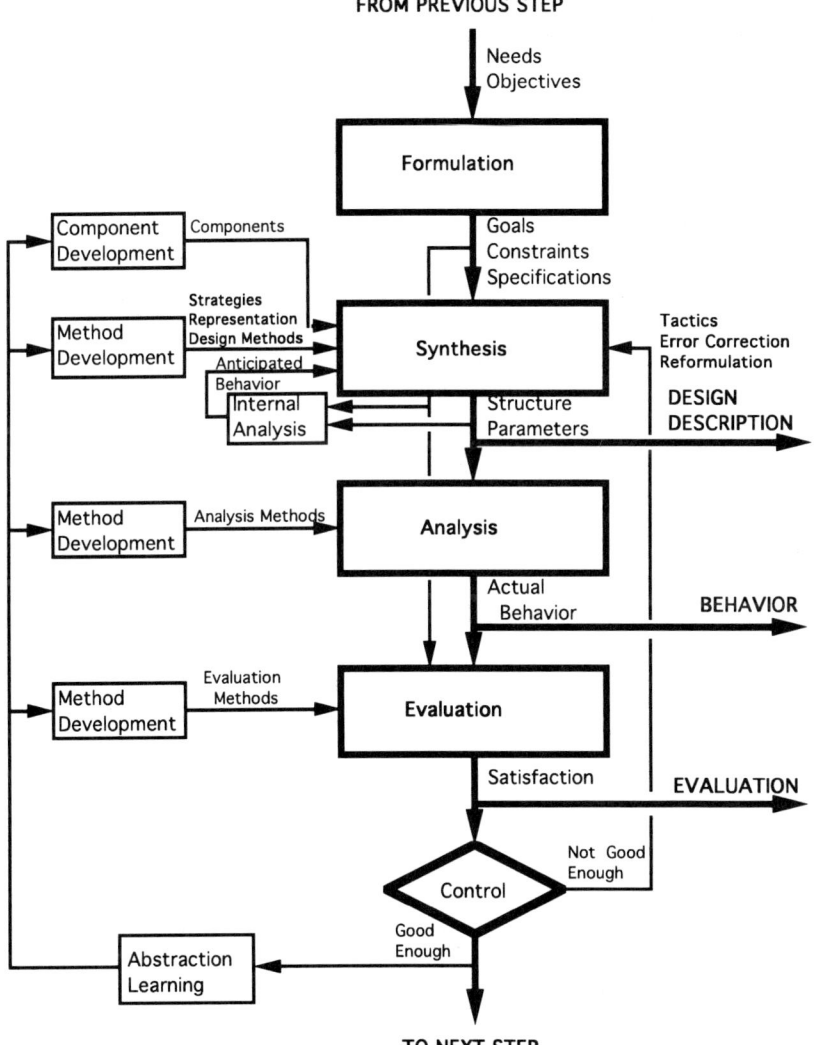

FIG. 3. General design paradigm.

alternative from available components), *analysis* (determination of the behavior of the alternative generated), and *evaluation* (comparison of the performance of the proposed alternative against the goals specified by the formulation block). If the performance of the proposed solution is judged to be satisfactory, the step is completed; otherwise, a new alternative must be generated, analyzed, and evaluated. In general, alternatives are not gen-

erated blindly, but are guided by some analysis internal to the synthesis step (sometimes encoded in the problem representation, sometimes encoded within synthesis rules of thumb, and sometimes performed by specialized analysis methods) so that alternatives are generated with anticipated behavior close to the desired behavior. In order to execute this general design strategy (which has variously been called *generate-and-test, propose–critique–modify,* and other names), components and formal or informal problem-specific alternative generation methods, analysis methods, and evaluation methods must be available. The development of such methods and tools for generating alternatives for the conceptual engineering steps of the innovation process is what much of process synthesis research is all about.
3. Because of the interacting nature of the innovation process stages, it is not efficient to make the attempt at generating a solution at the conceptual level particularly detailed. Rather, as additional information is developed in both previous and subsequent stages of the innovation process, a more *detailed* or *refined* solution can be attempted at a higher level of detail (see Fig. 2). The same basic four-block design procedure used at the conceptual level may be used in the refined level, but with different specific synthesis, analysis, and evaluation methods and tools capable of providing the required higher level of detail.
4. Finally, as yet more information becomes available, it may be necessary to repeat the refined level one *final* time using the same methods and tools that were used for the third pass.

B. A Journey through the Innovation Process

One possible strategy might to attack each stage of the innovation process in a linear manner, going from target level to the conceptual level to the refined level to a repeat of the refined level for a single stage of the innovation process, and then moving on to the next stage. An alternative strategy that also accomplishes an iteration among the innovation process stages is a diagonal approach wherein the execution of the conceptual level at one stage just precedes the execution of the detailed level of the previous stage, and just follows the execution of the targeting level of the following stage, etc. (Fig. 4). This back-and-forth approach in different innovation stages as well as in different levels of detail enables a certain amount of look-ahead and allows preliminary downstream results to have an impact on the final solutions developed at previous stages of the innovation process. This format also defines the types and detail of information that may reasonably be expected to be known as each step of each stage of the innovation process is approached.

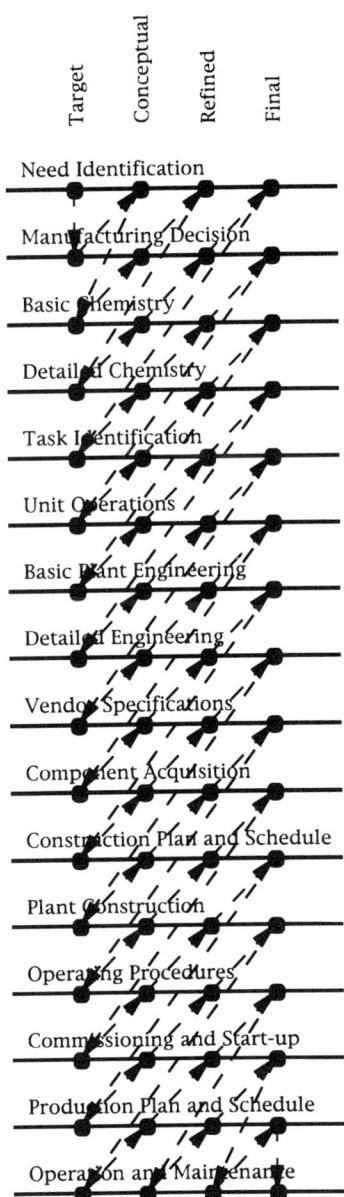

FIG. 4. Journey through the innovation sequence.

As an aside, this approach to the execution of the innovation process may also have an impact on the way industrial Research, Development, and Engineering functions might best be organized. A traditional organization might align specific innovation stages with particular company functions: for example, the need identification and manufacturing decision stages may be aligned with Business Development; basic and detailed chemistry stages with Research; task identification and unit operations stages with Development; plant and detailed engineering stages with engineering; and so forth (Fig. 5). Alternatively, the functional boundaries might be diagonal (Fig. 6), to more smoothly align with the iterative journey through the different levels of detail of the innovation process stages. This implies, for example, that in addition to the heavy chemistry emphasis usually associated with Research, it may also be appropriate for a research organization to maintain selected needs identification, business strategy, and conceptual process engineering functionalities at specific levels of detail.

III. Synthesis in Conceptual Process Engineering

Process synthesis is the invention of conceptual chemical process designs. Process synthesis is not just one step in the innovation process. Rather, it involves the generation of alternatives in all conceptual process engineering steps within the innovation process, including all the steps in the task identification stage and the less detailed steps of the unit operations stage. Because of the interacting nature of the innovation process stages, there is also some overlap with the preceding chemistry stages and the subsequent basic engineering stages.

The goal of conceptual process engineering is to develop a skeletal scheme for a material-making facility, as may be represented by a *flowsheet,* to implement the chemical conversion of available raw materials into desired product materials, fit for use, at the desired scale, safely, responsibly, economically, and on time. This flowsheet will be refined and optimized in subsequent basic and detailed engineering stages of the innovation process on the basis of additional experimental information, detailed calculations, standards, expert opinion, available equipment and construction capabilities, and other input. It is most desirable that this initial conceptual process scheme prove to be a better starting point for the remainder of the innovation process than other possible flowsheets. A flowsheet does not exist until it has been first synthesized.

Conceptual process engineering does not involve the invention of chemistry. However, it may very well help to select from among alternative candidate chemistries on the basis of what appears to be required to implement each chemical alternative in an industrial environment.

INDUSTRIAL APPLICATIONS OF CHEMICAL PROCESS SYNTHESIS

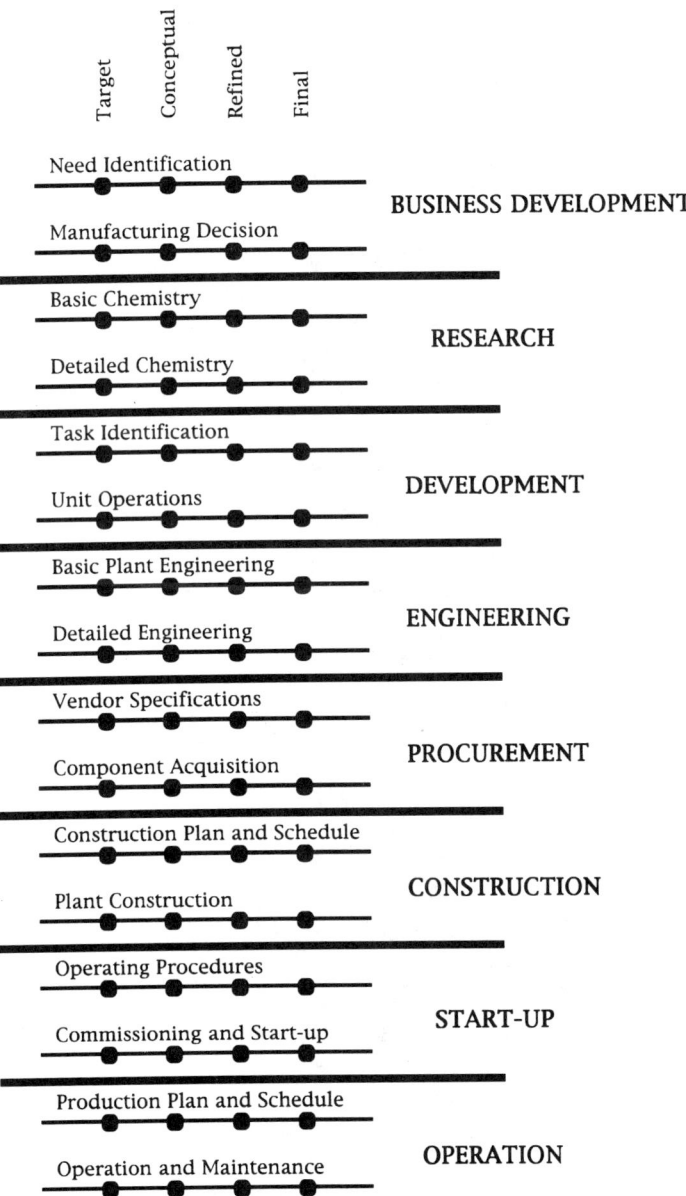

FIG. 5. Organization aligned with innovation stages.

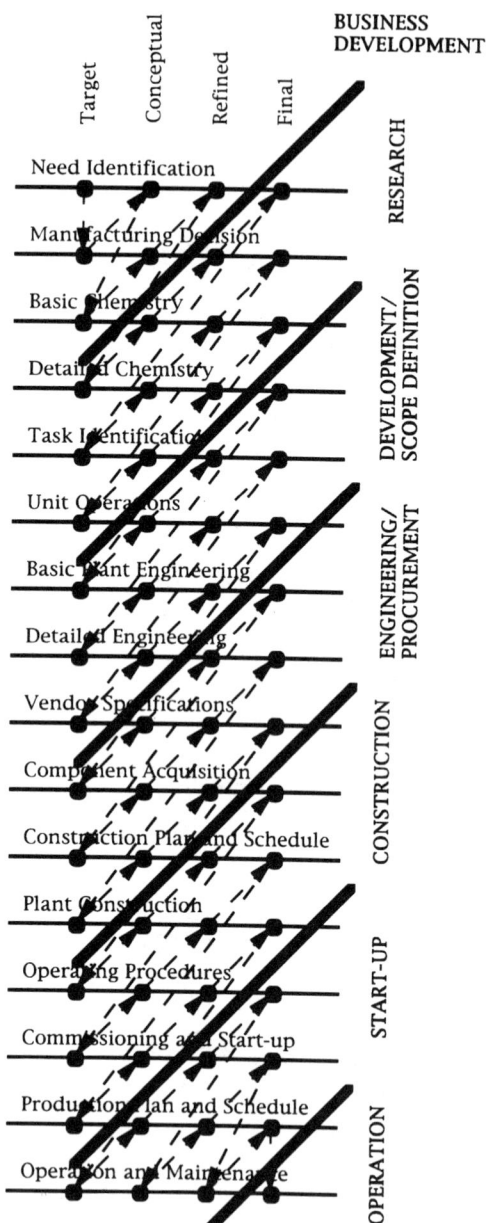

FIG. 6. Organization aligned with innovation journey.

There appears to be three fundamental approaches to the synthesis of chemical process flowsheets. The first, *systematic generation,* builds the flowsheet from smaller, more basic components strung together in such a way that raw materials eventually become transformed into the desired product. The second, *evolutionary modification,* starts with an existing flowsheet for the same or a similar product and then makes modifications as necessary to adopt the design to meet the objectives of the specific case at hand. The third, *superstructure optimization,* views synthesis as a mathematical optimization over structure; this approach starts with a larger superflowsheet that contains embedded within it many redundant alternatives and interconnections and then systematically strips the less desirable parts of the superstructure away.

Systematic generation from scratch sometimes leads to solutions that are already known. In the absence of the exhaustive generation of all alternatives, systematic generation cannot guarantee optimality. Likewise, the quality of solutions generated by evolutionary modification depends critically on the starting flowsheet as well as on the methods used to study and perturb it, thus rarely leading to clever designs. On the other hand, superstructure optimization offers the promise of simultaneous optimization of structural as well as other design parameters. However, it requires a starting superstructure from somewhere (which for some simple problems may be implicit in the formulation), as well as very extensive computational capability. Both *algorithmic* and *heuristic* methods have been proposed for the systematic generation and evolutionary modification approaches, whereas *mathematical programming,* especially mixed-integer nonlinear programming, is the obvious method for the superstructure optimization approach. Evolutionary modification is the approach traditionally used most frequently by conceptual design engineers. Systematic generation has been the basis of most of the academic process synthesis research results that so far have seen industrial application. Superstructure optimization, while not especially practical with current computing resources, offers tremendous potential for the future.

A. Systematic Generation: Means–Ends Analysis

Although much (perhaps most) conceptual process design is in fact done by the evolutionary modification of whole or partial existing flowsheets, it is sometimes necessary to generate a flowsheet from scratch.

In dealing with the manufacture of materials, a number of physical and chemical properties of materials or streams of materials are of particular interest. These properties include the *molecular identity* of the material; the *amount* involved; the *composition* or purity of the material; its thermodynamic *phase;* its *temperature* and *pressure;* and possibly size, shape, and other physical *form* characteristics. Available raw materials, intermediates, and the desired product materials are all characterized by these same parameters of identity, amount, composition, phase, temperature and pressure, form, and so on.

The raw materials may be considered an *initial state.* The desired product may be considered a *goal state.* If the value of a particular property of a raw material is different from the desired value of the corresponding property in the desired product, a *property difference* is detected. One of the purposes of an industrial chemical process is to apply technologies in sequence such that these property differences are systematically eliminated and the raw materials thereby become transformed into the desired products. The systematic detection of state differences and the elimination of such differences through the application of appropriate corrective *operators* is the essence of a general goal-directed problem-solving paradigm known as *means–ends analysis* (Simon, 1969). The detection of property differences and the identification of methods to eliminate such differences is also the goal of the task identification stage of the innovation process. The specification of actual equipment to implement these difference-elimination methods is the goal of the unit operations stage of the innovation process.

In many cases there are fairly obvious technologies (in some cases tautologies) for reducing or eliminating property differences. Familiar examples include *chemical reaction* to change molecular identity, *mixing* and *splitting* (and *purchase*) to change amount, *separation* to change concentration and purity, *enthalpy modification* to change phase, temperature, and pressure, etc. In some cases such close relationship exists between a difference-elimination method and an obvious type of equipment to implement the method that many designers think directly in terms of *equipment* when developing conceptual process designs, thereby translating property differences directly into unit operation solutions.

Generally, property-changing methods or tasks (the operators in means–ends analysis terminology) are applied to existing streams to produce new streams closer in properties to the desired product material. Most often operators are applied and the conceptual design is developed in a *synthetic, forward,* or *opportunistic* direction, in the same direction as material flow, from the source of raw material through any reactions and toward final products, with recycles added as an afterthought. These operators are applied until no further property differences remain. This is in contrast to the generally *retrosynthetic* (or *backward* from the desired product) approach taken by most chemists in the development of complex organic reaction sequences. In conceptual design, both the starting materials and the desired product are known, while in reaction sequence development sometimes only the desired product is known. Opportunistic approaches are possibly somewhat more comfortable because they are in a sense "anchored." At any point, the design generated is a feasible consequence of the raw materials and the operators chosen. Given the remaining goals, what must I do now? Where can I go from here?

One complication is that often property-changing operators can only be applied to a stream when certain other properties of the stream are within specified values, which may not be true at the time. For example, a method to select only crystals greater than a given size can be applied only if a stream contains solids. Similarly, a separation method expected to exploit relative volatility differences can be applied only if enthalpy conditions permit simultaneous liquid and vapor phases. If the *preconditions* for the immediate application of an operator believed to be useful are not met, a new design *subproblem* may be formulated whose objective is to reduce property differences between the initial stream and the conditions necessary for the application of the operator. This *recursive strategy* is a common feature of the means–ends analysis paradigm.

Alternative solutions are generated when more than one operator is identified that can reduce or eliminate a property difference. The decision of which operator to choose might be made on the basis of some evaluation at the time the operators are being examined. Alternatively, each may be chosen separately, then the consequences followed separately (leading to alternative design solutions) and each final solution evaluated. Yet another possibility is that all feasible alternative operators are selected and applied in parallel, leading to a redundant design or superstructure. At the end of the design process, the superstructure is reevaluated in its entirety, and the less economical redundant portions eliminated.

Another problem is that an operator may not completely eliminate a property difference. In such a case, another operator for the same property difference may need to be specified. For example, a drying task might follow a filtration task to completely remove a liquid contaminant from a desired solid. Furthermore, the application of an operator may not exclusively change a single property difference. The side effects of a difference-elimination method may change other properties as well. For example, the application of a concentration-changing separation method may also change the pressure or the temperature or the phase. In the worst case, the paradigm may not converge, but rather oscillate repeatedly, eliminating one difference but creating a second difference as a side effect. Other problems include the possibility of not finding a method in the repertoire that can eliminate a property difference. If this occurs in a recursive subproblem (that is, while attempting to meet the preconditions for the application of another operator), it may not be fatal if an alternative operator for the original problem is available. The paradigm fails, however, if no operator can be found to correct a property difference at the top level.

B. Hierarchical Approaches

Alternative designs may also result when more than one property difference exists between a stream and its goal and these differences are eliminated in different orders. However, a natural *hierarchy* among property differences seems to exist in the chemical process domain. The hierarchy is the same as the order in which the differences were previously mentioned: identity first, then amount, then concentration, then phase, then temperature and pressure, then form. The hierarchy arises because properties lower in the hierarchy are often more readily manipulated in order to satisfy the preconditions for the application of difference-elimination operators for properties higher in the hierarchy. Alternative solutions are still possible when multiple differences at the same level in the hierarchy exist: for example, when both temperature and pressure need to be changed or when a mixture of more than two components is to be separated and sent to different destinations.

Therefore, a natural approach to systematic generation of a conceptual design to a desired product given available raw materials using both the principles of hierarchical properties and the most obvious difference-elimination methods would be to attack identity differences first (resolved with reaction methods), then amount (resolved with mixing, splitting, or purchase methods), then composition (resolved with mixing or separation methods), then phase, temperature, and pressure (resolved with enthalpy-changing methods), then form properties (size, shape, etc.), as in Fig. 7. Preconditions are met by adjusting properties lower in the property hierarchy if necessary. The conceptual designer, thinking directly in terms of unit operations rather than tasks, would consider reactors first; then mixers, columns, decanters, filters, dryers, and the like; then heat exchangers, pumps, compressors, valves, etc., to satisfy goal properties for the desired product as well as the subgoals that arise as preconditions for the application of unit operations higher in the equipment hierarchy. As before, alternative designs result when alternative unit operations can be used to resolve the

Property Difference	Resolution Method
Molecular Identity	Reaction
Amount	Mixing, Splitting, Purchase
Composition	Mixing, Separation
Phase	Enthalpy Change
Temperature, Pressure	Enthalpy Change
Form	Various

FIG. 7. Property hierarchy and common difference-elimination methods.

same a property difference, or when multiple property differences at the same level in the property hierarchy are resolved in different orders.

The classical means–ends analysis paradigm, although recursive in nature, is nevertheless not iterative. However, for efficiency, many chemical processes exploit *recycle* of mass and energy. Recycle introduces a serious complication since even the existence, much less the properties, of the mass or energy to be recycled may not be known at the time property differences at the point of recycle were resolved. Sometimes the interactions are minimized by the way recycle tasks are implemented in equipment, as in indirect energy recovery with a heat exchanger. Other times the interactions are more severe, as in mass recycle by mixing. Recycles are generally considered at the end of one pass through the conceptual design. If the effects of recycle are severe, particularly with respect to structural validity, the design may need to be repeated from the beginning, this time assuming the existence of the potential recycle, until the design process converges.

C. THE ADAPTIVE INITIAL DESIGN SYNTHESIZER

A number of experimental chemical process synthesis procedures have been developed with a hierarchical architecture based on both the hierarchical structure of the chemical innovation process and the physical property hierarchy. Examples include PIP (Kirkwood *et al.*, 1988) and BALTAZAR (Mahalec and Motard, 1977). This hierarchical approach is also evident in the Pinch Technology Onion Diagram (Linnhoff and Ahmad, 1983). The hierarchical approach is also the foundation for the two early textbooks on process synthesis methodology (Rudd *et al.*, 1973; Douglas, 1988).

One such procedure, the Adaptive Initial DEsign Synthesizer (Siirola and Rudd, 1971), is instructive. The AIDES process synthesis hierarchy consists of *reaction path* (chemistry identification), *species allocation* (target mass flows among raw materials, reactions, products, and wastes including possible recycle of incompletely converted reactants), *task identification* (property difference detection and elimination method identification), *task integration* (association of tasks with actual unit operations, including combining adjacent tasks and also combining complementary tasks), *utilities system* (basic plant design and preliminary operating cost estimation), and *equipment design* (target level equipment design and preliminary capital cost estimation) (Fig. 8). With characteristics similar to the overall innovation process of which process synthesis is a part, the initial stages of the AIDES hierarchy are less well defined, but have a greater impact on the overall outcome of the design. Stages later in the procedure are more constrained and better defined, but have comparably less potential for impact. The stages clearly interact with each other. Decisions at one stage are

- **Reaction Path**
- **Species Allocation**
- **Task Identification**
- **Task Integration**
- **Utilities System**
- **Equipment Design**

FIG. 8. AIDES process synthesis hierarchy.

constrained by choices made at earlier stages, but optimal choices are influenced by how all the subsequent stages prove to be executed. Similar to the innovation process, passing one time through each stage sequentially does not generate exceptional results. Better are schemes to iterate among the stages possibly at different levels of detail. Better yet is the possibility of solving several stages simultaneously. The AIDES procedure employed the means–ends analysis paradigm. A key feature is that task identification, task integration, and equipment design are discrete and separate activities. In particular, and contrary to common conventional design practice, property differences are not resolved directly in terms of specific equipment.

The software version of the AIDES process synthesis procedure (Siirola, 1970) was one of the first applications of *artificial intelligence* in chemical engineering, exploiting dynamic storage allocation, linked data structures, and symbolic manipulation. More importantly, it employed a number of techniques at each stage of the synthesis hierarchy to anticipate certain downstream effects including a *linear programming* formulation for species allocation which sought to minimize the difficulty of the resulting separation problems that would arise because of the allocation of species in a source stream to different destinations. It also employed *multiobjective programming* to prioritize both the property differences to be addressed and the selection among competing property difference-elimination methods. It also used a *targeting* algorithm to maximize energy recovery and minimize utility requirements. Unlike the later PIP program and other expert systems approaches, however, it was not overtly rule-based. The automated means–ends analysis implementation was biased to propose obvious difference-elimination methods for each class of property differences, and preconditions were met, if necessary, by adjusting properties *lower* in the property hierarchy. For example, separations might be identified in order to get appropriately pure feed conditions for reaction tasks, and temperatures might be adjusted in order to get better conditions for separation tasks. However, AIDES would never consider changing the identity of a species to get better conditions for a separation, as that would have violated the property hierarchy. Multiple design alternatives were generated by applying different operators (for example, by exploiting different physical phenomena to effect a separation) and by attacking multiple differences at the same level of the property hierarchy in different orders. The AIDES program can interact as tightly or loosely with the conceptual designer as desired. It can accept advice or criticism, have its recommendations overruled, or formulate its own solution independently. The program is adaptive, and if feedback is provided it can *learn* from its performance (by adjusting internal multiobjective weighting parameters).

The AIDES program definitely has its limitations. The internal representation of solution thermodynamics is simplistic; it cannot, for example, handle azeotropes. Also, the species allocation, once proposed, is adhered to too rigidly and its task integration capability is somewhat weak. In industrial situations, its performance is judged to be good, but only slightly better than conceptual designers that resolve property differences directly in terms of common unit operations.

However sophomoric the AIDES computer implementation, the hierarchical process synthesis procedure itself offers three important lessons which have had a significant industrial impact—lessons that can be used even manually by conceptual process designers.

1. *Keep task identification distinct from task integration and equipment design.* Think carefully about what needs to be accomplished (tasks), then consider how it will be accomplished (equipment). In other words, remember the architect's maxim, "Form follows function."
2. *Do not adhere inflexibly to the property hierarchy.* The fact is, any property, not just those lower in the property hierarchy, might be worth adjusting as a precondition to the application of an operator, including properties at the same level or even at a higher level in the hierarchy. For example, identity change can sometimes be a route to resolving concentration differences, either because it turns an offending species into something else easier to separate—or better yet into something useful or at least innocuous so that it doesn't have to be separated at all.
3. *Be wary of convenience.* Much chemical laboratory practice is driven by the desire for high yields at ambient conditions. Thus there is a tendency toward irreversible procedures. For example, acid chlorides are used rather than acids for esterifications and amidations to avoid reaction equilibrium limitations. But such laboratory convenience comes at the price of more energetic and expensive reagents, and the production of very low-energy-state coproducts, such as volatile gases and salts, which are often of little economic value and large environmental impact. In a similar vein, it is more convenient in manufacturing operations to avoid the complications of material and energy recovery and recycle. Again, such luxury is often at the expense of greater raw material and energy requirements and costs, as well as greater thermal and material loading on the environment.

IV. A Total Flowsheet Example

Consider the following industrial example of a process for the production of methyl acetate, first synthesized in the conventional conceptual process design manner, then modified using evolutionary approaches, and finally resynthesized using the lessons from the hierarchical process synthesis procedure experience. For simplicity here, only identity, amount, and composition differences are resolved (that is, temperature and pressure changers will be ignored).

A. Conventional Conceptual Process Design

Methyl acetate may be produced by the elementary equilibrium-limited acid-catalyzed esterification reaction of methyl alcohol and acetic acid which also by-produces water. The reaction is nearly athermic, with an equilibrium constant on the order of unity.

Acetic acid + methanol ⇔ Methyl acetate + water

Acetates form azeotropes with both their corresponding alcohols and water, but the azeotrope with water is generally heterogeneous and lower-boiling. Therefore, acetates may be produced from the acetic acid and alcohol in a reactor fitted with a fractionating column which removes the acetate–water heterogeneous azeotrope overhead while forcing unreacted acetic acid and alcohol back to the reactor. The acetate and water are separated in a decanter and purified additionally as necessary (Witzeman and Agreda, 1993). However, this textbook flowsheet is not applicable to the production of methyl acetate because the methyl acetate–methanol azeotrope boils at a lower temperature than the methyl acetate–water azeotrope; furthermore, the latter azeotrope is unique among acetate–water azeotropes in that it is homogeneous. Therefore, a new conceptual process design must be generated.

If methanol and acetic acid are available as raw materials and methyl acetate is the desired product, according to the property-difference hierarchy, an identity difference is first detected between the desired product and each of the raw materials. A known chemical reaction operator, namely the esterification reaction, can be applied to a mixture of the raw materials brought to the proper conditions to produce methyl acetate and eliminate the identity difference between the reaction effluent and the desired product. Thinking directly in terms of equipment, this operator may be immediately implemented, for example, as a stirred tank reactor.

The effluent of the reactor contains product methyl acetate, by-product water, and significant unreacted methanol and acetic acid. The species allocation is trivial; methyl acetate is directed toward the product, water to wastewater treatment, and both methanol and acetic acid are recycled back to the reactor in order to improve feedstock usage efficiency. Since the reactor effluent does not meet the composition specifications of any desired destination, separation tasks need to be identified to eliminate composition differences between the reactor effluent and the various destinations of its various components.

The species are all liquids at ambient conditions. Normal boiling points vary from 57 to 118°C. It might be possible to exploit volatility differences—that is, distillation—to effect the required separations. Water and methyl acetate form a low-boiling azeotrope at 95 wt% methyl acetate boiling about 56°C. Water and methyl acetate are immiscible between about 23 and 92 wt% methyl acetate. Thus, as previously stated, this binary system is unusual in that it exhibits both azeotropy and heterogeneity, but the azeotrope is homogeneous. Also, methanol and methyl acetate form an azeotrope at 81 wt% methyl acetate boiling even lower at 54°C. There are no other binary or ternary azeotropes in the system, although water and acetic acid almost form an azeotrope exhibiting a tangent pinch on the water end.

How should the required separation system be synthesized? One common sequencing practice is to perform the easiest split first. Considering the boiling point distribution of the components and azeotropes and thinking in terms of equipment, an opportunistic first split is a distillation column in which all of the methyl acetate, all of the methanol, and as much water as azeotropes with the methyl acetate are taken as overhead. All of the acetic acid and the remaining water are taken as bottoms (Fig. 9). At first, this may not seem like much

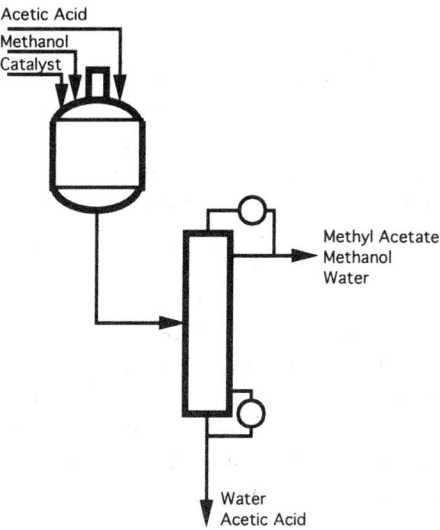

FIG. 9. Methyl acetate: Opportunistic first split.

progress since the composition of neither column effluent stream is directly suitable for any destination. However, the bottom stream is essentially a binary mixture of acetic acid and water, and the separation of that mixture is a problem that has been solved before.

The separation of acetic acid–water mixtures is a feature of a number of large-scale industrial processes including manufacture of cellulose acetate, terephthalic acid, and even acetic acid itself. Simple binary distillation is feasible but expensive because of the tangent pinch at the top of the column. The literature, however, contains a number of flowsheets for this separation. Most involve azeotropic distillation with an entrainer sometimes preceded by solvent extraction to circumvent the tangent pinch, but each is different because of the acetic acid feed concentration addressed and the specifics of the azeotropic entrainer or extraction solvent. One particular textbook flowsheet (Lodal, 1993) suitable for the acetic acid–water composition from the underflow of the splitting column in this example is shown in Fig. 10. The design consists of a solvent extractor, an azeo column and decanter, flash columns for solvent recovery from the extractor raffinate and the decanter aqueous layer, and a color column for purification of the azeo column underflow. All of the acetic acid appears in the color column distillate suitable for recycle back to the reactor. Water exits from the two flash columns suitable for discharge to the wastewater treatment system. Ethyl acetate in this case is used for both the extraction solvent and the azeo column entrainer, and is totally recycled within the flowsheet. Other flowsheets using different solvents or entrainers resulting in different configurations are also reported in the literature (Siirola, 1995).

This now leaves the overhead from the first splitting column. The homogeneous azeotropes in this mixture are a significant problem. However, a well-known pattern for breaking homogeneous azeotropes involves extractive distillation with a high-boiling solvent. Ethylene glycol is a suitable solvent in this case. A standard textbook configuration consisting of an extractive distillation column followed immediately by a solvent recovery column produces a pure methyl acetate product as distillate from the first column and a methanol–water mixture as distillate from the second. The methanol–water mixture may be separated in one final binary distillation column, producing methanol suitable for recycle back to the reactor and water suitable for discharge to the wastewater treatment system. Composition specifications of the product methyl acetate and the by-product water are met, and all unreacted methanol and acetic acid is recycled to the reactor. No additional identity, amount, or concentration differences exist and the conceptual design is now complete. The flowsheet includes one reactor, one extractor, one decanter, and eight distillation columns, and employs two different external mass separation agents (Fig. 11).

INDUSTRIAL APPLICATIONS OF CHEMICAL PROCESS SYNTHESIS 23

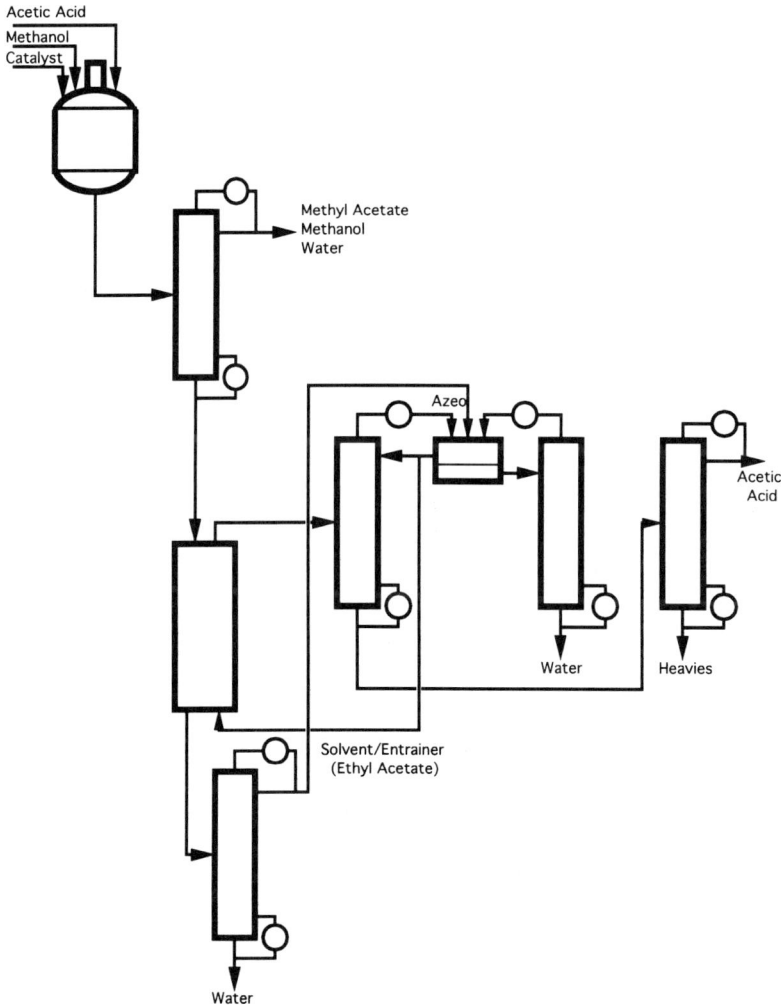

FIG. 10. Methyl acetate: Textbook acetic acid dehydration.

B. Evolutionary Modification

The classic flowsheet for the production of acetates was unsuitable for adoption for methyl acetate. However, the flowsheet generated by the conventional conceptual process design approach using literature schemes and standard patterns seems especially complicated for such simple chemistry involving so few components. This first flowsheet is a candidate for evolutionary modification.

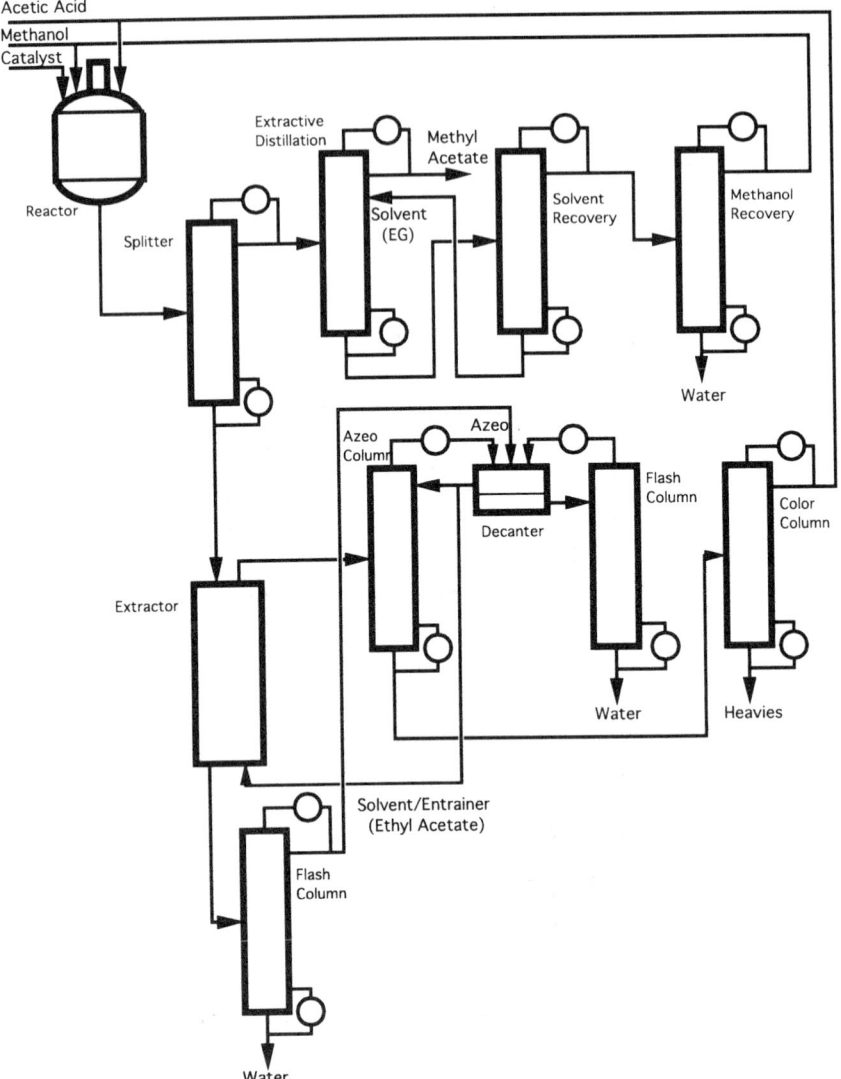

FIG. 11. Methyl acetate: Complete conventional flowsheet.

One evolutionary strategy is to search for thermodynamic inefficiencies to attack, for example, through exergy analysis. Another strategy is to search for structural inefficiencies including design redundancies. In our first flowsheet, both flash columns have similar feed compositions, both overheads go to the same decanter, and both underflows go to wastewater treatment. Therefore, these two columns can probably be combined. Second, the acetic acid color column, which was a feature of the textbook flowsheet copied for this part of the design, may not be required. Upon investigation, it is found that the original flowsheet was designed to handle dirty acetic acid–water mixtures containing minor unspecified high-boiling contaminants. In the present case, it is determined that the acetic acid composition in the bottoms of the azeo column is sufficiently pure to be recycled to the reactor directly, and thus another column is eliminated. Enthalpy-changing tasks were not considered in this example, but if they had been, a number of energy recovery opportunities might also be evaluated using heat integration techniques.

Another point of concern is the two external mass separating agents. Both were chosen from textbook patterns and flowsheets. Although both are readily recovered and recycled within their respective sections of the process, a common evolutionary heuristic suggests that if possible, mass separation agents should be chosen from among components already present within the system. First, consider the agent that must serve the dual role of extraction solvent and azeotropic entrainer. The only possible component in the original system that might work as an acetic acid extraction solvent is methyl acetate. However, as previously noted, the region of immiscibility is not large, and depending on the feed composition, it might not be a feasible extraction solvent at all. Moreover, its azeotrope with water is not heterogeneous, so the subsequent azeotropic distillate cannot be broken by decantation. Methyl acetate appears to be totally unsuitable as an internal mass separating agent for this problem.

On the other hand, considering the second mass separation agent, acetic acid is a possible candidate as an extractive distillation solvent to break the methyl acetate–water azeotrope. If acetic acid were substituted for the ethylene glycol, a methyl acetate–methanol mixture would be produced as distillate, and an acetic acid–water mixture would be produced in the underflow. The methyl acetate–methanol mixture (which also forms a homogeneous azeotrope) may be separated into methanol and sufficiently pure methyl acetate to meet the desired product specification either by binary distillation at reduced pressure (where the azeotrope disappears) or by pervaporation. The acetic acid–water mixture is sent directly to the previous part of the flowsheet with the underflow of the first splitting column, thus saving the solvent recovery column. Furthermore, the splitting column and the extractive distillation column might also be combined since one is fed from the other and both underflow to the same place. The final evolved design, Fig. 12, thus contains four fewer columns than the original flowsheet.

The relative performances of the original and evolved designs can be determined only by analysis. The evolved design has fewer columns, but has the added complication of either refrigeration or membranes, and the amount of acetic acid which must be separated from water is much increased. The results

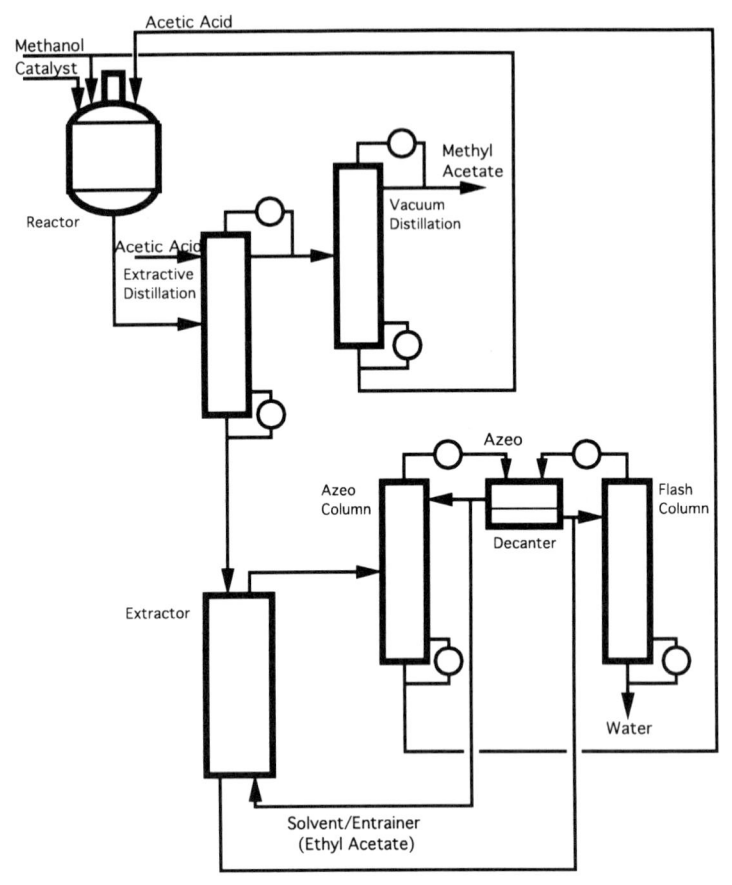

FIG. 12. Methyl acetate: Evolutionary modification.

of detailed simulations indicate in fact that combination of the flash columns and elimination of the color column is advantageous, but that using acetic acid as an extractive distillation mass separation agent in this case unfortunately is not.

C. The Hierarchical Process Synthesis Procedure

If the hierarchical means–ends analysis synthesis procedure is applied to the methyl acetate problem, the task identification, task integration, and equipment design stages are kept completely separate. Following the property-difference hierarchy, an identity-changing reaction task (Task A) is identified first, as before. When examining the differences between the result of this reaction task application and the product methyl acetate and by-product water destinations, two sets of composition differences are detected. For the methyl acetate destination, acetic acid, water, and methanol must all be essentially completely removed. Similarly for the water destination, methyl acetate, methanol, and acetic acid must all be essentially completely removed. One way to attempt to resolve these concentration differences is separation exploiting relative volatility. Considering the methyl acetate destination first, the separation of acetic acid should be relatively easy because of the large boiling point difference. However, removal of the water and methanol may be difficult because of homogeneous azeotropes. Turning to the water destination, removal of both methyl acetate and methanol should be easy because of the large boiling point differences (the methyl acetate–water azeotrope is not an issue where the goal is to produce pure water composition, although it can affect meeting amount goals), but removal of the acetic acid may be difficult because of the tangent pinch.

Relative volatility is not the only property that might be exploited to address composition differences. For example, if the property hierarchy is ignored, then an identity-changing esterification reaction task with acetic acid might be chosen to eliminate methanol from the methyl acetate stream (Task B). Similarly, an esterification reaction task with methanol might be chosen to eliminate acetic acid from the water stream (Task C). Furthermore, as in the previous example, solvent-enhanced distillative separation using acetic acid might be used to break any methyl acetate–water azeotrope (Task D), and conventional distillative separation might be used to separate both methyl acetate (Task E) and methanol (Task F) from water, and acetic acid from methyl acetate (Task G). What results then are three reaction tasks (one to produce the desired product and two to remove offending species from destination streams), one solvent-enhanced distillative separation task (using acetic acid), and three conventional distillative separation tasks (Fig. 13).

There are a number of ways in which these tasks may be integrated into actual processing equipment. It turns out in this case that considering hydraulics, material flows, and energy requirements, all seven tasks may be integrated into one single piece of equipment (Fig. 14). All three reaction tasks are integrated into the center of a column, with the main reaction Task A in the middle, the methanol-removing reaction Task B above, and the acetic acid-removing reaction Task C below. Extractive distillation Task D sits above the reaction tasks with extraction solvent knockdown Task G above that. Low-boiler separation from water Tasks E and F are combined and sit below the reaction zones. Proper operation requires that the extractive zone must be above and distinct from the reactive zones (otherwise the azeotrope of water and methyl acetate formed at the top of the reaction zone cannot be broken), which requires that the catalyst facilitating the reaction be either introduced between zones D and B (if homogeneous) or installed only below zone D (if heterogeneous). Although this single piece of equipment is admittedly complex, the resulting capital cost and operating cost of the single-column reactive–extractive distillation design for methyl

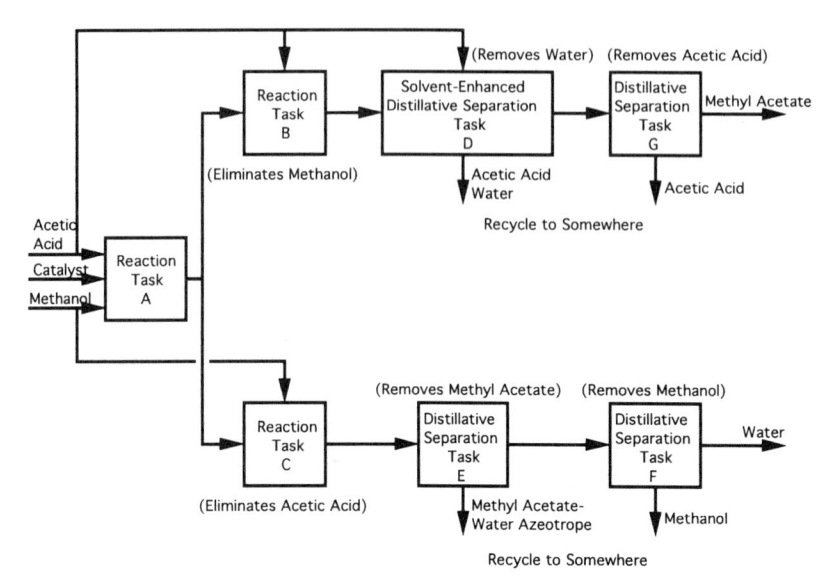

FIG. 13. Methyl acetate: Task identification.

acetate production are both just one-fifth of that for the optimized conventional flowsheet (Agreda and Partin, 1984). It is not clear how such a design might have been conceived without the explicit separation of task identification, task integration, and equipment design. The design is now in successful industrial operation in several plants.

To reiterate, most approaches to conceptual process design employ some kind of hierarchy to prioritize and order property-difference resolution. Sometimes the systematic generation of solutions from scratch is superior to copying existing patterns, tricks, and flowsheets. There may be some advantage to thinking of all properties as candidates for change to effect property-difference resolution, and selecting properties for change outside of the natural hierarchy sometimes leads to designs that are judged to be clever. Also useful is thinking primarily in terms of tasks rather than unit operations or equipment, being open to unique ways in which multiple tasks may be integrated into actual equipment, and being especially suspicious of laboratory convenience.

V. Energy Efficient Distillation Synthesis

A. HEAT-INTEGRATED DISTILLATION TRAINS

Heat integration, or the synthesis of heat-exchanger networks, is a classic problem of integrating complementary tasks at the task integration stage of the

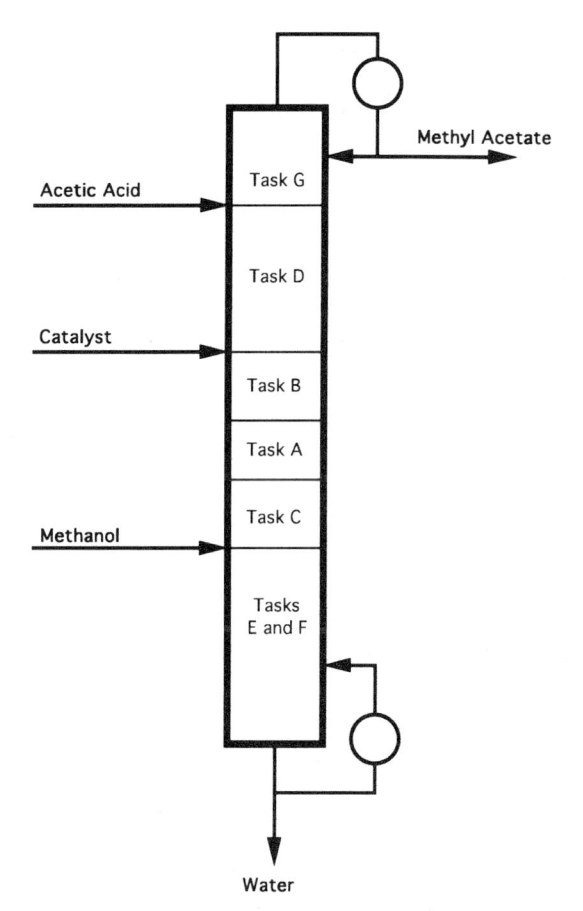

FIG. 14. Methyl acetate: Reactive extractive distillation process.

process synthesis hierarchy. The tradeoffs between capital and operating costs are well understood. Excellent procedures exist for targeting the expected amount of energy recovery, the remaining utilities requirements, and the expected capital requirements under a number of constraints even before any heat-exchanger networks have been actually designed. The *pinch* concept (Linnhoff and Hindmarsh, 1983) provides further guidance on restrictions on the placement of exchangers, the use of heat pumps, and determining situations where stream splitting and parallel heat-exchange designs are especially advantageous. It also introduced important visual representations of the relevant thermodynamics including *composite curves* and *grand composite curves*.

Procedures for the actual synthesis of heat-exchanger networks are somewhat less well developed; however, all of the standard process synthesis approaches—including systematic generation, evolutionary modification, and superstructure optimization using all the standard methods and tools including heuristics, expert systems, deterministic algorithms, and mathematical programming—have been discussed in the literature. Aspects of network flexibility, resiliency, and controllability have also been investigated. Available academic and commercial heat-integration software includes HEXTRAN, ADVENT, SUPERTARGET, MAGNETS, SYNHEAT, HEATNET, and others, as well as several proprietary codes.

Heat integration, which has been the most studied problem in the process synthesis literature, has also been the most widely applied synthesis technique in industry. The problem is well defined and highly constrained, being near the bottom of the synthesis hierarchy. Heat-integration technology has been applied to large industrial problems, including sections of individual processes, whole process plants, and even whole plant sites. The straightforward application of heat-integration technology to a typical chemical process design generally reduces the net present cost (NPC) by on the order of 10%, due largely to sensible heat recovery.

It is also well appreciated that greater savings may be possible if energy recovery is considered during, not just after, the task identification stage while the magnitudes and temperatures of the various heat sources and sinks are actually being determined. The following example considers the effect of closely coupling the conceptual detail level of the task identification stage of a distillation-based multicomponent separation problem with the target detail level of the subsequent enthalpy task integration stage.

The components involved in this example are proprietary, but the results are general (Siirola, 1981). During the species allocation stage of the process synthesis procedure, it was determined that each species of a particular four-component stream was required to be relatively pure at four different destinations. The components are liquids at ambient temperatures, have about equal relative volatility differences, and form no azeotropes. Distillative separation methods were selected to resolve all composition property differences. The feed stream composition was dominated (about 70%) by the heaviest component (D).

From exhaustive application of alternative simple distillation operators to all possible separations for this four-component system, or from application of ranked-list–based separations synthesis methods, it is easily shown that there are five different separation train structures for this four-component problem. Each can be generated systematically; or since this pattern of solutions is already well known, each can be written down immediately or design heuristics can be used to generate one or more of the structures expected to be most suitable. After each structure is synthesized, its performance can be analyzed and evaluated with a flowsheet simulator.

Probably the most common industrial solution to a problem of this type is a train of three columns, all operated at essentially atmospheric pressure, and each

Relative NPC = 1.00

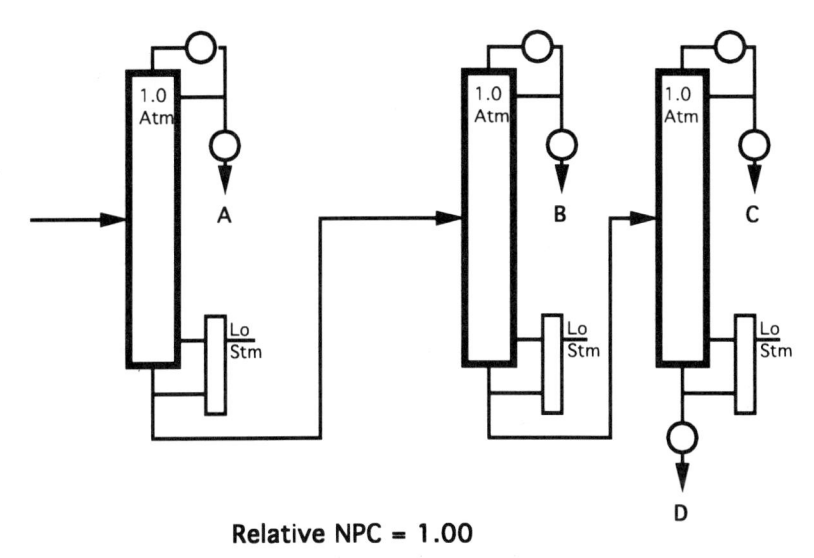

FIG. 15. Conventional sequence: Atmospheric operation.

operated to remove component relatively pure overhead one at a time (Fig. 15). In this example, this three-column train is the reference design to which we assign a relative NPC (including both capital and operating costs) of unity.

It turns out that, when all five sequences are analyzed, a different scheme has about 5% lower NPC than the reference case (Fig. 16). In this case, the

Relative NPC = 0.94

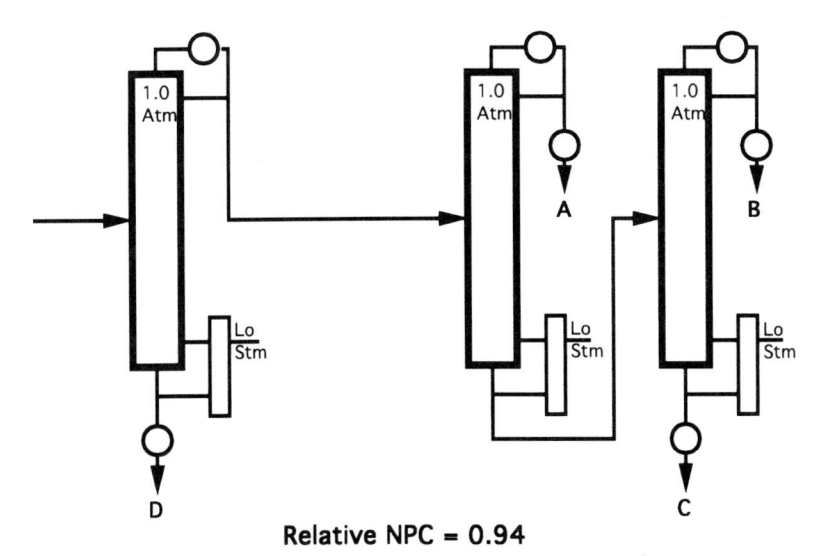

FIG. 16. Most plentiful first: Atmospheric operation.

dominant heavy species D is removed as the underflow in the first column, and then the remaining components are removed one at a time overhead. This is also the scheme that would have been generated by application of the most common distillation train sequencing heuristics (Seader and Westerberg, 1977).

Each of the distillative separation tasks is implemented as a separate standard single-feed distillation column with reboil and reflux. The thermal quality of each column feed might be adjusted externally to saturated liquid (although the size of the reboiler or condenser may be altered instead). There are also four product cooling (for storage) tasks. Altogether, therefore, there are potentially 13 enthalpy-changing tasks associated with this separations system (3 reboilers, 3 condensers, 3 feed conditioners, and 4 product coolers). Some of these tasks may be resolved with utilities and some integrated into a heat-exchanger network (see Fig. 17 for the separation sequence of Fig. 16). (In this example, there were no other significant enthalpy-changing tasks outside of the separation train to be simultaneously considered.) The NPC improvement from heat integration of the best separation scheme is on the order of 10% (15% reduction total from the reference case), more or less as expected. It should be noted that all of the heat-integration opportunities involve sensible heat.

Now one might return to the task identification stage, but at a higher level of detail to consider the possibility of separator designs not necessarily constrained to atmospheric pressure. In this example, the solution thermodynamics are such that the A/B split is favored by higher pressures and the C/D split is favored by lower pressures. In the best unintegrated case, the A/BC column is increased in pressure until just below where a more expensive reboiling utility would be required, while the ABC/D column is decreased in pressure until just

Relative NPC = 0.85

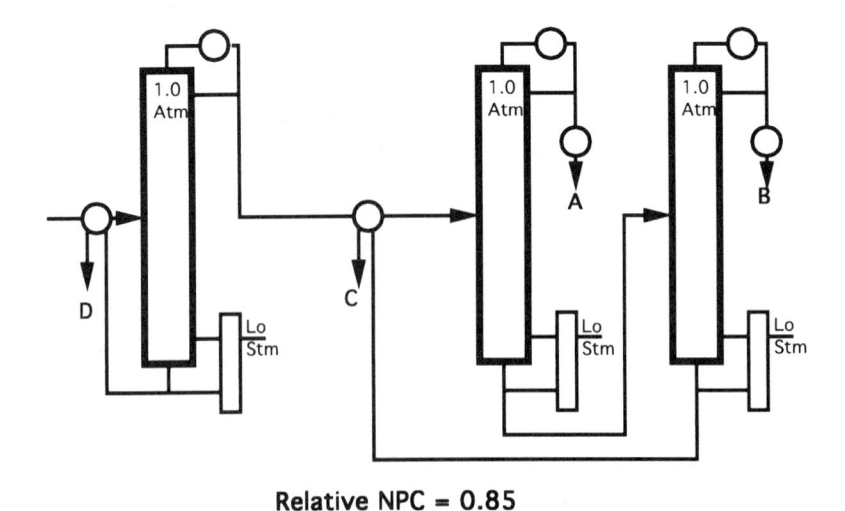

Fig. 17. Most plentiful first: Atmospheric operation, heat-integrated.

Relative NPC = 0.84

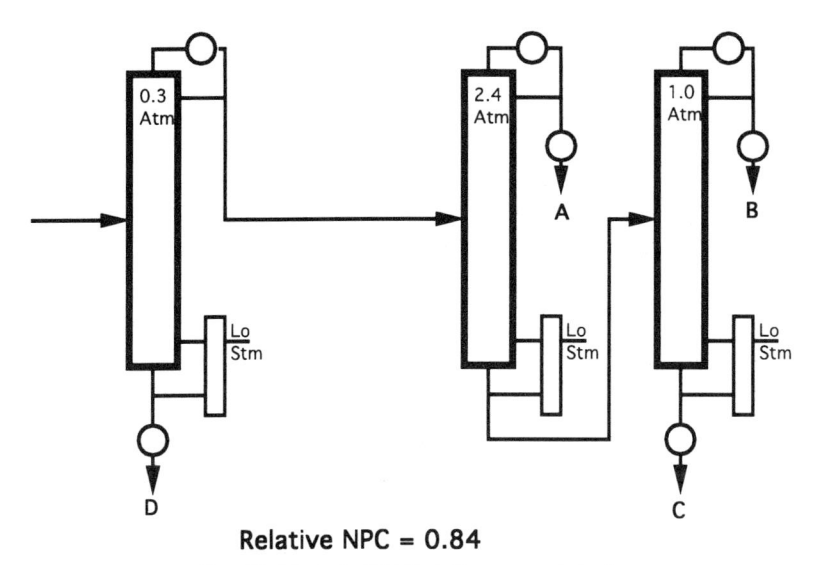

FIG. 18. Most plentiful first: Pressure-optimized.

above where a more expensive condensing utility would be required (Fig. 18). This pressure-optimized design has a 10% NPC improvement over the reference case due to flowsheet structure and column size minimization.

Once again the task integration stage may be revisited and a heat-exchanger network designed for the 13 enthalpy-changing tasks of the pressure-optimized design (Fig. 19). The result is a cost improvement similar to that for heat integrating all-atmospheric designs, and again represents only sensible heat recovery.

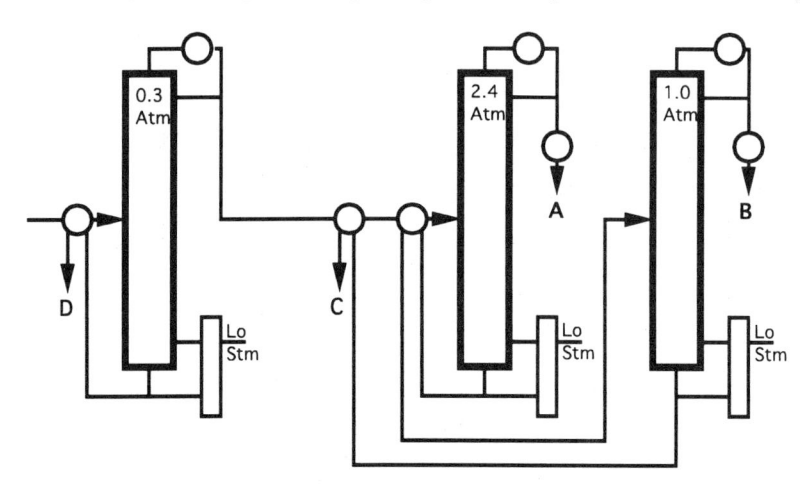

Relative NPC = 0.79

FIG. 19. Most plentiful first: Pressure-optimized, heat-integrated.

So much for the results of iterating between the task identification and task integration stages of the process synthesis procedure. What if the column pressure selection in the task identification stage were more tightly coupled with the heat-integration network generation of the task integration stage? A conceptual design level heat-integration procedure (Siirola, 1974) was made fast enough to be included within the objective function of the column-pressure optimization algorithm enabling such a coupling. Each objective function evaluation involved the design and costing of three columns; the determination of 13 enthalpy-changing tasks; the design and sizing of an associated heat-integrated energy recovery network including all reboilers, condensers, and other heat exchangers; and determination of the remaining utility requirements and costs. The result for this example of such a simultaneous column-pressure optimization with heat-exchanger network synthesis is shown in Fig. 20. The pressure of the ABC/D column is still low. However, the pressures of the remaining columns are both raised to a point where their reboilers require a higher cost utility and just below the point where they would require an even higher cost utility. However, latent heat from both of their condensers is recovered in reboilers for the first column. Operation of the two columns at these pressures does not make sense without the simultaneous latent heat integration. The NPC savings for the resultant design is a significant 35% reduction from the original reference case. The utility requirement was reduced by 50%.

The integration of enthalpy-changing tasks among different parts of a process can introduce interactions and control difficulties similar to other recycle

Relative NPC = 0.65

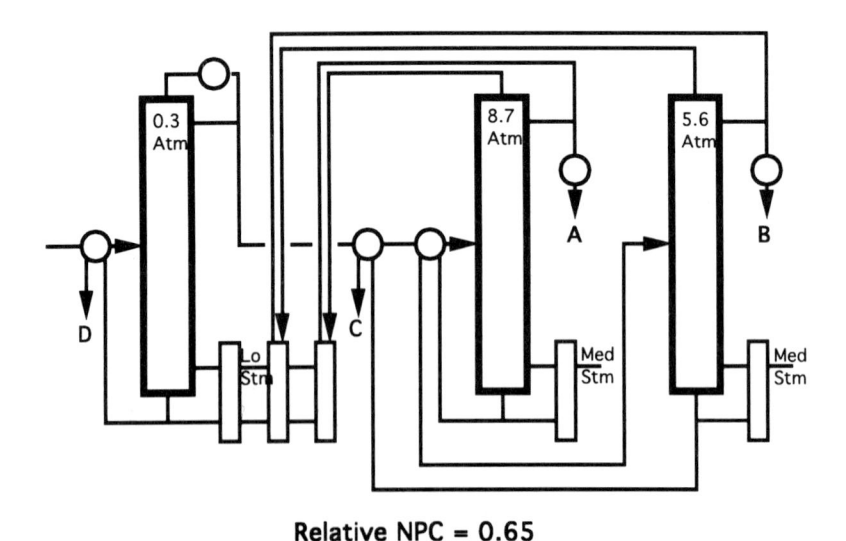

Fig. 20. Most plentiful first: Simultaneous pressure optimization and heat integration.

schemes. Sometimes these interactions may be decoupled through the use of an intermediate heat-transfer fluid between tasks, such as an intermediate-pressure steam, an imbalance of which could be made up from a mixture of standard utilities. Similarly, in a large plant, it may also be possible to effectively do heat integration mostly with the utility system, rather than within the process. This is especially true when there exists a large utility network with energy sources and sinks available at many levels of temperature. In such cases, it may be worthwhile to consider steam generation as a cooling utility. In the present example, simultaneously optimizing column pressures and heat integration while allowing the option of steam generation for credit produced the design in Fig. 21. The best separation train in this case turned out to be the more conventional single-species-at-a-time overhead arrangement. Higher pressure steam was used in all reboilers, and steam was generated in the condensers of two, for return to the utility network. There was only minimal sensible heat integration (for the fairly high-temperature first column). The NPC was again 35% below the original reference case. Because of perceived ease of operation, this was the design that was actually implemented industrially.

Utility requirement savings in the neighborhood of 50% and NPC savings in the neighborhood of 35% for distillation trains are typical of what can be expected from the tight interaction of the process synthesis task identification and task integration stages involving simultaneous distillation sequence synthesis, operating-parameter optimization, and heat-integration network synthesis.

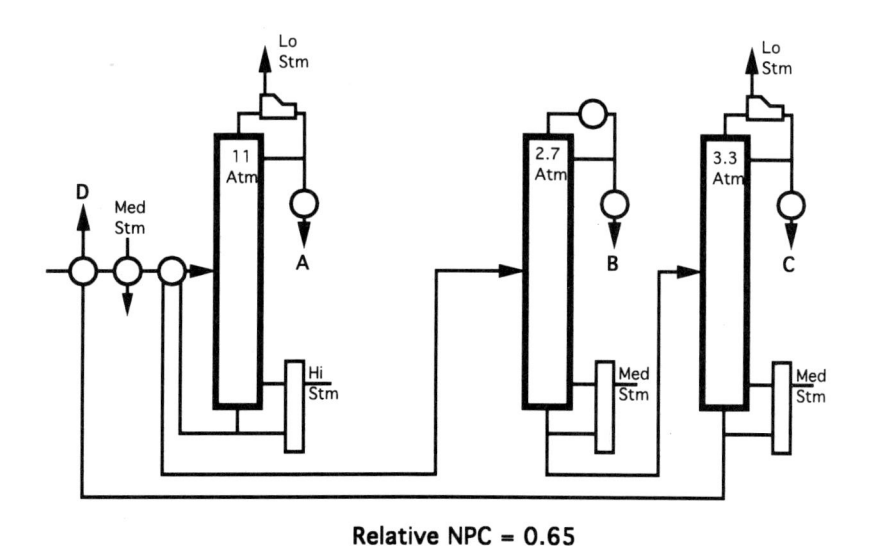

Relative NPC = 0.65

FIG. 21. Simultaneous pressure optimization and steam generation.

B. COMPLEX COLUMNS FOR BINARY SEPARATIONS

The previous section indicated that there can be a significant benefit to heat-integrated distillation sequences, with energy recovery either within the sequence or with the utility system. For processes that contain a single or a particularly dominant distillation, significant energy recovery may not be possible. Pinch analysis would suggest the possible use of a heat pump or vapor recompression in such cases. However, another possibility is to split the implementation of the distillative separation task into two columns in such a way that energy rejected from one might be recycled to the other, approximately halving the net utility requirement. The question is, given a separation that can be accomplished perfectly feasibly in a single simple distillation column, under what conditions is it more advantageous to implement the separation in two interlinked columns?

The motive for multiple-effect distillation for a single separation task is net energy reduction, generally at the expense of greater capital cost and operating complexity. For this to be accomplished, the condenser of one column must be at a higher temperature than the reboiler of the other. This may be done by operating the columns at different pressures, possibly assisted by the degree of separation (and hence temperature difference from bottom to top) designed for each column.

In one parametric study, six different designs (Fig. 22) were compared (Blakely, 1984). Design 1 was the single-column reference case. Design 2 consisted of a high-pressure and a low-pressure column. The feed was split and fed to both columns in parallel. Both columns performed the same sharp separation, producing pure products. The condenser of the high-pressure column was the reboiler of the low-pressure column.

In Design 3, the feed was sent to the high-pressure column, which produced a pure high-boiler bottom product but a mixed overhead product (a sloppy separation decreasing the temperature difference across the high-pressure column). The mixed distillate was then completely separated in the low-pressure column. Again, the condenser of the high-pressure column was the reboiler of the low-pressure column (heat integration in the same direction as flow).

In Design 4, the feed was sent to the low-pressure column, which produced a pure low-boiler distillate but a mixed underflow (again, a sloppy separation decreasing the temperature difference across the low-pressure column). The mixed bottoms was then completely separated in the high-pressure column. The condenser of the high-pressure column was the reboiler of the low-pressure column (heat integration in the opposite direction as flow).

In Design 5, both columns were operated at the same pressure. However, the first column produced a pure low-boiler distillate but a mixed underflow, thus allowing the use of a less expensive utility. The mixed bottoms was then com-

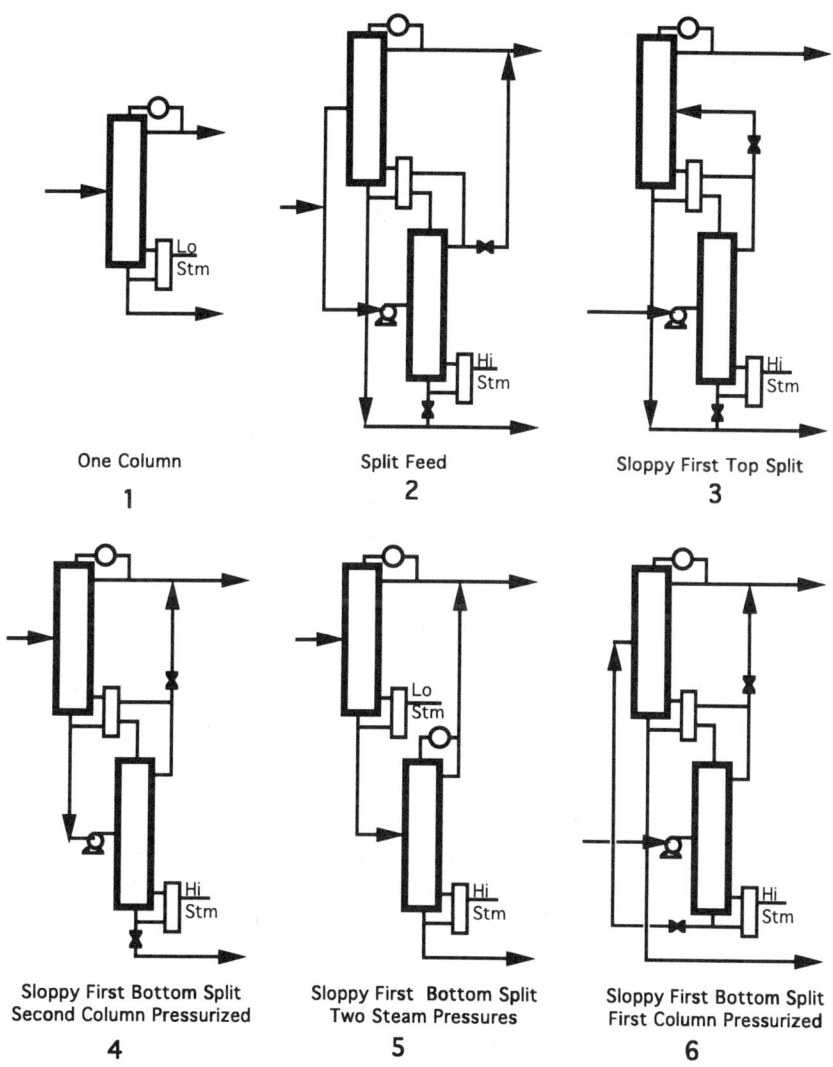

FIG. 22. Multiple-effect distillation configurations.

pletely separated in the second column, producing pure top and pure bottom products using a more expensive utility. No heat integration was involved.

In Design 6, the feed was sent to the high-pressure column, which produced a pure low-boiler distillate but a mixed underflow (again, a sloppy separation). The mixed bottoms was then completely separated in the low-pressure column.

The condenser of the high-pressure column was the reboiler of the low-pressure column (heat integration in the same direction as flow).

(The last obvious combination, with the feed sent to the low-pressure column producing a pure high-boiler bottoms product, with a mixed overhead completely separated in the high-pressure column, was not considered because it always requires a higher temperature utility than the other schemes.)

In the study, the pressure in the low-pressure column and the boiling point of the more volatile component were fixed. The composition of the mixed stream was chosen so that the heat duty of the high-pressure condenser was equal to that of the low-pressure reboiler. A parametric study on feed flowrate, feed composition, and relative volatility between the binary pair was conducted with a detailed process simulator. For each set of conditions, the NPC of each design was compared to that of the single-column reference case.

The specific results depend on the costs and temperatures of the utilities actually used in the study as well as on the capital cost correlations, and are not given here. However, the general trends are instructive. At very low flowrates, the economics of the single-column reference design was generally superior to all of the multiple-effect designs largely because at those scales the capital costs of the columns dominate the energy costs. However, at higher flowrates, there were conditions of feed composition and relative volatility for which the multiple-effect designs (but not the multiple-utility design) exhibited lower NPC than the single column. The shaded areas in Fig. 23 indicate the regions of feed composition and relative volatility in which the costs of each of the designs were at least 10% lower than the corresponding cost of the single-column design. There are still regions of high relative volatility and heavier feed composition where the single column is superior. But as feed flowrate increases, there are surprisingly more and more combinations of relative volatility and feed composition for which one or another of the multiple-effect schemes (each reducing the energy requirement by 50%) is worthy of examination. For some cases, the NPCs of the complex schemes were as much as 35% lower than the single-column reference design, savings similar in magnitude to typical savings in other multiple-column heat-integrated distillation designs.

In the production of methanol from synthesis gas, some water is by-produced from which the methanol product must be separated. This is the only major distillative separation task in the process, and because methanol has a high latent heat, it is very energy-intensive. For the feed conditions for this separation, the charts from this study indicate that Designs 2, 4, and 6 should be better than a single column, with Design 6 somewhat superior to the others. Steady-state simulation confirmed this prediction, and dynamic simulation confirmed the controllability of the tightly coupled column pair. Design 6 has been implemented industrially for this service with neither trim condensers nor reboilers (although multiple exchanger shells are required because of the low approach temperatures).

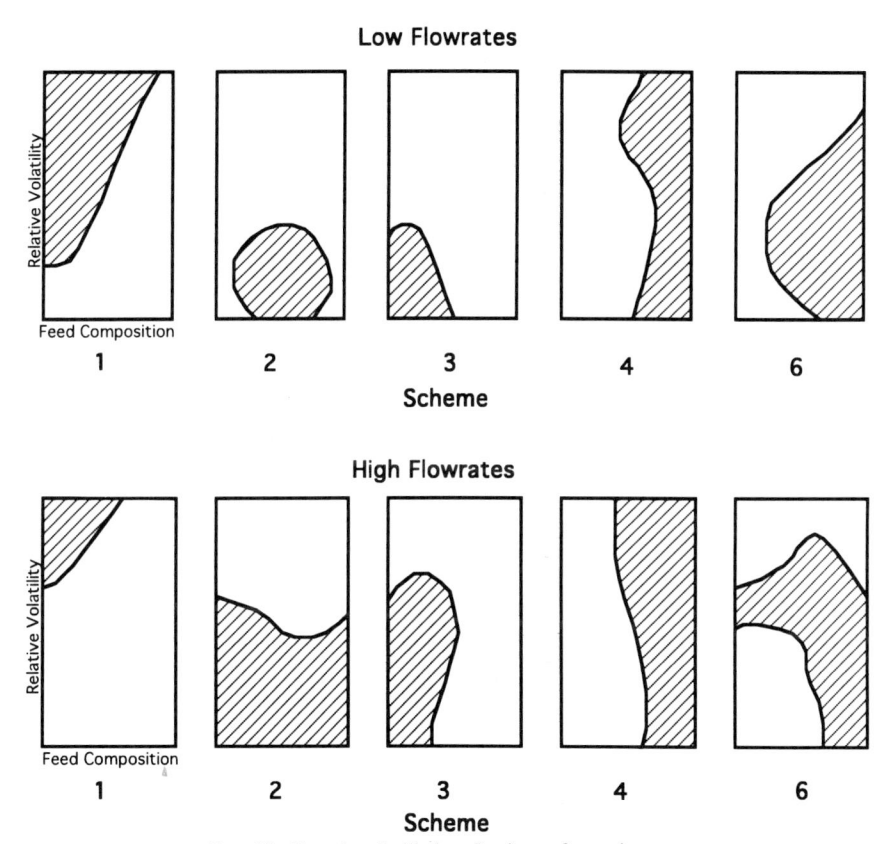

FIG. 23. Complex distillation: Regions of superior cost.

VI. Residue Curve Maps and the Separation of Azeotropic Mixtures

The synthesis tools required to propose property-difference resolution methods at the task identification stage of the process synthesis hierarchy depend in part on the nature of the system. In the case of distillative separations for resolving compositional differences, if the solution thermodynamics of the system are relatively ideal, a simple ranked-list volatility ordering may be sufficient to represent the thermodynamics for the purpose of separations flowsheet synthesis. Alternative separation trains can be synthesized at the targeting level of detail using list-processing techniques. However, if the system exhibits more nonideal thermodynamics involving azeotropes and regions of immiscibility, a more detailed representation of the thermodynamics may be necessary for the synthesis of separation schemes even at the targeting level.

Residue curve maps (RCMs) are phase equilibrium representations for systems involving azeotropes (Foucher et al., 1991). RCMs are derived from an analysis of a single-stage batch still, but they may be applied to the understanding of the thermodynamic behavior of solutions and the behavior of continuous distillative separation operations on those systems. For a three-component system, the analysis can be plotted on familiar triangular composition diagrams. However, the analysis may be performed on systems having any number of components. Residue curves trace the composition of the stillpot of a single-stage flash in time as the vapor is slowly removed. Curves may be generated from any starting composition. As time proceeds, the curves move toward the composition of the last drop remaining in the stillpot. For some mixtures, this composition is the highest-boiling pure component in the system. For other systems, this composition may be a high-boiling azeotrope. For yet other systems, the final composition varies depending on the initial composition in the still, sometimes ending at one composition, sometimes ending at another. Residue curves may also be mathematically extrapolated backward in time to give the most volatile composition whose residue curve would pass through the specified feed composition. When extrapolated both forward and backward in time, all residue curves that originate and terminate at the same two compositions are said to define a *region*. The demarcation between regions in which adjacent residue curves diverge and terminate at different compositions is called a *separatrix*. Separatrices are related to the existence of azeotropes in the system.

Residue curve maps are being employed in an industrially useful method for the synthesis of separation schemes for azeotropic systems. To a first approximation, the compositions of the overhead and underflow of a single-feed continuous distillation column lie on the same residue curve; therefore, it is not possible for the composition profile of a continuous distillation column to cross into different regions. The boundary that a distillation composition profile can never cross is called a *distillation boundary*. The precise location of distillation boundaries is a function of reflux ratio, but distillation boundaries closely approximate the RCM separatrices. Distillation boundaries connect azeotropic and pure component compositions just as separatrices do. If an RCM separatrix exists, a corresponding distillation boundary will also exist.

Distillation boundaries together with the composition diagram boundaries define *distillation regions* which closely approximate residue curve regions. All pure components and azeotropes in a system lie on the boundaries of such regions. The most volatile composition on the boundary of a distillation region (either a pure component or a minimum-boiling azeotrope) is called the *low-boiling node* (the origin of all residue curves within the region). The least volatile composition on the boundary of a region (either a pure component or a maximum-boiling azeotrope) is called the *high-boiling node* (the terminus of all residue curves in the region). All other pure components and azeotropes on the region boundary are called *saddles* and are never exactly reached by any residue curve. Saddle ternary azeotropes are particularly interesting because they are less obvious to determine experimentally (being neither minimum-boiling nor maximum-boiling), and are not generally recorded in the literature; however, their existence implies distillation boundaries that may have an important impact on separation system design.

The thermodynamic nature of nonideal systems places limitations on the exploitation of volatility to effect separation tasks, and on what may be accomplished with simple distillation. However, if the resolution of a property difference requires that a composition be changed from one distillation region to another, a number of possibilities still exist. In some cases, some new stream may be mixed with the initial stream to produce a composition on the other side of the distillation boundary (although care must be taken to determine where this new stream came from and how it might be continuously regenerated). In other cases, the system thermodynamics and hence the distillation boundary are pressure-sensitive. If the system pressure is changed, a stream initially on one side of a boundary may find itself on the other side of the shifted boundary and in a different distillation region. Finally, other phenomena, unrelated to volatility, may be exploited to change the composition of a stream. One common phenomenon often used in this context is liquid–liquid immiscibility. Others are based on solid–liquid equilibria, such as crystallization and adsorption, and various kinetic phenomena, such as diffusion through membranes.

A. Dehydration of Ethanol

The traditional textbook flowsheet for the dehydration of ethanol involves three distillation columns. However, some industrial implementations for exactly the same separation use only two columns with a corresponding savings in capital cost (Doherty and Knapp, 1993). How can these two alternative designs be systematically understood?

The ethanol–water system has a well known minimum-boiling azeotrope at about 96 wt% ethanol. As ethanol and water are completely miscible in all proportions, the ethanol–water azeotrope in homogeneous. If pure ethanol is desired from a feed richer in water than the azeotropic composition, the azeotrope must be broken.

There is a standard flowsheet pattern for breaking heterogeneous binary azeotropes, but that, of course, is not applicable here. It is possible to break homogeneous azeotropes by pressure shifting the azeotrope or by extractive distillation with a suitable high-boiling solvent, in this case for example, ethylene glycol. Here we wish to consider a third alternative, the use of an entrainer and the exploitation of a heterogeneous ternary azeotrope. A number of specific entrainers will work, including benzene, cyclohexane, and hexane among many others.

Consider a typical entrainer that forms binary minimum-boiling azeotropes with both ethanol and water as well as a minimum-boiling ternary azeotrope. The RCM for such a system indicates that all residue curves originate at the ternary azeotrope (the lowest-boiling composition in the system). However, some residue curves terminate at pure water, some at pure ethanol, and some at pure entrainer, thus forming three distillation regions, with distillation boundaries connecting each of the binary azeotropes with the ternary azeotrope (Fig. 24). The binary feed and the pure water product are in one region, and the pure ethanol product is in another region. Also, as shown in Fig. 25, the ternary system contains a region of liquid–liquid immiscibility that includes the entrainer–water azeotrope and the ternary azeotrope.

The distillation boundary separating the desired ethanol pure product from the feed and the desired water pure product may be called a thermodynamic *critical feature*. In all solutions to this problem, this boundary must be crossed somehow. Overcoming this critical feature is so important that a hierarchical

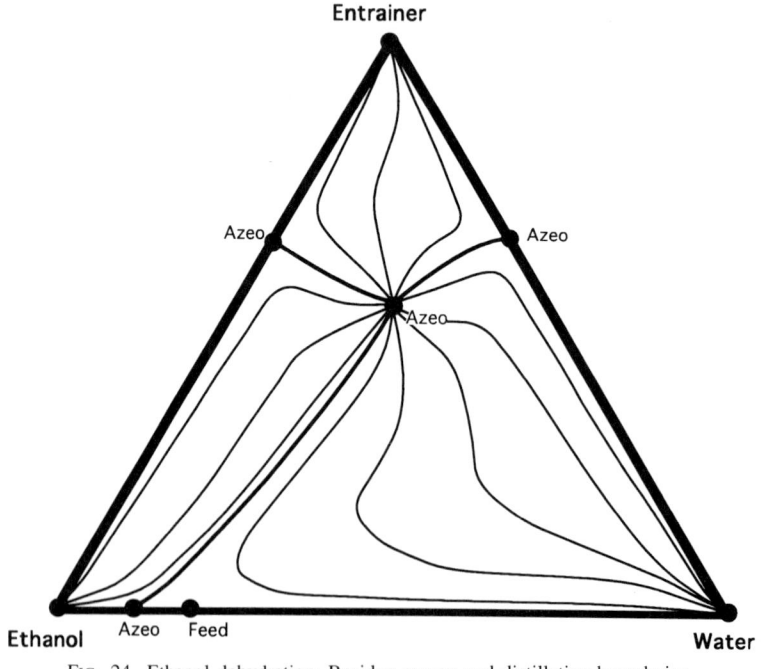

FIG. 24. Ethanol dehydration: Residue curves and distillation boundaries.

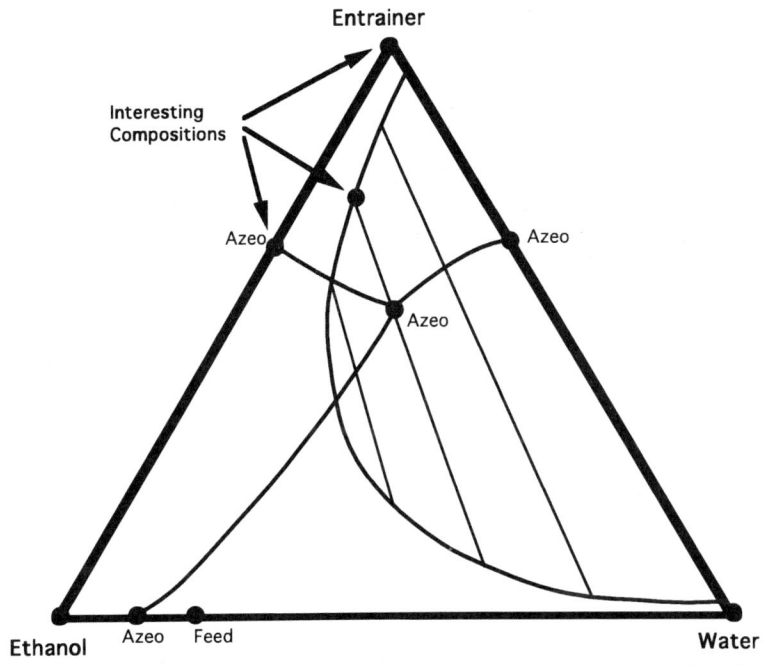

FIG. 25. Ethanol dehydration: Distillation regions and heterogeneous liquid–liquid region.

separations synthesis approach might be followed to concentrate on resolving this strategic problem first, rather than a straightforward opportunistic approach starting from the original feed stream. In other examples, avoiding a tangent pinch might be another type of critical feature that should be addressed first.

Assuming the distillation boundary does not shift significantly with pressure, at least two boundary-crossing methods are possible. The first is exploitation of that portion of the liquid–liquid immiscibility region having tie lines crossing the distillation boundary. The second is mixing with some new stream such that the resulting composition has crossed the boundary. The first method requires that somehow the feed composition be shifted onto one of the desired tie lines. It is not obvious, however, how that might be accomplished.

The second method requires that some composition be chosen so that when mixed with the original feed, the resulting mixture has crossed into the other distillation region. Criteria for selecting such a composition for mixing include the requirement that the composition be easily regenerated, for example, by simple distillation or decantation of a composition that may be generated by distillation. A material recycle is anticipated. From Fig. 25, it is seen that possible candidate compositions for this recycle include the pure entrainer, the ethanol–entrainer azeotrope, and the organic-rich end of the liquid–liquid tie line passing through the ternary azeotrope. Each choice will lead to a different design, which will be affected in part by how difficult the regeneration of the selected composition proves to be.

Let us choose for this example the organic-rich end of the tie line through the ternary azeotrope (as shown later, the ternary azeotrope, being the low-boiling node in all three distillation regions, is readily reached by distillation from any ternary composition in the system). This composition (which we will now call the *azeotropic entrainer*) may easily be mixed with the feed to produce a new composition in the distillation region which includes pure ethanol (Fig. 26). In this region pure ethanol in the highest-boiling composition (high-boiling node), while the ternary azeotrope is the lowest-boiling composition (low-boiling node). In fact, the amount of azeotropic entrainer to be mixed with the feed might just be selected such that the net composition is exactly on the straight line between pure ethanol and the ternary azeotrope.

A distillative separation task with this net feed may then be specified to simultaneously produce a pure ethanol underflow and the ternary azeotrope overhead. This is feasible because the net feed, distillate, and bottoms compositions are collinear by mass balance, and because the distillate and bottoms are on the same residue curve (since all the curves in that region originate at the ternary

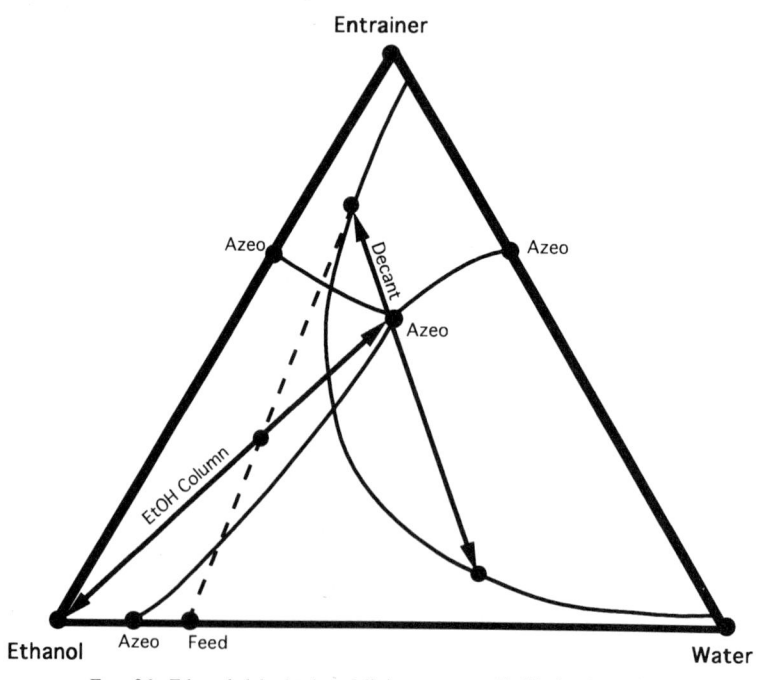

FIG. 26. Ethanol dehydration: Mixing to cross distillation boundary.

azeotrope and terminate at pure ethanol). In the integration of these mixing and separation tasks into actual equipment, the azeotropic entrainer may be directly mixed with the ethanol–water feed and introduced together into the middle of the column as a single feed, which would then be refluxed with condensed distillate in the conventional manner. Alternatively, since the azeotropic entrainer in this case has a lower boiling point than the ethanol–water feed, it may better be introduced separately as a second feed in the upper part of the column. Yet again, depending upon the staging and reflux requirements, the amount of azeotropic entrainer may enough to provide most or even all of the required column reflux at the very top of the column. In this last case, the staging may be adjusted so that the amount of azeotropic entrainer required by mass balance considerations (so that the net feed composition lies directly between pure ethanol and the ternary azeotrope) is also the amount of reflux appropriate for the number of stages to produce the desired distillate and bottoms purities.

Since the overhead ternary azeotrope is heterogeneous, it may be opportunistically decanted, producing the exact composition in the organic layer previously assumed for the azeotropic entrainer as well as an aqueous layer as determined by the liquid–liquid tie line through the ternary azeotrope (Fig. 26). By mass balance it can be shown that the total amount of azeotropic entrainer required to be mixed into the column has not yet been generated by the decant, so the recycle problem is not quite completely solved.

Now the decant aqueous layer must be processed. In any distillation region a number of overhead and underflow compositions are achievable from a given feed through simple single-feed distillative separation depending on the staging, reflux, distillate-to-feed ratio, and constraints of the region boundaries (Fig. 27). The determination of exactly all reachable compositions by distillative separation of a given feed is somewhat complex (Wahnschafft *et al.*, 1992). However, in practice, just two alternatives are generally of most interest: production of the most volatile composition in the region (the low-boiling node) as distillate (with the bottoms determined by mass balance constrained by the region boundaries), or production of the least volatile composition in the region (the high-boiling node) as bottoms (with the distillate determined by mass balance constrained by the region boundaries). Unless the feed, high-boiling node, and low-boiling node are collinear (as they were in the previous column and as they also are for binary systems), these two different choices lead to two different separation flowsheet alternatives.

By far, the most common prejudice (sometimes completely overlooking the other alternative) is to propose the most volatile composition (low-boiling node) as distillate (Fig. 28). In this region containing the decant aqueous layer, the lightest composition is the ternary azeotrope. With enough stages, all of the azeotropic composition in the decant aqueous layer may be recovered, and the underflow will contain only a binary ethanol–water mixture. If the distillate of

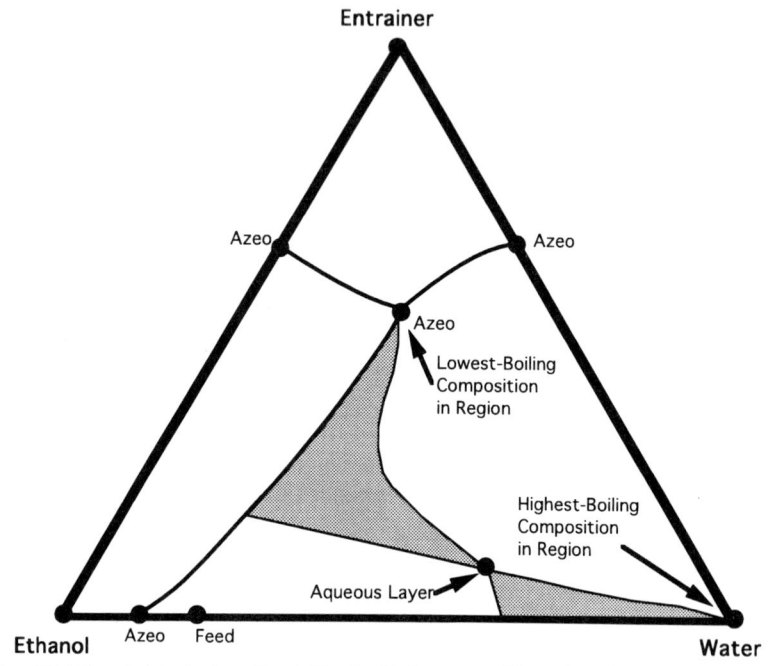

FIG. 27. Ethanol dehydration: Reachable distillative compositions from decant aqueous layer.

this second distillative separations task, whose composition is nearly identical to the distillate from the first column, is recycled to the decant task, mass balance now indicates that the full amount of azeotropic entrainer assumed to be available for the boundary-crossing mixing with the original feed has now been regenerated.

As a final step, the binary ethanol–water underflow from the second column may be opportunistically separated to produce a pure water underflow and a composition close to the original feed to which it also may be recycled. The amount of extractive entrainer must now be readjusted because of the increased amount of feed to the first column, but iteration shows that the flowsheet structure generated remains unchanged and has converged. The tasks specified accomplish the composition goals, producing pure water and pure ethanol. The entrainer remains totally within the system. The subsequent integration into equipment involves three columns and one decanter (Fig. 29).

Returning for a moment to the decant aqueous layer, the other separation alternative is to produce the least volatile composition in the region (water, the high-boiling node) as bottoms. Given enough staging, an amount of water can be removed such that the overhead composition is constrained by the distillation region boundary. This may appear not to be useful, since the resulting overhead

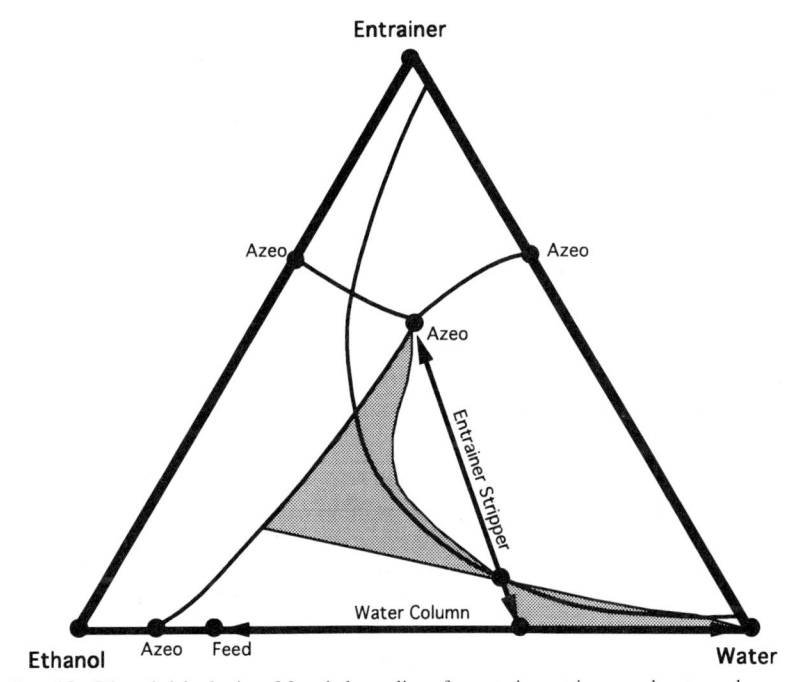

INDUSTRIAL APPLICATIONS OF CHEMICAL PROCESS SYNTHESIS   47

FIG. 28. Ethanol dehydration: Mass balance lines for entrainer stripper and water column.

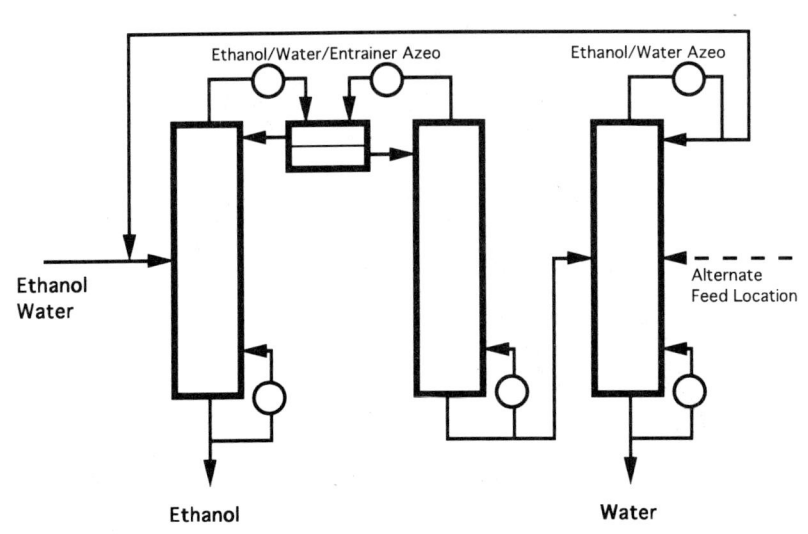

FIG. 29. Ethanol dehydration: Three-column flowsheet.

composition contains entrainer, ethanol, and water and can be separated no further. However, in this case, the composition of this overhead is similar to the mass balance line of the first mixing task, and thus could be recycled to the first column (Fig. 30). This changes the composition and amount of that overall feed from what was originally assumed when that part of the flowsheet was first addressed, but iteration confirms that the structure of the design is not affected. Further, analysis confirms that this alternative flowsheet again produces pure ethanol and pure water and completely recovers all entrainer, but requires only two columns (Fig. 31). Both flowsheets in this example have been implemented industrially, although this second two-column alternative saves about 20% in capital costs (although the energy costs are approximately equivalent). In this example, the original feed illustrated was fairly close to the ethanol–water azeotrope. It turns out that both alternative flowsheets synthesized in this example dehydrate ethanol–water mixtures of any composition, although the feed may have to be introduced more appropriately into a different column than shown.

In the preceding example, the feed composition in the first column was adjusted such that simultaneously the most volatile composition in the region was distillate while the least volatile composition was bottoms, thereby reducing the distillative separation task to a single alternative (with apparently maximum

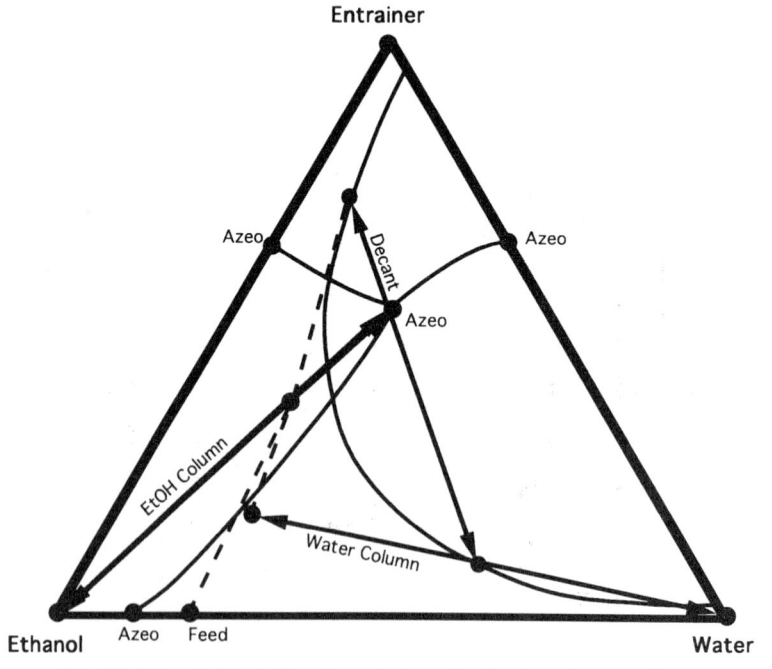

FIG. 30. Ethanol dehydration: Second alternative for decant aqueous layer.

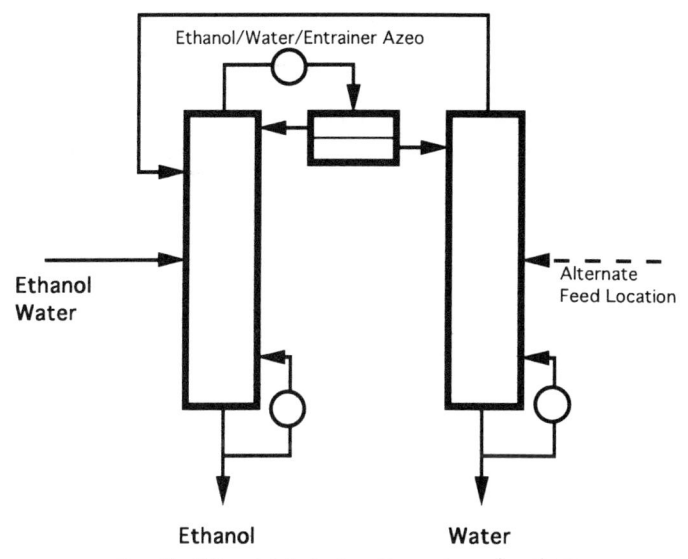

INDUSTRIAL APPLICATIONS OF CHEMICAL PROCESS SYNTHESIS 49

Fig. 31. Ethanol dehydration: Two-column flowsheet.

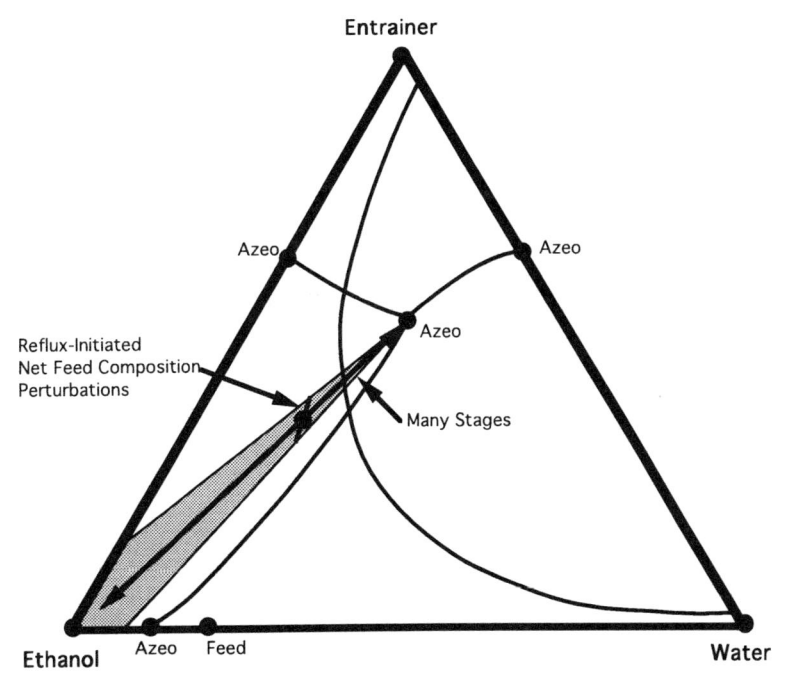

Fig. 32. Ethanol dehydration: Failure to meet ethanol product specification with excessive stages in upper part of column.

separation potential, that is, maximum volatility difference and maximum temperature difference across the column). For multicomponent systems, this is not necessarily as clever as it may appear. Since all residue curves in the region originate at the desired overhead composition and terminate at the desired underflow composition, many designs very near the desired one will be feasible, and slight perturbations in reflux conditions might result in operation associated with very different residue curves. This can be a particular problem in designs that reflux compositions different from the overhead (as in this case), as slight changes in reflux lead to overall feed composition shifts in a direction orthogonal to the column mass balance line (which does not happen with columns that reboil the underflow and reflux the overhead). In this case, slight shifts in the mass balance line can result in major shifts in the actual column composition profile, so much so that the major contaminant in the ethanol product may change from water to entrainer or vice versa. If there are excessive stages in the top of the column, the distillate composition will be pegged to the ternary azeotrope. If there are reflux perturbations that result in overall feed composition shifts, the bottoms composition may swing from the ethanol–water face of the composition triangle to the ethanol–entrainer face, and the ethanol purity specification may never be met (Fig. 32). It may appear that the problem can be resolved by relatively increasing the staging in the bottom of the column, pegging the bottoms composition to pure ethanol and letting the top composition float along the distillation region boundaries converging on the ternary azeotrope. As long as the overhead composition remains in the heterogeneous region, the flowsheet pretty much works as designed (Fig. 33). However, even in such a case, if the mass balance line moves from one side of the ternary azeotrope to another, the column composition profile will also shift dramatically and at the bottom of the column will approach the ethanol vertex from one side of the composition triangle or the other. Although the ethanol purity specification may still be met, the identity of the next plentiful contaminant will change from water to entrainer or vice versa, which may be entirely unacceptable (Fig. 34).

B. Separations System Synthesis Method for Nonideal Mixtures

In general, the steps of this separations system synthesis method for nonideal mixtures involving azeotropes include examination of the RCM representation (overlaid with vapor–liquid equilibria (VLE) pinch information, liquid–liquid equilibria (LLE) binodal curves and tie lines, and solid–liquid equilibria (SLE) phase diagrams if appropriate); determination of the critical thermodynamic features to be avoided (e.g., pinched regions), overcome (e.g., necessary distillation

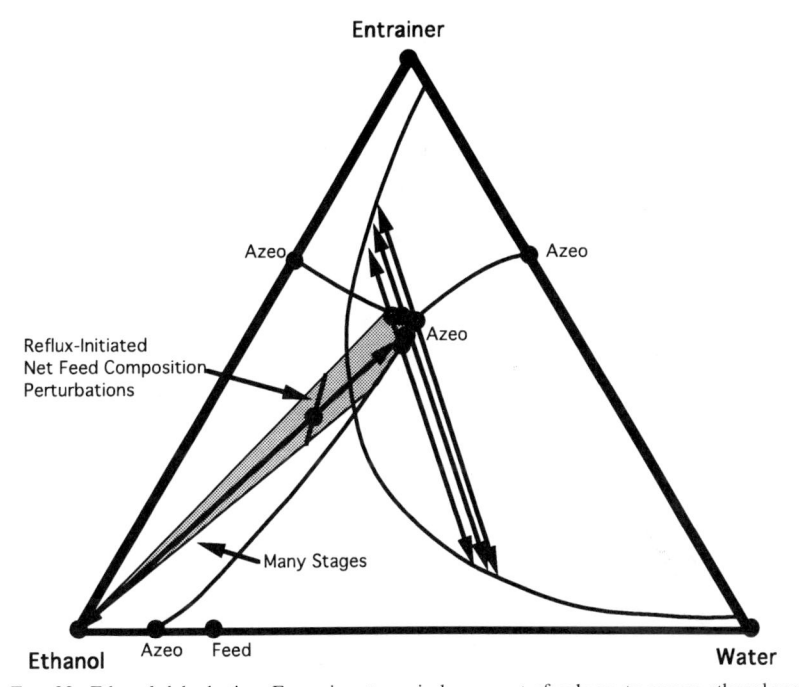

FIG. 33. Ethanol dehydration: Excessive stages in lower part of column to ensure ethanol product purity.

boundary crossings), or exploited (e.g., decant, extraction, or crystallization opportunities); strategically addressing issues raised by the critical features first; pursuing multiple interesting compositions (mass separation agents and other compositions which are useful for mixing but which must be regenerated and recycled); opportunistic resolution of remaining concentration property differences (with the desired products or unanchored critical feature strategic tasks, guided by constraints represented by the RCM); pursuing both possible distillative alternatives when different (heaviest possible underflow in addition to lightest possible overhead); wariness of reflux (reboil) composition different from the distillate (bottoms); avoiding separating multicomponent mixtures simultaneously into lightest possible overhead and heaviest possible underflow; recycling opportunistically to similar compositions already processed; analyzing anticipated and opportunistic recycle feasibility (particularly mass balances); elimination of redundant or duplicate tasks if any; and final integration of tasks into actual equipment. These steps are briefly outlined in Fig. 35.

There are numerous other features of the separations system synthesis method based on the RCM representation that space permits only brief mention. Residue

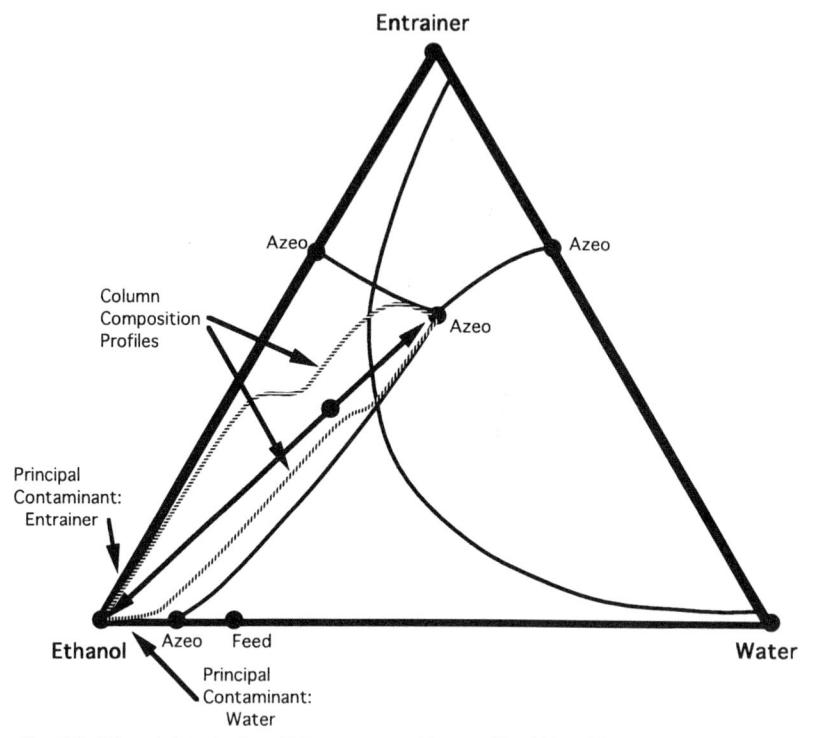

FIG. 34. Ethanol dehydration: Column composition profile shifts with small changes in organic phase reflux.

curves never quite pass through saddle compositions. As a result, the reachable composition analysis for a distillative separation usually does not include saddles; and in the systems where it comes acceptably close, the required distillate-to-feed ratio goes through a minimum at the saddle, resulting in difficult control and potential multiple solutions. Saddle products can also be reached by extractive distillation, a unique form of two-feed distillative separation in which a less volatile solvent is introduced above a more volatile feed with the objective of altering the liquid phase activity coefficients in the upper part of the column ways not otherwise possible in single-feed columns. (Azeotropic columns also typically introduce the entrainer above the feed, but they would also work if the entrainer were introduced with the feed; extractive columns will not work if the solvent is introduced only with the feed.) The behavior of extractive distillation does not follow the usual rules to determine reachable compositions. However, which saddle will come overhead in extractive distillation (generally two are possible) can be determined from the shape of the residue curves in the vicinity of the feed composition. Yet another way to reach a saddle composition is to

- Define Goal Specifications
- Construct RCM Representation
- Identify Critical Features
- Identify Interesting Compositions
- Address Critical Features First
- Resolve Remaining Differences Opportunistically
- Pursue All Resolution Alternatives
- Pursue Both Distillative Separation Alternatives
- Avoid Separating Ternary Mixtures Simultaneously into Lightest Overhead and Heaviest Underflow
- Beware of Reflux Composition Different from Overhead
- Recycle Opportunistically
- Check Recycle Mass Balance
- Special Consideration for Pressure-Shifting Boundaries, Exploiting Boundary Curvature, Reaching Saddle Compositions, and Solvent-Enhanced Separations
- Eliminate Redundant Tasks
- Integrate Tasks into Equipment
- Evaluate Each Design Alternative

FIG. 35. Separations system synthesis method.

reduce the number of components in the system. Compositions are identified as nodes and saddles, depending on the boiling points of all of the components and azeotropes in the system (Barnicki and Siirola, 1996). A pure component that is a saddle vertex of a ternary distillation region will be a node in either binary system that does not contain the other third component. Removing a component may turn saddles into nodes in the resulting reduced-order system, which might then be reached by distillation.

Although it is not possible to cross a distillation boundary in a column, the actual restriction is that the entire composition profile be in a single distillation region. Since the column feed need not necessarily (and in general does not) lie on the composition profile (it only needs to be collinear with the distillate and bottoms compositions), it is possible to operate a column in a region with a sufficiently curved boundary, even though the feed composition is in the adjacent region on the concave side of the boundary. Exploitation of distillation-boundary curvature is similar to exploiting pressure shifting of boundaries. Such schemes generally involve a great deal of recycle for mass balance and turn out not to be energetically very practical for boundaries that involve minimum-boiling azeotropes (because the large recycles are all taken overhead), but have been exploited industrially to break maximum-boiling azeotropes (where the recycles are underflows, as in the well-known example of breaking the nitric acid–water azeotrope with sulfuric acid). In other systems in which boundaries terminate in a pure component vertex, curvature can also be exploited not to break an azeotrope, but to produce an acceptably pure product (analogous to pressure-shifting an azeotrope almost to one pure component).

There are, in addition, techniques for avoiding pinched regions, avoiding solidification in distillative separations, and for exploiting crystallization and other SLE and LLE phenomena as well as kinetic-based phenomena. Residue curves with inflections, common near heterogeneous azeotropes, can lead to multiple solutions for some distillation designs, which can result in control problems. RCMs have also been useful for the evaluation of the consistency of solution thermodynamic data and for determining the impact of potential (often hidden) saddle ternary azeotropes. They are also useful in troubleshooting column malfunctions, operational sensitivities, control problems, and even simulation convergence problems. Much RCM analysis and distillation column design technology was developed in the experimental MAYFLOWER system, and is becoming commercially available in such specialized process synthesis software as HYCON and SPLIT.

C. Production of Diethoxymethane

As one last example of the separations system synthesis method, a process was being developed to produce the chemical diethoxymethane (DEM), the ethanol acetal of formaldehyde. The raw materials are paraformaldehyde and ethanol, and the reaction by-produces water. Diethoxymethane, ethanol, and water form a homogeneous ternary azeotrope that is readily removed from the reaction mass (the paraformaldehyde being relatively nonvolatile). The species allocation is such that DEM is the desired product, water is waste, and ethanol is to be recycled to the reactor. A conceptual separations system to produce pure DEM, pure water, and pure ethanol from the reactor effluent is to be developed.

The problem here is that the ternary azeotrope is homogeneous. Using traditional patterns, one might consider adding yet another component to the system to form some heterogeneous ternary or possibly quaternary azeotrope that can be broken, similar to the breaking of the homogeneous ethanol–water azeotrope of the previous example. Such a system turns out to require four distillation columns. Can an alternative system not requiring the use of an external component be devised?

The DEM–ethanol–water system has three binary minimum-boiling azeotropes and one minimum-boiling ternary azeotrope, and an RCM (Fig. 36) similar to the previous ethanol–water–entrainer example. Distillation boundaries connect the ternary azeotrope with each binary azeotrope. The desired products, DEM, ethanol, and water, are each high-boiling nodes in their respective distillation regions. The system does exhibit a small region of immiscibility (Fig. 37), but both the reaction mass and the ternary azeotrope are in the single-phase region.

The reaction mass is in the same distillation region as water. Since DEM is in a different region, the critical feature distillation boundary between these two

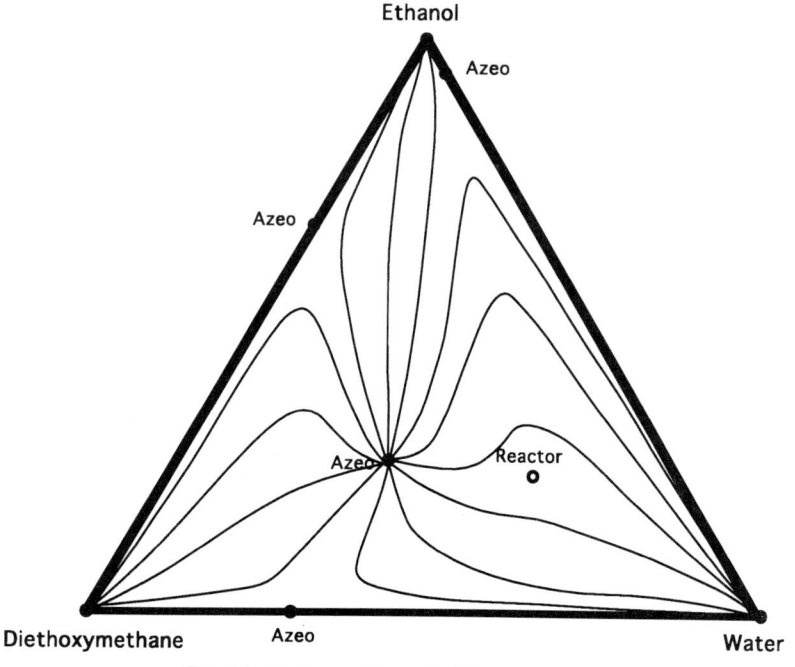

FIG. 36. Diethoxymethane: Residue curve map.

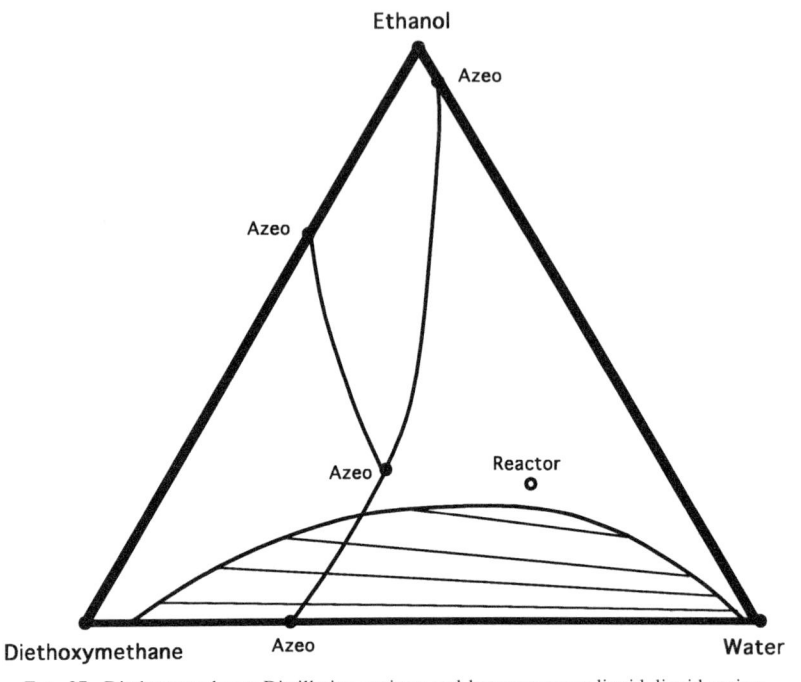

FIG. 37. Diethoxymethane: Distillation regions and heterogeneous liquid–liquid region.

regions must be crossed. Attacking this strategic problem first, it is noticed that some liquid–liquid tie lines span the distillation boundary. If the composition from the reaction can be moved into the lower part of the two-phase region, the distillation boundary may be crossed.

Two possible interesting compositions exist for mixing to a composition in the heterogeneous region: the DEM product and the water product. The DEM choice might be preferable for direct mixing with the reaction mass. The water choice is possible if the composition from the reactor is first somewhat concentrated in DEM (otherwise, two phases will not result). Following the second option (which has the additional benefit of retaining the nonvolatile paraformaldehyde in the reactor), again two options are possible for a first opportunistic distillative separation to concentrate DEM from the reaction mass: distill the lowest-boiling composition (the ternary azeotrope) overhead, producing a DEM-free ethanol–water mixture as an underflow, or distill the highest-boiling composition (water) as underflow, producing a composition constrained by a distillation boundary overhead. The former is chosen for the first design. The ternary azeotropic composition so produced may now be mixed with a portion of the water product. The result is two liquid phases which split into an organic-rich phase layer in the new distillation region, while the water-rich layer has a composition close to the underflow of the first column (Fig. 38).

The organic-rich layer may be opportunistically distilled to produce either the lowest-boiling ternary azeotrope overhead and a DEM–water heterogeneous mixture as underflow, or the highest-boiling DEM product as bottoms with a composition constrained by the distillation boundary close to the ternary azeotrope overhead. The latter is chosen first, producing the desired DEM product and an overhead composition close enough to be opportunistically recycled with the ternary azeo to the decant task.

The water-rich layer and the first column underflow may be combined and distilled. Again, either the lowest-boiling ternary azeotrope may be taken overhead, or the highest-boiling product water taken as underflow. If the latter is chosen, the overhead composition will lie on the distillation boundary close to the ethanol–water azeotrope (Fig. 38).

Two of the three composition goals, pure DEM and pure water, have been met; and by mass balance, sufficient water has been regenerated to serve as the mixing agent for dragging the ternary azeotrope into the two-phase region (actually, the extra water is first put in the system, then continually recycled as is the entrainer in the previous example). A third composition near the ethanol–

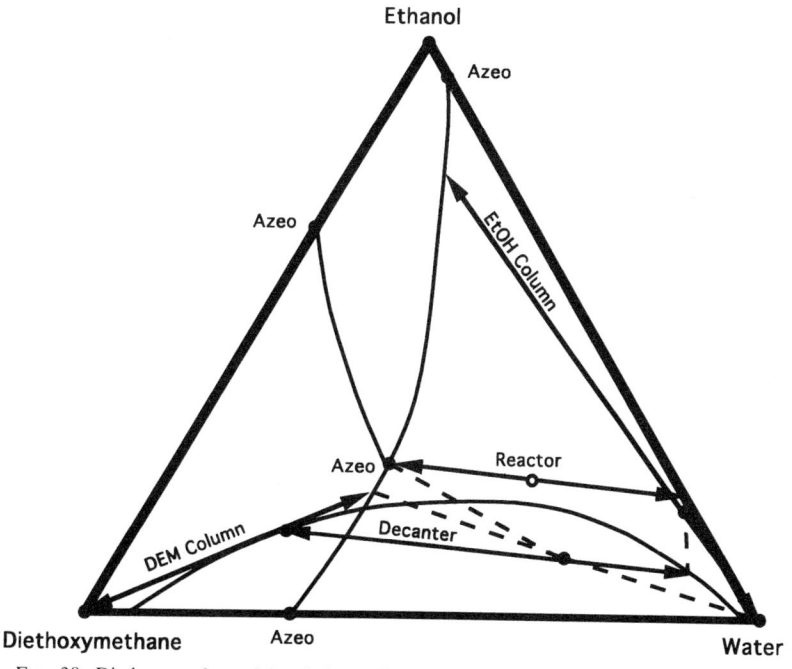

FIG. 38. Diethoxymethane: Mass balance lines plotted on distillation boundary diagram.

water azeotrope has also been generated, but this does not meet the original product specification of pure ethanol.

Getting pure ethanol will be difficult. Ethanol is in the third distillation region, and no obvious means exist to cross the relevant distillation boundaries. A new entrainer might be added, as was the case for the previous example. However, since the ethanol is only going to be recycled to the reactor, the required purity specification might be questioned. What is the effect of small amounts of contaminants DEM and water in the recycled ethanol? The acetal formation reaction is equilibrium-controlled, so there may be some deleterious effect of including products with a reactant. However, additional analysis indicated that the effect is small, and so the ethanol composition requirement was relaxed and the overhead composition recycled to the reactor directly as produced.

Following each of the other design alternatives identified above also leads to feasible solutions, but all involve more separation tasks. The tasks for the present case may be integrated among themselves and with the reaction task in a number of different ways, some of which involve fewer columns but passing more water through the reaction mass.

The conceptual design actually implemented industrially involved three columns and a decanter in addition to the reactor (Fig. 39). The savings over the straightforward scheme to form a heterogeneous azeotrope with an external entrainer was approximately 30% reduced capital and 30% reduced energy. In later refinement of the conceptual design, an extractor was substituted for the decanter, requiring even less water for mixing. The process was patented and the RCM representation was used in the patent application to help explain the process, particularly the method for breaking the homogeneous ternary azeotrope without the addition of an external component (Martin and Raynolds, 1988).

VII. Challenges for the Future

Many of the methods and tools being developed in the conceptual process engineering research community are still more concerned with analysis (targeting, representation, operability, etc.) than with synthesis of alternatives. More effort needs to be directed specifically to alternative generation for all conceptual design stages at all levels of detail.

Most of the applications discussed here were based on algorithmic systematic generation approaches to process synthesis. There also exist a number of heuristic rule-based or expert-systems–based approaches including PIP for total flowsheet synthesis, SPLIT for separation schemes (Wahnschafft et al., 1991),

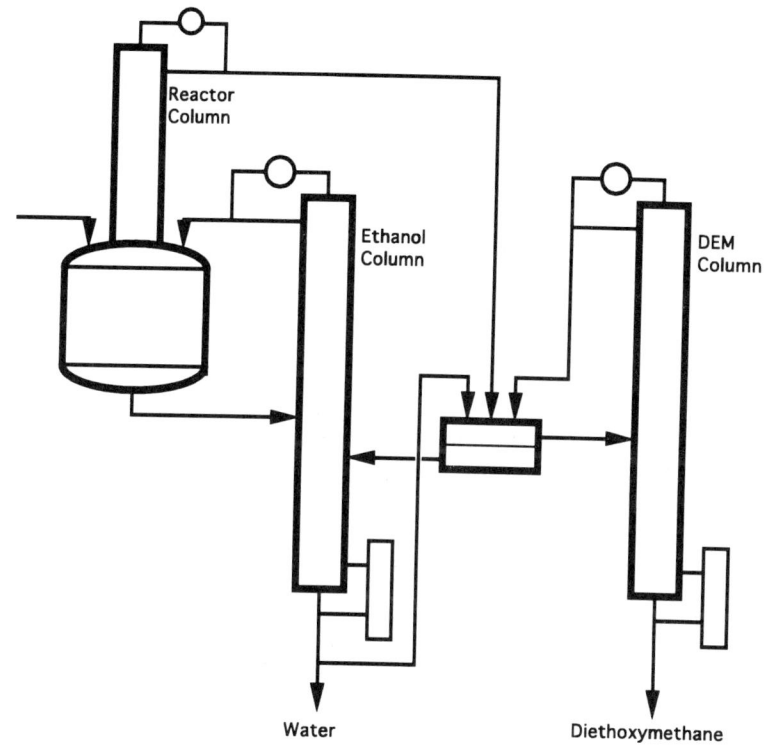

FIG. 39. Diethoxymethane: Process flowsheet.

and most of the informal methods used by conceptual design engineers. Because of combinatorial difficulties, neither the algorithmic nor the heuristic methods carry any guarantees of optimality.

The superstructure optimization approach, on the other hand, does hold out the promise of structural optimality. Given a superstructure, however, there are significant challenges remaining to improve computational efficiency to overcome combinatorial difficulties and to develop global optimization strategies for the generally nonconvex mixed-integer nonlinear problems involved. Steady progress continues to be made on these fronts (Grossmann and Daichendt, 1994). There still remains, however, the general problem of generating appropriate superstructures in the first place.

One possibility might be to employ algorithmic and/or heuristic systematic generation methods not for the synthesis of a single alternative, but for the synthesis of an entire superstructure of alternatives, for example, by following and including all alternatives encountered at all decision points during the procedure. The resulting superstructure could still depend upon the order in which various differences were attacked. Therefore, this superstructure might be subjected to evolutionary modification methods, possibly to trim away demonstrably inferior segments, but more likely to add patterns to one part of the superstructure similar to that generated for other parts. Finally, the superstructure so generated and refined could be in principle optimized to produce structurally and parametrically superior solutions.

However, no matter how completely the objective function for the optimization is formulated, there will always be additional important design criteria not included in the optimization. Examples include safety, controllability, flexibility, constructibility, maintainability, and many, many others. On these other criteria the several best solutions selected by the optimization procedure might be critiqued, possibly subjectively, before a final design is selected. The process synthesis paradigm of the future might very well then combine all of the existing approaches (systematic generation, evolutionary modification, and structural optimization), exploiting all of the existing methods (algorithmic, heuristic, and mathematical programming). The result would be a new process synthesis strategy that might be called *generate–evolve–optimize–critique* (Fig. 40).

GENERATE — Algorithmic and/or Heuristic Systematic Generation of Flowsheet Superstructure

EVOLVE — Algorithmic and/or Heuristic Evolutionary Modification of Flowsheet Superstructure

OPTIMIZE — Simultaneous Mathematical Optimization of both Structural and Design Parameters

CRITIQUE — Evaluate Best Flowsheet Alternatives on the Basis of Criteria not in the Optimization Objective Function

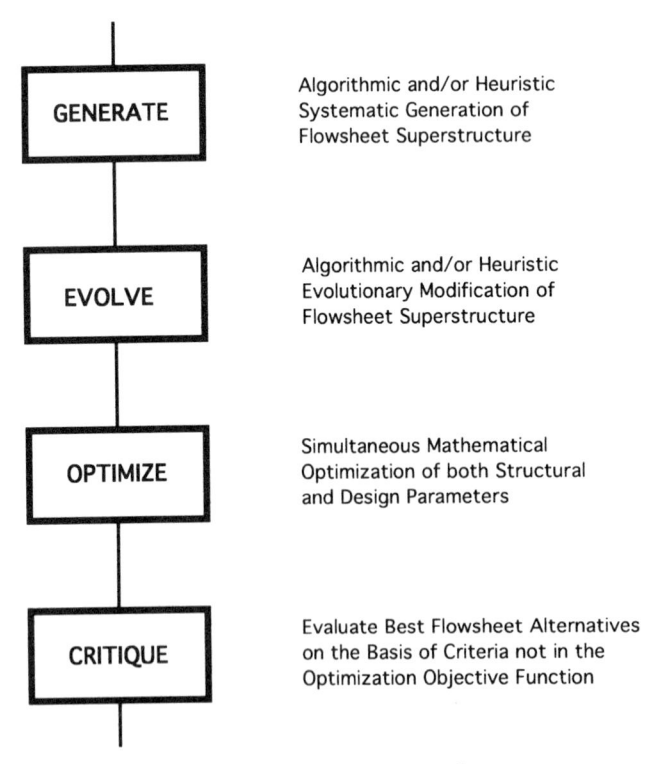

FIG. 40. Future process synthesis paradigm.

VIII. Conclusions

Systematic approaches, both to the material-making innovation process and to the process synthesis steps of the conceptual engineering stages within that innovation process, have begun to have measurable industrial impact. With these techniques—some automated and some not—an increased number of higher value, lower energy, lower environmental impact, and sometimes even novel design alternatives have been synthesized and actually implemented. Energy reductions of 50% and net present cost reductions of 35% are typically achievable.

Certain features of systematic approaches to process synthesis appear to have special merit. These include architectures that are hierarchical in scope and iterative in level of detail, targeting as a useful bounding exercise, means–ends analysis as a recursive problem-solving paradigm with hierarchical strategic as well as opportunistic goals implemented by an iterative formulation–synthesis–analysis–evaluation design strategy, thinking specifically in terms of tasks to be accomplished before equipment to be employed, thinking of nonconventional properties to be exploited to accomplish identified tasks, recognizing the importance of representations to encapsulate analysis within synthesis activities, and exploring the advantages of solving related synthesis problems iteratively or even simultaneously.

Advances in problem formulation and in computational hardware and software capability offer the promise of a new generation of practical process synthesis techniques based directly on structural optimization. Soon the goal of synthesizing provably unbeatable conceptual process flowsheets may be at hand.

Acknowledgments

I would like to thank my colleagues S. D. Barnicki, R. D. Colberg, D. L. Martin, E. J. Peterson, and M. R. Shelton for their help with the preparation of this manuscript. I would also like to thank Eastman Chemical Company for allowing these experiences to be shared.

References

Agreda, V. H., and Partin, L. R. U.S. Patent 4,435,595 (to Eastman Kodak Company) (1984).
Barnicki, S. D., and Siirola, J. J. "Enhanced Distillation," *in* "Perry's Chemical Engineering Handbook," 7th ed. McGraw-Hill, New York, 1996.

Blakely, D. M. "Cost Savings in Binary Distillation through Two-Column Designs," M. S. Dissertation, Clemson University, Clemson, SC (1984).
Doherty, M. F., and Knapp, J. P. "Azeotropic and Extractive Distillation," *Kirk-Othmer Encycl. Chem. Technol. 4th Ed.* **8**, 358 (1993).
Douglas, J. M. "Conceptual Design of Chemical Processes." McGraw-Hill, New York, 1988.
Foucher, E. R., Doherty, M. F., and Malone, M. F. "Automatic Screening of Entrainers for Homogeneous Azeotropic Distillation," *Ind. Eng. Chem. Res.* **30**, 760 (1991).
Grossmann, I. E., and Daichendt, M. M. "New Trends in Optimization-Based Approaches to Process Synthesis," *Comput. Chem. Eng.* (1996).
Kirkwood, R. L., Locke, M. H., and Douglas, J. M. "A Prototype Expert System for Synthesizing Chemical Process Flowsheets," *Comput. Chem. Eng.* **12**, 329 (1988).
Linnhoff, B., and Ahmad, S. "Towards Total Process Synthesis," Paper 26d, AIChE Meeting, Washington, DC (1983).
Linnhoff, B., and Hindmarsh, E. "The Pinch Design Method of Heat Exchanger Networks," *Chem. Eng. Sci.* **38**, 745 (1983).
Lodal, P. N. "Production Economics," *in* "Acetic Acid and its Derivatives" (V. H. Agreda and J. R. Zoeller, eds.), p. 61. Dekker, New York, 1993.
Mahalec, V., and Motard, R. L. "Procedures for the Initial Design of Chemical Processing Systems," *Comput. Chem. Eng.* **1**, 57 (1977).
Martin, D. L., and Raynolds, P. W. U.S. Patent 4,740,273 (to Eastman Kodak Company) (1988).
Rudd, D. F., Powers, G. J., and Siirola, J. J. "Process Synthesis." Prentice-Hall, Englewood Cliffs, NJ, 1973.
Seader, J. D., and Westerberg, A. W. "A Combined Heuristic and Evolutionary Strategy for Synthesis of Simple Separation Sequences," *AIChE J.* **23**, 951 (1977).
Siirola, J. J. "The Computer-Aided Synthesis of Chemical Process Designs," Ph.D. Dissertation, University of Wisconsin, Madison (1970).
Siirola, J. J. "Status of Heat Exchange Network Synthesis," Paper 42a, AIChE Meeting, Tulsa, OK (1974).
Siirola, J. J. "Energy Integration in Separation Processes," *in* "Foundations of Computer-Aided Process Design" (R. S. H. Mah and W. D. Seider, eds.), vol. 2, p. 573. Engineering Foundation, New York, 1981.
Siirola, J. J. "An Industrial Perspective on Process Synthesis," *in* "Foundations of Computer-Aided Process Design" (M. F. Doherty and L. T. Biegler, eds.), p. 222. Cache Corporation, Austin, TX, 1995.
Siirola, J. J., and Rudd, D. F. "Computer-Aided Synthesis of Chemical Process Designs," *Ind. Eng. Chem. Fundam.* **10**, 353 (1971).
Simon, H. A. "Science of the Artificial." MIT Press, Cambridge, MA, 1969.
Wahnschafft, O. M., Jurian, T. P., and Westerberg, A. W. "SPLIT: A Separation Process Designer," *Comput. Chem. Eng.* **15**, 565 (1991).
Wahnschafft, O. M., Koehler, J. W., Blass, E., and Westerberg, A. W. "The Product Regions of Single-Feed Azeotropic Distillation Columns," *Ind. Eng. Chem. Res.* **31**, 2345 (1992).
Witzeman, J. S., and Agreda, V. H. "Alcohol Acetates," *in* "Acetic Acid and its Derivatives" (V. H. Agreda and J. R. Zoeller, eds.), p. 257. Dekker, New York, 1993.

SYNTHESIS OF DISTILLATION-BASED SEPARATION SYSTEMS

Arthur W. Westerberg and Oliver Wahnschafft*

Department of Chemical Engineering and the Engineering Design Research Center
Carnegie Mellon University
Pittsburgh, Pennsylvania

I. Introduction	64
II. The Richness of the Solution Space	66
III. Assessing the Behavior of a Mixture	69
A. Azeotropic Behavior: Infinite-Dilution K-Values	69
B. Liquid/Liquid Behavior: Infinite-Dilution Activity Coefficients	73
IV. Separating Nearly Ideal Systems	75
A. Analysis	75
B. System Synthesis for Nearly Ideal Systems	80
V. Separating Highly Nonideal Mixtures	90
A. Azeotropic Separation: Example 1	91
VI. Synthesis Discussion	94
A. Analysis-Driven Synthesis	94
B. Impact of Number of Species on Representation, Analysis, and Synthesis Methods	94
C. In Summary	97
VII. Pre-analysis Methods	98
A. Equilibrium-Phase Behavior	98
B. Distillation Column Behavior	105
VIII. Synthesis Method for Nonideal Mixtures	107
A. Azeotropic Separation: Example 2	108
B. Azeotropic Separation: Example 3	121
IX. More Advanced Pre-analysis Methods	131
A. Species Behavior	131
B. Limiting Simple Distillation Column Behavior	140
C. Extractive Distillation	157
X. Post-analysis Methods: Column Design Calculations	166
Acknowledgments	167
References	167

*Currently as AspenTech, Cambridge, Massachusetts.

This tutorial paper is a review of recent advances in the synthesis of ideal and nonideal distillation-based separation systems. We start by showing that the space of alternative separation processes is enormous. We discuss simple methods to classify a mixture either as nearly ideal or as nonideal, in which case it displays azeotropic and possibly liquid/liquid behavior.

For nearly ideal mixtures, insights based on marginal vapor flows permit the development of a simple screening criterion computed using only relative volatilities and component feed flowrates to find the better column sequences from among the many possible. This criterion explains several of the traditional heuristics.

We ask how one can invent alternative structures to separate nonideal mixtures. We present and illustrate an approach with three examples: separating n-*butanol and water; separating acetone, chloroform, and benzene; and separating* n-*pentane, acetone, methanol, and water. We find that these processes always contain recycles because we are unable to obtain the sharp separations possible for ideal mixtures.*

Next, we explore more advanced methods to assess the behavior of complex mixtures. We discuss two algorithms to find all azeotropes for a mixture; we also discuss the problem of finding the regions for liquid/liquid behavior.

Example problems are included to highlight the need to estimate the entire set of products that can be reached for a given feed when using a particular type of separation unit. We show that readily computed distillation curves and pinch point curves allow us to identify the entire reachable region for simple and extractive distillation for ternary mixtures. This analysis proves that finite reflux often permits increased separation; we can compute exactly how far we can cross so-called "distillation boundaries." For extractive distillation, we illustrate how to find minimum solvent rates, minimum reflux ratios, and, interestingly, maximum reflux ratios.

I. Introduction

The goal of this paper is to discuss a methodology for the *preliminary design* of separation processes for liquid mixtures using distillation- and extraction-based technologies. We define the preliminary design step to be the one in which we *discover* the alternative overall system structures that might be reasonable for solving the problem. Choosing which alternative to use among those proposed in this step and adjusting the operating conditions to their best values is a follow-up step not covered by the preliminary design step; it will not be covered here to any significant extent. If the vapor/liquid equilibrium behavior of the species is reasonably ideal, we shall, however, present a simple method to screen among the distillation-based alternatives to find the likely better ones. The selection of the better sequences for relatively ideal mixtures is the subject of several past reviews (Westerberg, 1980, 1985; Nishida *et al.,* 1981; Hlavacek, 1978; Hendry *et al.,* 1973).

We use the word "discover" in the previous paragraph very carefully. In artificial intelligence the discovery step is the one in which we identify the building blocks we have to use when solving a problem. For well behaved mixtures, discovery is simple: we can quickly sketch likely alternatives built from such building blocks. It is the search among those alternatives—of which there can be a large number—that is the problem. Often, especially when the species display azeotropic behavior, the problem is to discover which types of separation steps can be used—a process that requires us to conjecture a method and then to carry out significant computer calculations (e.g., column or flash unit simulations) to find if the conjectured method is useful. Proposing any solution for these harder problems, much less enumerating a number of alternatives, is a difficult task.

The following illustrates the type of problem we would like to be able to solve:

Design a separation process to split a (liquid) mixture of 25% methanol, 40% water, and 35% ethanol into the three relatively pure products of methanol, water, and ethanol.

This particular problem is not a simple one to solve because water and ethanol form an azeotrope. In this paper, we concentrate first on selecting the better alternative processes for ideal (or near ideal) mixtures, and then we present a prototypical method for generating alternatives for nonideal mixtures such as the one above.

Distillation processes are large consumers of hot and cold utilities. It is often useful to consider their heat integration (Rathore *et al.,* 1974; Andrecovich and Westerberg, 1985) where the heat expelled from one column supplies part or all of the heat needed by another. While important, we shall not consider such issues here.

This paper assumes the reader is familiar with the standard textbook presentation for staged processes such as that in McCabe and Smith (1976). More extensive texts on distillation include King (1980), Henley and Seader (1981), and Holland (1981). We shall build on the standard background to develop needed insights for designing separation systems and to understand less conventional single-unit configurations which are often a part of such designs.

II. The Richness of the Solution Space

As a first step in design, we must be aware that many more alternatives may be available for solving a problem than at first seem likely (Westerberg, 1985), even for well-behaved mixtures. To confirm this statement, we show that many alternative structures are available first for a single distillation step and then for a system of steps. The intent of this section is simply to expand our thinking about this problem.

We start by sketching a simple two-product distillation column, as shown in Fig. 1. This column is the unit operation most likely to be considered as a building block for these problems. The two-product distillation column splits its feed into two essentially disjoint product sets. The distillate contains species A, B, and C (plus, of course, a small amount of D and generally only traces of the other species) while the bottom product contains predominantly D and E. Heat is injected at the bottom into the reboiler, the hottest point in the column, and removed from the condenser at the top, the coldest point in the column. Liquid flows downward against a flow of vapor upward, with the more volatile species enriching as one moves up the column.

Typically, we put a specification onto such a column of the following form: recover 99.8% of the species C in the distillate while recovering 99% of species D in the bottoms. From a practical point of view, then, the *light* and *heavy keys* for this split are C and D, respectively. In this case, 0.2% of C will make it to the bottom product and will be the lightest species to make it to the bottom in more than trace amounts. Similarly, 1% of D will make it to the top along with at most a trace of E.

Columns can be run differently from the one shown above. For example, we may remove heat from or add it to trays within a column. We can add heat by removing part of the liquid from a tray, vaporizing it, and returning the vapor

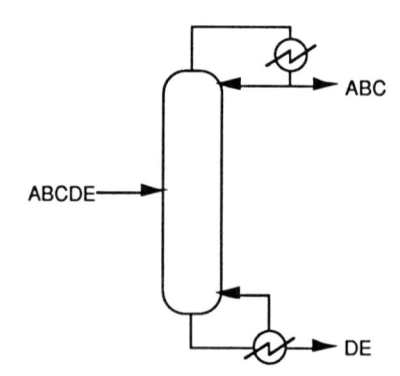

FIG. 1. A simple two-product column.

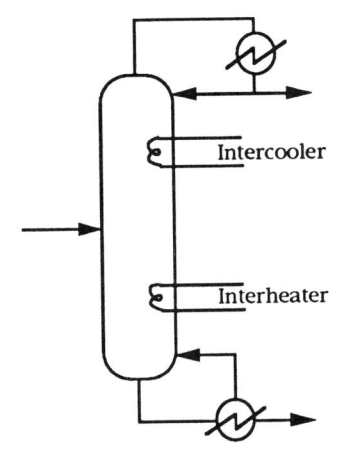

FIG. 2. Columns with interheating and/or intercooling.

just below the tray from which it was removed; or we can remove heat by withdrawing some vapor, condensing it, and placing it back into the column. Such a configuration is called an intercooled and/or interheated column, as shown in Fig. 2. When there are three or more species in the feed, we can alter the set of product compositions a column can produce by using interheating or intercooling, as we shall note later.

Another commonly used configuration is a column with a side enricher or a side stripper. As illustrated in Fig. 3, the column with a side stripper can separate

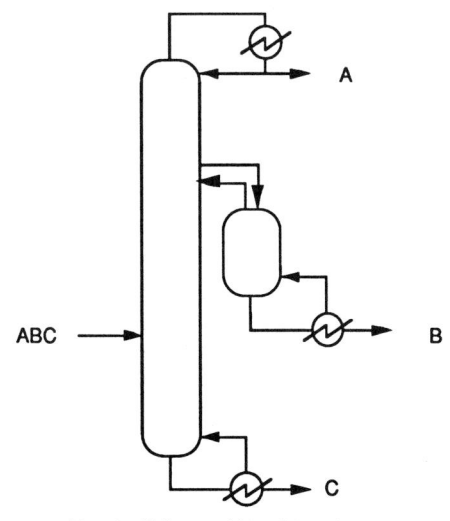

FIG. 3. Column with a side stripper.

three species, each to a desired purity. This column has one condenser and two reboilers. At first glance, it looks as if it is the equivalent of one and one-half columns; but it is really more like two.

We next consider separating a mixture of ethane, propane, n-butane, and n-pentane. Two evident alternative solutions are shown in Fig. 4. Three more, equally "evident" solutions exist, as well as many more less obvious solutions. There is, however, a problem with these two (and the remaining three) sequences. The top of the column in which ethane is a product is very cold. The column is pressurized to elevate its temperature, but, even so, it cannot be brought up to anywhere near ambient temperatures. Thus, we will have to cool the condenser using refrigeration, which is very expensive.

Is there any way we might reduce the amount of refrigeration required for the sequence on the left of Fig. 4? One possibility is to recycle some of the n-pentane produced back to the top of the first column, thus providing part or all of the liquid required for reflux instead of condensing ethane. The normal boiling point for n-pentane is about 310 K, which means it can be liquefied easily at ambient temperatures and very slightly elevated pressures. Running the column at several atmospheres pressure can reduce but cannot eliminate the loss of n-pentane with the ethane. If this loss cannot be allowed, we can add a few trays above where n-pentane is fed into the column and use a much reduced amount of ethane as reflux, significantly reducing rather than eliminating the need for refrigeration. Economics will, of course, indicate whether such designs are a good idea.

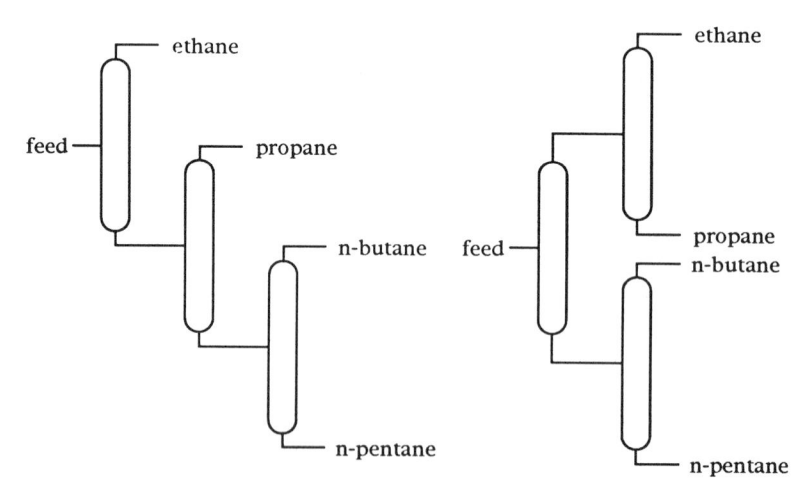

FIG. 4. Two alternative separation sequences for separating a mixture of four species.

III. Assessing the Behavior of a Mixture

It does little good to assume ideal vapor/liquid equilibrium behavior for a mixture that is highly nonideal. The design proposed assuming ideal behavior has little to do with the design actually required. There is, in fact, no reason even to assume it represents a lower cost bound. For example, suppose we wish to separate water from toluene. Assuming ideal behavior would lead us to design a distillation column. But in fact, toluene and water "hate" each other, and can be separated by using a simple decanter. Therefore, the first step in designing a separation process is often to assess the vapor/liquid/liquid equilibrium behavior of the species to be separated. We start with an approach that is useful for this activity, given today's plethora of excellent computer-based physical-properties packages. Having experimental data is the only assured way to know the behavior, and many such data are currently available. Horsley (1973), for example, has compiled a listing of known azeotropes.

Assume we can have a reliable flash computation available to us and that the physical property options to be used for the species can be reasonably selected (a step for which help from an expert consultant within the company may be necessary). Then the following is a first step we can use to discover if the mixture displays highly nonideal behavior.

A. AZEOTROPIC BEHAVIOR: INFINITE-DILUTION K-VALUES

We will examine a quick method based on performing two flash computations to determine the existence of azeotropic behavior for binary mixtures. The hydrogen-bonding classes for the species in a mixture are also a clue that the mixture might exhibit liquid/liquid behavior. Indeed, we have used these classes to find mixtures that display nonideal behavior as illustrative examples.

1. Infinite-Dilution K-Values

We can predict azeotropic behavior as follows from infinite-dilution K-values. Using a flowsheeting system, we perform a bubble-point calculation for each species in the mixture. Assuming a mixture contains the species A, B, C, and D, we wish to compute the infinite-dilution K-values for three of the species in the remaining one. For example, we perform a flash calculation where A is dominant and B, C, and D are in trace amounts, using something like a feed composition of 0.99999, 0.000003333, 0.000003333, 0.000003334. It does not matter what type of flash computations we do: bubble point, dew point, or a 50/50 split. We do these at the pressure intended for the separation device to be considered and repeat for all species. Then from each we tabulate the K-values for the trace species. The K-value for the abundant species is always unity at infinite dilution. The ratio of the vapor composition to the liquid composition for each of the trace species gives its infinite-dilution K-value. As we shall discuss later when considering liquid/liquid behavior, these same flash computations should also allow us to retrieve infinite-dilution activity coefficients.

Given infinite-dilution K-values, we want next to examine each species pair where one is plentiful and the other is in trace amount. As an example, Fig. 5 shows how K-values vary for a binary mixture of acetone and chloroform versus composition. We see that the K-value for a drop of chloroform (far left) is less than unity. The vapor composition y_C is less than that for the liquid, x_C. The K-value for a drop of acetone in chloroform is also less than unity. The mixture displays a maximum-boiling azeotrope.

An interpretation of these K-values is as follows:

$$\text{For} \quad K_{12}^\infty = \frac{y_1}{x_1}\bigg]_{1 \text{ in } 2}; \quad K_{21}^\infty = \frac{y_2}{x_2}\bigg]_{2 \text{ in } 1}$$

Maximum-boiling azeotrope: $K_{12}^\infty < 1 \wedge K_{21}^\infty < 1$

Minimum-boiling azeotrope: $K_{12}^\infty > 1 \wedge K_{21}^\infty > 1$

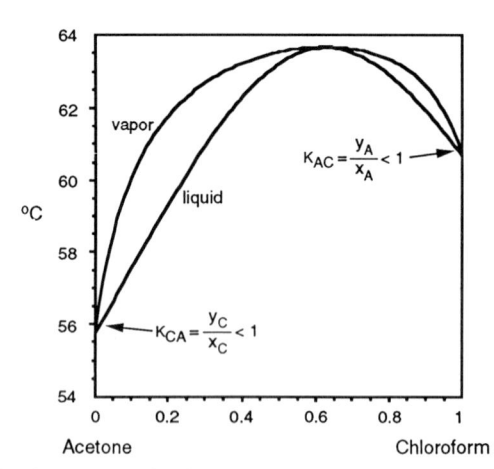

FIG. 5. Liquid and vapor mole fractions vs T at equilibrium for acetone and chloroform. Pressure is 1 atm.

Example. Table I (each row comes from one of the above flash computations) lists infinite-dilution K-values that were computed for a mixture of acetone, chloroform, and benzene. The infinite-dilution K-value for acetone in chloroform is 0.6 and for chloroform in acetone is 0.4. Both are less than 1; a maximum-boiling azeotrope must exist. For chloroform and benzene, the values are 1.5 and 0.4. These values do not suggest the existence of an azeotrope. We assume normal behavior. A similar conclusion is obtained for acetone and benzene where the K-values are 3.0 and 0.7, respectively.

2. Hydrogen-Bonding Guidelines

Berg (1969), in a paper on selecting agents for extractive distillation, classifies species into hydrogen-bonding classes. The deviation from ideality is then predicted depending on the classes represented in the mixture. Quoting from Berg, the classes are as follows:

Class I: Liquids capable of forming 3-dimensional networks of strong hydrogen bonds—e.g., water, glycol, glycerol, amino alcohols, hydroxylamine, hydroxyacids, polyphenols, amides, etc.

Class II: Other liquids composed of molecules containing both active hydrogen atoms and donor atoms (oxygen, nitrogen, and fluorine)—e.g., alcohols, acids, phenols, primary and secondary amines, oximes, nitro compounds with alpha-hydrogen atoms, nitriles with alpha hydrogen atoms, ammonia, hydrazine, HF, HCN (plus nitromethane, acetonitrile even though these form 3-dimensional networks; they have weaker bonds than —OH and —NH bonds in class I)

Class III: Liquids composed of molecules containing donor atoms but no active hydrogen atoms—e.g., ethers, ketones, aldehydes, esters, tertiary amines (including pyridine type), nitro compounds and nitriles without alpha-hydrogen atoms

Class IV: Liquids composed of molecules containing active hydrogen atoms but no donor atoms—e.g., chlorinated hydrocarbons with two or three chlorines per carbon ($CHCl_3$, CH_2Cl_2, CH_3CHCl_2, CH_2Cl—CH_2Cl, CH_2Cl—$CHCl$—CH_2Cl, CH_2Cl—$CHCl_2$)

TABLE I
INFINITE-DILUTION K-VALUES FOR MIXTURE

		K^∞		
		Acetone	Chloroform	Benzene
In:	Acetone	1.0	0.4 (max)	0.7 (normal)
	Chloroform	0.6	1.0	0.4 (normal)
	Benzene	3.0	1.5	1.0

Class V: All other liquids—i.e., liquids having no hydrogen-bonding capabilities—e.g., hydrocarbons, CS_2, sulfides, mercaptans, halohydrocarbons not in class IV, nonmetallic elements such as iodine, phosphorus, sulfur.

Quoting again, their expected deviation from Raoult's Law are shown in Table II.

We can consider our example again. The classes for these species are as follows:

Acetone: class III
Benzene: class V
Chloroform: class IV

The binary-pair behaviors suggested by this article are as follows:

Acetone, benzene	III+V	d: Quasi-ideal
Acetone, chloroform	III+IV	b: Always − behavior
Chloroform, benzene	IV+V	d: Quasi-ideal

As before, we see that the problem in this mixture is the acetone, chloroform pair. As we have already seen, they have a maximum boiling azeotrope.

TABLE II
Expected Deviations from Raolt's Law
(from Berg, 1969)

	I	II	III	IV	V
I	c	c	c	a	a
II	—	c	c	a	a
III	—	—	d	b	d
IV	—	—	—	d	d
V	—	—	—	—	d

a. Always + deviations, frequently limited solubility (min boiling azeotropes if any). Hydrogen bonds broken only.
b. Always − deviations (tendency for max boiling azeotropes). Hydrogren bonds formed only.
c. Usually + deviations; some very complicated situations. Some will give maximum azeotropes (from negative deviations). Hydrogen bonds both formed and broken, but dissocation of class I or II liquid is more important effect.
d. Quasi-ideal systems, always + deviations or ideal. Minimum azeotropes only if any. No hydrogen bonds involved.

B. Liquid/Liquid Behavior: Infinite-Dilution Activity Coefficients

Mixtures may also form two or more liquid phases at equilibrium. For example, a 50/50 mol% liquid mixture of toluene in water will partition into a water-rich liquid phase and a toluene-rich liquid phase. We just used infinite-dilution K-values as a means to predict azeotropic behavior. We can argue that we should use infinite-dilution liquid activity coefficients to alert us to the potential for liquid/liquid behavior. We do so as follows.

For a liquid mixture at constant temperature and pressure, an equilibrium state is one that minimizes the total Gibbs free energy for the system (Smith and Van Ness, 1987). Figure 6 shows a plot of the total Gibbs free energy of a binary mixture where we form the total from three terms: one that mole-fraction averages the Gibbs free energy for the two pure species, one that computes the ideal Gibbs free energy of mixing, and one that estimates the excess Gibbs free energy. We can model this last term by using an empirical relationship such as the Margules equation. We have rescaled the ordinate for this plot by dividing all terms by RT, where R is the universal gas constant and T is the absolute temperature.

Suppose the rescaled Gibbs free energy, G_1/RT, for species 1 (left side Fig. 6) is 0 while for pure species 2 it is 0.5. Then, if the Gibbs free energy simply mixed with no effect of mixing nor any effect from nonidealities, the line joining 0 on the right to 0.5 on the left would give us the mixture Gibbs free energy:

$$\frac{G_{avg}}{RT} = x_1 \frac{G_1}{RT} + x_2 \frac{G_2}{RT} \quad \text{where} \quad x_1 + x_2 = 1$$

However, even for an ideal mixture, there is an effect on the Gibbs free energy from the entropy of mixing, namely,

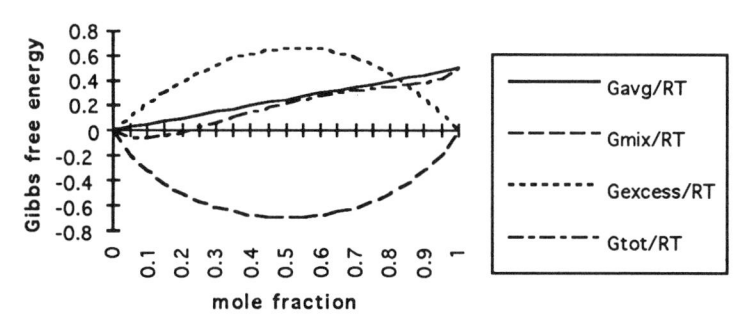

FIG. 6. Gibbs free energy for binary mixture.

$$\frac{G_{\text{mix}}}{RT} = x_1 \ln x_1 + x_2 \ln x_2$$

Note it always has this exact shape—that of an upward-opening "convex" curve that passes through zero at 0 and 1. If the mixture were ideal, this curve plus that averaging the pure species Gibbs free energies would assume this upward-opening convex shape.

The following Margules equation is one form in which we can approximately represent nonideal behavior, allowing us to estimate the excess Gibbs free energy as a function of composition:

$$\frac{G_{\text{excess}}}{RT} = x_1 x_2 (A_{21} x_1 + A_{12} x_2)$$

We would find, when taking compositions to their limiting values, that

$$A_{21} = \ln(\gamma_1^\infty); \quad A_{12} = \ln(\gamma_2^\infty)$$

where A_{12} and A_{21} are constants for this equation and γ_i^∞ is the infinite-dilution activity coefficient of species i in the other species. This curve can assume all sorts of shapes as it is cubic in mole fraction. It too must be zero at both ends.

With a downward-opening *concave* shape as illustrated here, it starts by canceling only a part of the effect of ideal mixing; then it more than cancels this effect and takes over in making the G_{tot}/RT curve switch from convex-upward to concave-downward. It is this switch in shape that indicates liquid/liquid behavior. The mixing term approaches its endpoints with an infinite slope so the G_{tot}/RT curve always starts out in the downward direction, no matter what model we use to estimate the nonideal behavior.

If our total curve switches to a concave-downward appearance anywhere along it, as it does here between approximately 10% and 90% B in A, any mixture with compositions between these two points will break into two liquid phases at equilibrium. Suppose we compute the total Gibbs free energy for a 50/50 mixture on the G_{tot}/RT curve. We can get a lower total Gibbs free energy by breaking the mixture into two mixtures, one at approximately 10% and the other at approximately 90%. Their total Gibbs free energy is along a straight line connecting their individual Gibbs free energies. The lowest possible value would be along a support line that just touches the total curve from below. The value on this line is below that predicted for the mixture; thus the system can reduce its total Gibbs free energy by breaking into these two phases.

By carrying out numerical studies, we find that the Margules equation predicts the onset of liquid/liquid behavior if either of the following is (approximately) true:

- if either $\gamma_{1 \text{ in } 2}^\infty$ or $\gamma_{2 \text{ in } 1}^\infty$ is greater than 9;
- for $\gamma_{j \text{ in } k}^\infty > \gamma_{k \text{ in } j}^\infty$, if $\gamma_{j \text{ in } k}^\infty > 9(\gamma_{k \text{ in } j}^\infty)^{1/3}$.

For example, if $\gamma_{1\text{ in }2}^{\infty}$ is 0.001, then we need to worry about liquid/liquid behavior if $\gamma_{2\text{ in }1}^{\infty}$ is greater than about $9(0.001)^{1/3} = 0.9$.

We can propose to use this guideline to alert us to the potential for liquid/liquid behavior. For example, we might consider the need to check more thoroughly for liquid/liquid behavior if we replace the 9 by a 6 and either of these test passes.

We used infinite-dilution activity coefficients of 10 and 20 to create Fig. 6. Both are greater than 9, so we should expect the Margules equations to predict liquid/liquid behavior. Water and toluene have infinite-dilution activity coefficients in the thousands. They really dislike each other and break into relatively pure phases. If we examine the total Gibbs free energy curve, we gain the impression that the curve is totally convex-upward; however, there is a slight downward move at the extremes because of the infinite downward slope of the mixing term at the extreme compositions. The two liquid phases are almost, but not quite pure.

1. Relating Infinite-Dilution K-Values and Activity Coefficients

If we have evaluated infinite-dilution K-values to test for the existence of azeotropes, we can use those numbers to get a quick estimate of the corresponding activity coefficients by noting that

$$\gamma_{i\text{ in }j}^{\infty} = \frac{\phi_i P}{f_i^0} K_{i\text{ in }j}^{\infty} \approx \frac{P_j^{\text{sat}}(T)}{P_i^{\text{sat}}(T)} K_{i\text{ in }j}^{\infty}$$

If the mixture is at 1 atm, T is the normal boiling point for the plentiful species j.

IV. Separating Nearly Ideal Systems

A. ANALYSIS

Here, we review some techniques that are useful for analyzing nearly ideally behaving distillation columns—i.e., for predicting how these columns might perform, given certain specifications. The first topic will be minimum reflux calculations so we can determine the required internal flowrates in a column. However, to really understand this topic, we should first examine the concept of a *pinch point* in a column. Using our understanding of a pinch point on a McCabe–Thiele diagram, we shall see that a pinch point occurs when the compositions passing each other between two trays (thus satisfying the operating line equations) are also in equilibrium with each other (King, 1980).

1. Pinch

We follow the development in Terranova and Westerberg (1989) to explain a pinch point. The material balance and equilibrium relationships for the top section of a column are as shown in Fig. 7.

The species material balance is

$$Vy_i = Lx_i + Dx_{D,i}$$

We can express equilibrium as follows:

$$y_i = K_i x_i \equiv \frac{\alpha_{ik}}{\sum_j \alpha_{jk} x_j} x_i = \frac{\alpha_{ik}}{\bar{\alpha}_k} x_i$$

where α_{ik} is the *relative volatility* of species i relative to an arbitrarily selected key species k—e.g., the heaviest species in the mixture or the most plentiful (or the lightest, etc.). Note that

$$\bar{\alpha}_k \equiv \sum_j \alpha_{jk} x_j$$

is a *mole-fraction–averaged relative volatility*. It will lie somewhere between the largest and the smallest relative volatility for the mixture.

Using Raoult's law

$$y_i = K_i x_i \approx \frac{P_i^{\text{sat}}(T)}{P} x_i$$

we can estimate relative volatility as

$$\alpha_{ik} \equiv \frac{K_i}{K_k} \approx \frac{P_i^{\text{sat}}(T)/P}{P_k^{\text{sat}}(T)/P} = \frac{P_i^{\text{sat}}(T)}{P_k^{\text{sat}}(T)}$$

The ratio of K-values is an exact definition for relative volatility. The ratio of vapor pressures is an approximate one that assumes Raoult's law holds, thereby assuming that the K-values are not composition-dependent. *The advan-*

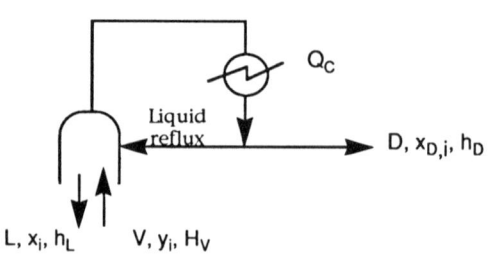

FIG. 7. Top section of a column showing flows and compositions.

tage to relative volatilities are that they are much less temperature- and pressure-sensitive than are K-values.

A pinch point occurs when the vapor and liquid compositions for the operating relationships are also in equilibrium with each other. For such a situation, the operating point of streams flowing against each other between stages is also on the equilibrium surface for the problem. To step through a pinch point requires an infinite number of stages. Substituting equilibrium into the species material balance equation for species i, we get

$$\left(\frac{\alpha_{ik}V}{\overline{\alpha}_k} - L\right)x_i = Dx_{D,i}$$

which, when solved for x_i, gives

$$x_i = \frac{Dx_{D,i}}{\dfrac{\alpha_{ik}V}{\overline{\alpha}_k} - L}$$

We multiply both sides by the relative volatility α_{ik} and sum over all species to get

$$\sum_i \alpha_{ik} x_i = \overline{\alpha}_k = \sum_i \frac{Dx_{D,i}}{\dfrac{\alpha_{ik}V}{\overline{\alpha}_k} - L}$$

If we are given

- all the relative volatilities,
- the mole fractions and total flow for the distillate product, and
- the vapor and liquid flows in the column at the pinch point (i.e., the reflux ratio defined at the pinch point),

this equation is a single equation in the one unknown $\overline{\alpha}_k$. It has to be solved numerically, using something like Newton's method. Once we have $\overline{\alpha}_k$, we can use the previous equation to solve for each of the mole fractions x_i at the pinch point.

If the relative volatilities are composition-, temperature-, and pressure-dependent, we can use this composition as input to a bubble point computation. When solved, we will have new estimates for the relative volatilities. Then we can iterate the computation until all the numbers are consistent.

We can also compute the condenser duty with a heat balance around this part of the column:

$$Q_C = H_V V - (h_L L + h_D D),$$

where H_V, h_L, and h_D are molar enthalpies for the given mixtures and Q_C is the condenser heat duty. Note that there is nothing approximate about this computation. We can compute the molar enthalpies given the temperature, pressure, and stream composition using a rigorous physical property package and obtain as accurate a number as the property computations allow.

2. Underwood's Method

It is possible to derive—roughly—the equations underlying the Underwood method (Underwood, 1946) from the above. The variable R represents the reflux ratio defined in terms of the liquid flow *at the pinch point* relative to the distillate top product flow; i.e.,

$$R \equiv L/D.$$

We again write the equation giving x_i at the pinch point, and then do some rearrangements and variable transformations.

$$x_i = \frac{x_{D,i} D}{\dfrac{\alpha_{ik} V}{\overline{\alpha}_k} - L} = \frac{x_{D,i}}{\dfrac{\alpha_{ik} V/D}{\overline{\alpha}_k} - \dfrac{L}{D}} = \frac{x_{D,i}}{\dfrac{\alpha_{ik}(R_{\min} + 1)}{\overline{\alpha}_k} - R_{\min}}$$

$$= \frac{\dfrac{1}{R_{\min}} \dfrac{R_{\min}\overline{\alpha}_k}{R_{\min} + 1} x_{D,i}}{\alpha_{ik} - \dfrac{R_{\min}\overline{\alpha}_k}{R_{\min} + 1}} = \frac{\dfrac{1}{R_{\min}} \phi x_{D,i}}{\alpha_{ik} - \phi}$$

Summing over all species, we get

$$R_{\min} = \sum_i \frac{\phi x_{D,i}}{\alpha_{ik} - \phi}$$

Adding $1 = \sum_i x_{D,i}$ to both sides gives

$$R_{\min} + 1 = \sum_i \frac{\phi + \alpha_{ik} - \phi}{\alpha_{ik} - \phi} x_{D,i} = \sum_i \frac{\alpha_{ik}}{\alpha_{ik} - \phi} x_{D,i}$$

Finally, multiplying both sides by D gives the form

$$(R_{\min} + 1)D = \sum_i \frac{\alpha_{ik}}{\alpha_{ik} - \phi} x_{D,i} D = \sum_i \frac{\alpha_{ik}}{\alpha_{ik} - \phi} d_i = V_{\min} \qquad (1)$$

which is one of the familiar equations from the Underwood method. A similar relationship exists for the bottom of the column (note the minus sign):

$$\bar{R}_{min}B = -\sum_i \frac{\alpha_{ik}}{\alpha_{ik} - \phi} x_{B,i} B = -\sum_i \frac{\alpha_{ik}}{\alpha_{ik} - \phi} b_i = \bar{V}_{min} \qquad (2)$$

We can now write a relationship between the vapor flows in the top and bottom of the column:

$$V = \bar{V} + (1 - q)F$$

Then assuming the roots ϕ for the top and the bottom are both the same and using this equation, we derive the remaining Underwood equation:

$$\sum_i \frac{\alpha_{ik}}{\alpha_{ik} - \phi} d_i + \sum_i \frac{\alpha_{ik}}{\alpha_{ik} - \phi} b_i = \sum_i \frac{\alpha_{ik}}{\alpha_{ik} - \phi} f_i = (1 - q)F \qquad (3)$$

The assumption that the ϕ values are the same is quite an assumption. It takes a good deal of arguing to make that plausible. We shall not go into that here, however.

Use of Underwood's Method. Table III presents an example that illustrates a computation we might do to compute the minimum reflux flows for a column. In this example, species C distributes between the top and bottom product in the column. Underwood's method permits us to compute how it distributes. The approach for using Underwood's equations to compute minimum reflux is as follows:

- Write Eq. (3) for each of the roots ϕ lying between the species that appear in both the top and bottom products—here B, C, and D. So we write it twice. There will be a root between α_{BE} and α_{CE} and between α_{CE} and α_{DE}. Call them ϕ_{BC} and ϕ_{CD} (for the relative volatilities, E is arbitrarily selected to be the key species).

TABLE III
PROBLEM FOR ILLUSTRATING UNDERWOOD'S METHOD

Species	Relative volatility	Feed (kmol/s)	Top product (kmol/s)
A	4	1	1
B (lk)	3	1	0.98
C	2	1	?
D (hk)	1.5	1	0.03
E	1	1	0

$$\frac{4 \cdot 1}{4 - \phi_{BC}} + \frac{3 \cdot 1}{3 - \phi_{BC}} + \frac{2 \cdot 1}{2 - \phi_{BC}} + \frac{1.5 \cdot 1}{1.5 - \phi_{BC}} + \frac{1 \cdot 1}{1 - \phi_{BC}}$$
$$= (1 - 1)4 = 0$$

$$\frac{4 \cdot 1}{4 - \phi_{CD}} + \frac{3 \cdot 1}{3 - \phi_{CD}} + \frac{2 \cdot 1}{2 - \phi_{CD}} + \frac{1.5 \cdot 1}{1.5 - \phi_{CD}} + \frac{1 \cdot 1}{1 - \phi_{CD}}$$
$$= (1 - 1)4 = 0$$

which gives $\phi_{BC} = 1.673825$ and $\phi_{CD} = 2.395209$.

- Write Eq. (1) twice, one for each root, and solve these two equations for the two unknowns d_C ($=x_{D,C}D$) and R_{min}. (Solved for these two variables, the equations will be linear.) Once d_C is known, D can be computed and then the mole fractions for the top product:

$$\frac{4 \cdot 1}{4 - 1.673825} + \frac{3 \cdot 0.98}{3 - 1.673825} + \frac{2d_C}{2 - 1.673825} + \frac{1.5 \cdot 0.03}{1.5 - 1.673825}$$
$$+ \frac{1 \cdot 0}{1 - 1.673825} = V_{min}$$

which gives $V_{min} = 5.664$ kmol/s and $d_C = 0.324$ kmol/s.

B. System Synthesis for Nearly Ideal Systems

We shall present our ideas in this section by example. Consider the following separations problem, for which the species should all behave relatively ideally when in a mixture.

1. Example: Separation of Five Alcohols

Table IV gives the species flows for a five-species alcohol mixture. Design a system of distillation columns to separate them.

TABLE IV
Five-Alcohol Separation

Species i	$f(i)$ (kmol/h)
Isobutanol	5
1-pentanol	10
1-hexanol	20
1-heptanol	50
1-octanol	15

SYNTHESIS OF DISTILLATION-BASED SEPARATION SYSTEMS 81

We first need to assess how many different sequences we might actually invent for this problem. Using only simple columns, we can construct the alternative sequences shown in Fig. 8, where each column does a fairly sharp split between adjacent key species. As we can see, there are 14 sequences. The third sequence has two binary separations at the end for it in this "tree" of alternatives. Note that both are required, so only one sequence results in the counting.

Thompson and King (1972) developed a formula for predicting the number of simple sequences for such a problem:

$$\text{No. Seq.} = \frac{[2(N-1)]!}{[N-1]!N!} = \frac{[2(5-1)]!}{[5-1]!5!} = \frac{8!}{4!5!} = 14$$

where N is the number of species in the original mixture. We note that for a problem where the vapor/liquid equilibrium behavior of the species is relatively ideal, discovery of the alternative simple sequences is straightforward. The number of sequences grows to over 290,000 for a ten-species mixture. Separating a ten-species mixture into ten relatively pure single species products is a rather large separation problem. However, if the analysis is simple, this number of alternative sequences is not too large a problem to be investigated using a computer. We would want to do it efficiently, none the less.

2. Separation Selection Using Marginal Costs

In this section we present a very simple method based on marginal cost (Modi and Westerberg, 1992) to compare the different sequences. As it is a very ap-

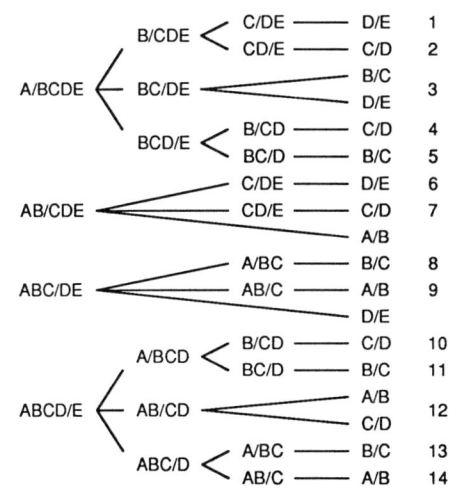

FIG. 8. All 14 "simple" separation sequences for separating a mixture of five species.

proximate method, it should be used only to remove sequences from consideration, leaving one with a (hopefully) much reduced set to investigate with more accurate methods.

We start by noting two heuristics presented a number of years ago by Hendry et al. (1973) that we should consider before proceeding very far into the selection of the sequences.

Dangerous Species Heuristic: *Remove dangerous and corrosive species early.*

We would wish to handle dangerous (deadly if released, toxic, carcinogenic, explosive) and corrosive species in as few pieces of equipment as possible. (Best of all, avoid handling them completely by not designing a process with such species present.) They should be removed first. If such a species is either the most or the least volatile, then we can remove it in the first column. Otherwise, it will take at least two columns (a split just above and a split just below—in either order) to remove it.

Final Product Heuristic: *Produce final products as distillate products, not as bottoms products.*

All mixtures will contain contaminants, such as heavy organics, which tend to discolor the yield if they are removed from the process with final products. That is, heavy species will exit a column with the bottoms product, contaminating it. Customers will pay more for a cleaner looking product (even when it really may not matter). However, this heuristic should not be followed blindly. Many times it is not cost-effective to follow this heuristic; at other times it may be impossible to abide by it—for example, when the heaviest species is the product to be sold. Moreover, it is possible to clean up a product in other ways: e.g., by vaporizing and then condensing the material, leaving the contaminant as a heavy residual, or by passing the product through a bed of activated carbon.

Assuming we have considered these two heuristics, we need to select among the many alternative sequences that may remain for a problem. One criterion is to select the one with the lowest annual cost in units like $/year (£/year, DM/year, etc.). Annual cost is the total of the costs per year to operate the plant and the annualized cost of the investment required to build the plant. Investments are measured in dollars. Annualizing investments means that we convert an investment cost in dollars ($) into an equivalent expense in dollars per year ($/yr).

A very simple method for annualization is to divide investment cost by the number of years over which the company wants that investment paid back by the earnings of the process; e.g., we divide by 3 years if we want the investment back in three years. Many other ways exist, such as establishing a set of equal yearly payments over a specified lifetime of the process, say 15 years, that would have the same present value as the investment. Present value requires us to set an acceptable interest rate over the inflation rate, say 10–15% per year.

We propose finding the sequence with lowest costs by using *marginal costs*. We compute marginal costs by devising a base cost that all sequences have as a minimum, then estimating only the added costs that distinguish one sequence from another. One appealing base cost is the cost to carry out all the needed separations—such as A/B, B/C, C/D, and D/E for our earlier example—as if they were done with no other species present. Then the *incremental cost* of a task is the cost to carry out those separations with the other species present, a cost that differs from sequence to sequence. For nonideal systems, the cost may actually go down when other species are present. However, for ideal systems, the costs increase.

Returning to our previous five-species example, the sequence {A/BCDE, B/CDE, C/DE, D/E} differs from the sequence {AB/CDE, A/B, CD/E, C/D} as shown in Table V. The entries in the second and third columns indicate the extra species present for each binary separation when accomplished using these two different sequences. The binary separation B/C is done with species D and E present in the first sequence, while it is done with species A, D, and E for the second.

The particular *cost-related* quantity we shall consider here is the *marginal vapor flow* rather than the actual annual cost. It is a quantity we can more readily estimate. In fact, for nearly ideal systems, we shall show a very easy way to approximate it.

We define the symbol

$\Delta V(i/j, \text{list})$ = Marginal vapor flow for column having i and j as light and heavy key species, respectively. The list contains species other than i and j which are present in the feed.

For example, $\Delta V(\text{B/C,ADE})$ is for the task AB/CDE: This column is splitting B from C, and the other species present are A, D, and E.

TABLE V
EXTRA SPECIES AS A FUNCTION OF THE COLUMN SEQUENCE USED

Binary Separation	Sequence 1	Sequence 2
A/B	CDE	none
B/C	DE	ADE
C/D	E	none
D/E	none	C

Marginal vapor flow is the added vapor flow required in the column because the other species—i.e., those on the list—are present. The vapor flow in a column is an indicator of the cost of purchasing and operating the column. A difficult separation will have a large vapor flow because it will require a large reflux ratio. Also, refluxed material has to be vaporized and condensed, which directly affects the utility costs for operating the column. Therefore, it makes sense to try to minimize the total of the vapor flows for a system of columns.

Formally, we can define marginal vapor flow as

$$\Delta V(i/j, \text{list}) = V(i/j, \text{list}) - V(i/j)$$

where the last term is the vapor flow in a column to split i from j with no other species present (the list is empty).

We should note that all sequences to separate a mixture will have the same set of binary splits. For example, each alternative sequence for the separation of ABCDE into five single-species pure products will have a split between A and B, another between B and C, etc. The difference among the alternative sequences is the presence or absence of other species when carrying out each of these binary splits. The total of the vapor flows for a sequence is the base set of vapor flows $V(i/j)$, where i and j are A/B, B/C, C/D, and D/E, plus its marginal vapor flows. Thus, the difference in marginal vapor flows is the difference in total vapor flows among the sequences. The sequence with the minimum marginal vapor flows is the sequence with the minimum total vapor flows.

How can we estimate a marginal vapor flow for a column? One approach is to estimate the minimum reflux required using any method that is appropriate. If the separation is among species that are acting nearly ideally, we can use Underwood's method.

3. Five-Alcohol Example Continued

For our five-alcohol example, let species A be *n*-butanol, B be 1-pentanol, etc. Now, let us consider the column that accomplishes the separation AB/CDE. To estimate the internal vapor flows in the column, we will have to assume recoveries for the species. Here, we shall assume that 99% of the key species go to their respective products, while everything lighter than the light key goes to the distillate and everything heavier than the heavy key ends up in the bottoms.

Applying Underwood's method gives us a minimum vapor flow of 72.5 kmol/h for a column accomplishing the separation of AB/CDE. Without species A, D, and E present, the minimum vapor flow is computed to be 44.5 kmol/h. The marginal vapor rate is therefore 38.0 kmol/h.

Let us look more closely at the Underwood equations to see if we can quickly compute an approximate answer. The minimum vapor flow for the split AB/CDE is given by

$$V_{min} = \frac{\alpha_{AC} d_A}{\alpha_{AC} - \phi_{BC}} + \frac{\alpha_{BC} d_B}{\alpha_{BC} - \phi_{BC}} + \frac{\alpha_{CC} d_C}{\alpha_{CC} - \phi_{BC}}$$

but it is also related to the bottoms vapor flow as follows:

$$V_{min} = \overline{V}_{min} + (1 - q)F$$

$$= -\frac{\alpha_{BC} b_B}{\alpha_{BC} - \phi_{BC}} - \frac{\alpha_{CC} b_C}{\alpha_{CC} - \phi_{BC}}$$

$$- \frac{\alpha_{DC} b_D}{\alpha_{DC} - \phi_{BC}} - \frac{\alpha_{EC} b_E}{\alpha_{EC} - \phi_{BC}} + (1 - q)F$$

Let us assume the root ϕ_{BC} does not move all that much if we add other species. Then these two equations would suggest that V_{min} is increased by an amount

$$\frac{\alpha_{AC} d_A}{\alpha_{AC} - \phi_{BC}}$$

because species A is present and by an amount

$$-\frac{\alpha_{DC} b_D}{\alpha_{DC} - \phi_{BC}} - \frac{\alpha_{EC} b_E}{\alpha_{EC} - \phi_{BC}}$$

because species D and E are present. The extra species are those lighter than the light key (species A) and heavier than the heavy key (species D and E). They will be essentially fully recovered in their respective products, so we should be able to substitute the feed flows for the product flows for each of them. Also in the terms for the species D and E, the denominator of the expressions is negative, so the terms are positive.

We can thus estimate the added vapor flow for each as the absolute value for the appropriate term, i.e.,

$$\left| \frac{\alpha_{ik} f_i}{\alpha_{ik} - \phi_{lk,hk}} \right|$$

where species i is in the feed when separating species lk from hk.

Figure 9 is a sketch of the relative sizes of the numerator and denominator for terms such as this for our example problem.

The amount of added vapor flow because A is present is proportional to α_{AC} divided by $\alpha_{AC} - \phi_{BC}$. We do not need a particularly accurate value for ϕ_{BC}

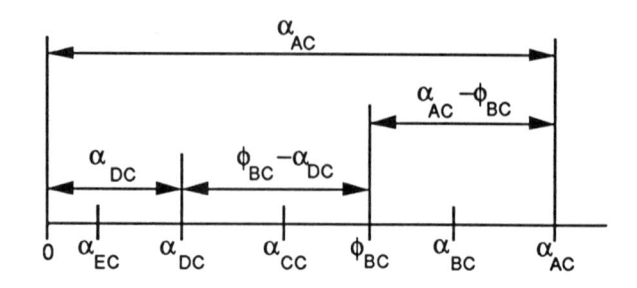

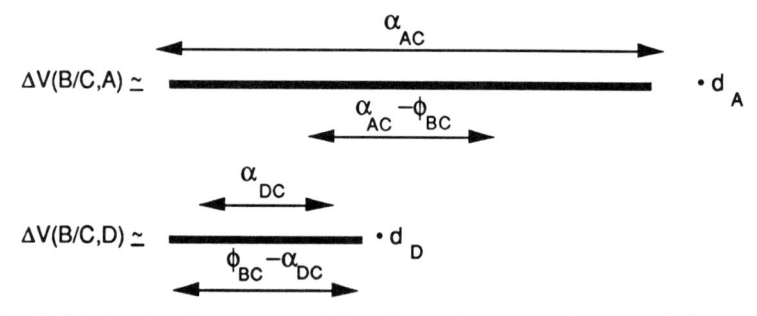

FIG. 9. Relative size of numerators and denominators in term to estimate marginal vapor flow: (a) terms relative to each other; (b) ratios.

to get a reasonable value for this term. A similar argument holds for the term when D is present and, although not shown, when E is present, too. We could, for example, let ϕ_{BC} be the average of the two surrounding relative volatilities. Our final term to approximate the increase in internal vapor flow for a column is thus

$$\left| \frac{\alpha_{ik} f_i}{\alpha_{ik} - \left(\frac{\alpha_{lk} + \alpha_{hk}}{2} \right)} \right|$$

We see that the term in front of the flow for the species in its respective product is much larger for a light species than for a heavy one. Thus, Figure 9 alerts us to the heuristic that the vapor flow is more sensitive to the presence of extra light species than the presence of extra heavy ones. This heuristic is often stated as follows:

Direct Sequence Heuristic: *Prefer removing the most volatile species first.*

Repeated application of this heuristic gives what is called the direct sequence. For example, the direct sequence for our problem is A/BCDE, B/CDE, C/DE, and D/E. (The indirect sequence—the other one with a special name—is ABCD/E, ABC/D, AB/C, A/B.)

4. Five-Alcohol Example Continued Further

The calculations were done using vapor pressure data available in Reid *et al.* (1987) for each of the species; Table VI gives the results. The temperature selected for evaluating the vapor pressure is the bubble point at 1 atm for the feed mixture, i.e., at 434.21 K. Using the relative volatilities and feed flows shown in Table VI, we can estimate the marginal vapor rates shown in Table VII using the equation

$$\left| \frac{\alpha_{ik} f_i}{\alpha_{ik} - \left[\dfrac{\alpha_{lk} + \alpha_{hk}}{2}\right]} \right|$$

For example, with C present for the A/B split, we get

$$\Delta V(A/B,C) = \left| \frac{1.0000 \cdot 20 \text{ kmol/h}}{1.0000 - \left(\dfrac{3.3199 + 1.7735}{2}\right)} \right| = 12.9 \text{ kmol/h}$$

To read Table VII, the marginal rate for AB/CDE, $\Delta V(B/C, ADE)$, is computed by adding the terms for the B/C split (second row) with A, D, and E

TABLE VI
VAPOR PRESSURE AND RELATIVE VOLATILIES FOR EXAMPLE PROBLEM

Species	Vapor pressure (bars)	Relative volatility	$f(i)$ (kmol/h)
Isobutanol	3.8279	3.3199	5
1-pentanol	2.0449	1.7735	10
1-hexanol	1.1530	1.0000	20
1-heptanol	0.6407	0.5557	50
1-octanol	0.3542	0.3072	15

TABLE VII
ESTIMATED MARGINAL VAPOR RATES[a]

	A	B	C	D	E
A/B	****	****	12.9	14.0	2.1
B/C	8.6	****	****	33.4	4.3
C/D	6.5	17.8	****	****	9.8
D/E	5.7	13.2	35.2	****	****

[a] Key species are listed along the left side, extra species across top.

added: namely, 8.6 + 33.4 + 4.3 = 46.3 kmol/h. This number should be compared to the more exact number, 38.0 kmol/h computed earlier. We should not expect it to be any more accurate than these numbers indicate.

We note there are two rather large numbers: one for which D is present with the B/C split and the other for which C is present with the D/E split. No doubt these should be avoided—as they can be if we split ABC/DE at the start. Then the "cost" to have A present for the AB/C split (8.6) compared to that of having C present in the A/BC split (12.9) suggests we should probably split C off next. That fixes the sequence as

ABC/DE, AB/C, A/B, D/E

The only other sequence that avoids these two high costs is A/BCDE followed by BC/DE.

We can search over all possible sequences quite quickly as follows. We sum up the marginal costs for each split possible in the problem and organize them in the following way.

		B/CDE	37.7		
A/BCDE	28.9	BC/DE	27.6		
		BCD/E	48.4	C/DE	9.8
				CD/E	35.2
AB/CDE	46.3				
				B/CD	33.4
				BC/D	17.8
ABC/DE	34.1				
				A/BC	12.9
				AB/C	8.6
		A/BCD	26.9		
		AB/CD	42.0		
ABCD/E	54.1	ABC/D	19.0		

To develop a solution using the least cost, we start with the total feed and compare the costs for the first splits. Here, the costs range from a low of 28.9 to a high of 54.1 kmol/h, so we select A/BCDE first. We now must complete the sequence; i.e., separate BCDE. The least-cost next step is 27.6 corresponding to BC/DE, which gives us a total of 56.5 kmol/h. That completes this sequence because the marginal costs for the two remaining splits, B/C and D/E, are zero (there are no extra species present for them). Note that this is the second of the two solutions we spotted above. Can we find a cheaper solution? Observe that neither of the other first splits for BCDE is preferable: each has a higher-cost first step (and each requires another split, with an added cost, to finish them).

We try ABC/DE next at a first-step cost of 34.1. The AB/C split adds 8.6 for a cost of 42.7. This sequence is the first one we found intuitively above, and it is better. No sequence starting with AB/CDE (first-step cost of 46.3) or ABCD/E (54.1) can possibly be better. We are done.

The approach we just used to completely search this space is a form of branch and bound. We branched off the least expensive first steps and slowly eliminated the need to look at others as they each had a first-step cost (bound) that was too high to win.

Because errors in these numbers can occur, we might want to look at anything within 20% of the best with a more accurate analysis. If none exists, we are done with a relatively cheap analysis. If some exist, we could use Underwood's method (i.e., find the roots) to estimate the minimum reflux ratios as we did above for one of the columns.

5. Distillation Heuristics

The marginal cost approach explains (at least partially) many of the commonly published heuristics used to select the better sequences. We have already seen how it explains the direct sequence heuristic, which is stated as follows:

Direct Sequence Heuristic: *All other aspects of the problem being equal, remove the most volatile species first.*

For all other things to be equal, the amounts of the species must be the same and the relative volatility between all adjacent pairs of species must also be the same. For example, in the case of a four-species feed having equal amounts of each species (e.g., 1 kmol/h each) and relative volatilities of $\alpha_{AD} = 1.2^3 = 1.728$, $\alpha_{BD} = 1.2^2 = 1.44$, $\alpha_{CD} = 1.2^1 = 1.2$, and $\alpha_{DD} = 1.2^0 = 1$, "all other aspects of the problem [would be] equal." For each binary split (e.g., A/B or B/C), the McCabe–Thiele diagram would be drawn with the equilibrium curve based on a relative volatility of 1.2.

In carrying out the above search, we become aware that different sequences are associated with different numbers of extra species contributions. For example, in the sequence

{A/BCDE, B/CDE, C/DE, and D/E}

CDE, DE, and E show up as six instances of extra species, while in the sequence

{ABC/DE, A/BC, B/C, D/E}

only four instances, ABE and C, show up. The more we split the problem into halves, the fewer the number of extra species contributions we will add into the objective for the above search. We can use this observation to justify partially the following commonly stated heuristic:

> **50/50 Split Heuristic:** *Separate the mixture into roughly equal amounts of products.*

Using marginal vapor flows, we can also explain the following heuristic fairly straightforwardly, as the approximation for the added vapor flow for a species in any mixture in which it appears is proportional to its flow in the feed:

Major Species Heuristic: *Have the major species in as few splits as possible.*

One other heuristic very commonly stated is

> *Save the difficult splits for last.*

In other words, *do the easy splits first.* A split is easy if the relative volatility between the two key species is large. The (plausible) argument is that the hard splits should be done when no other species are present. Since the marginal vapor flow computation neither supports nor rejects this heuristic, we might draw the conclusion that this heuristic is not valid for the problem as we have posed it above. But there is some justification for this heuristic when we consider the energy integration of columns (e.g., using the heat expelled from the condenser of one column as the heat input into the reboiler of another).

All of these heuristics depend on amounts and relative volatilities for the species in the mixture. Based on precisely these quantities, the simple computation

$$\left| \frac{\alpha_{ik} f_i}{\alpha_{ik} - \left[\frac{\alpha_{lk} + \alpha_{hk}}{2} \right]} \right|$$

would seem to quantify their relative importance.

V. Separating Highly Nonideal Mixtures

We shall now look at the synthesis of separation systems for liquid mixtures of species that display highly nonideal behavior. We reference a small sampling of the work on this topic: Ewell and Welch (1945), the classical text by Hoffman (1964), early work on classifying behavior by Berg (1969), a scheme for classifying vapor–liquid behavior for ternary mixtures by Matsuyama (1975), many papers on homogeneous continuous azeotropic distillation by Doherty and coworkers (Doherty and Perkins, (1978); Doherty, (1985); Doherty and Caldarola, (1985); Levy *et al.*, (1985); Van Dongen and Doherty, (1985); Knight and Doherty, (1989); Julka and Doherty, (1990); Foucher *et al.*, (1991); Fidkowski *et al.*, (1993); work by Petlyuk (1978), Stichlmair *et al.*, (1989), an expert system described by Barnicki and Fair (1990), work on multiple steady states by Bekiaris *et al.*, (1993), work by Bossen *et al.*, (1993), and our own work. Poellmann and Blass (1994) presented a review and new analysis methods for azeotropic mixtures.

Distillation remains a likely option for separating a complicated liquid mixture except for the fact that we cannot readily predict what the products will be if we do distill a given mixture. Extractive distillation is often a viable alternative. If the species display liquid/liquid behavior (e.g., hydrocarbons and water), we can consider simple decantation or liquid/liquid extraction, too.

In general, we will try simple distillation; but, when we do, we often discover that significant amounts of almost all of the species show up in either or both of the distillate and bottoms products, no matter how we run the column. The inability to effect sharp splits gives rise to the recycling of streams within the separation process itself—something we did not require earlier when we looked at the separation of ideally behaving mixtures.

A. Azeotropic Separation: Example 1

Consider the separation of water from *n*-butanol. The phase behavior for this mixture is quite complex. There is a minimum-boiling azeotrope formed as well as a liquid–liquid phase separation when the liquid is cooled enough. Figure 10 is a sketch of the general shape of the phase behavior for this system. (It is not an accurately drawn phase diagram.)

Several textbooks and reference books (e.g., the third edition of Perry's *Handbook,* 1950) use this example to illustrate azeotropic distillation. They show the solution sketched in Fig. 11 for a feed whose composition is about 60% *n*-butanol. If we analyze this configuration, we see that it separates the mixture. The same textbooks suggest that for a feed below the azeotropic composition, one should use the same configuration but put the feed into the decanter. The question that occurs immediately is: How was this configuration selected? Was it a trial-and-error procedure, or is there some way to find it directly?

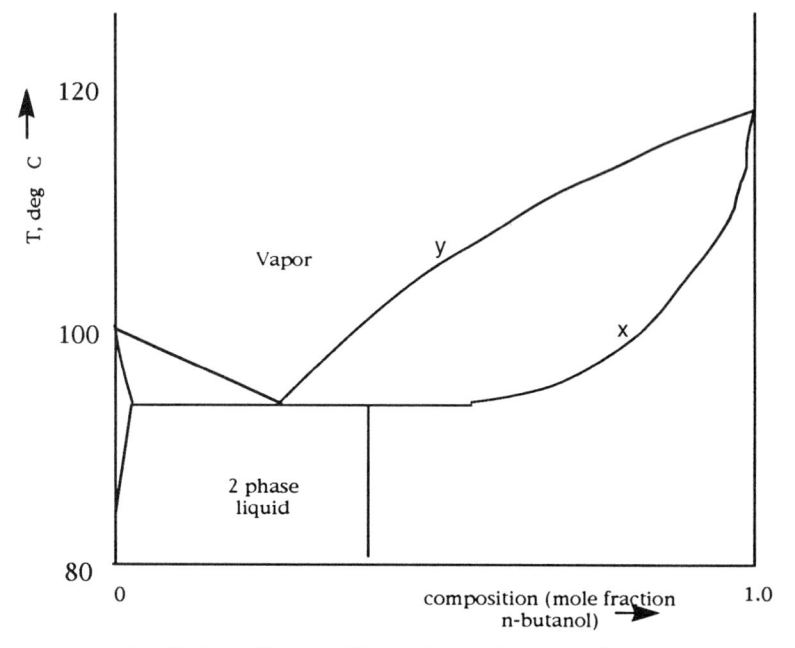

FIG. 10. Vapor/liquid equilibrium diagram for n-butanol/water.

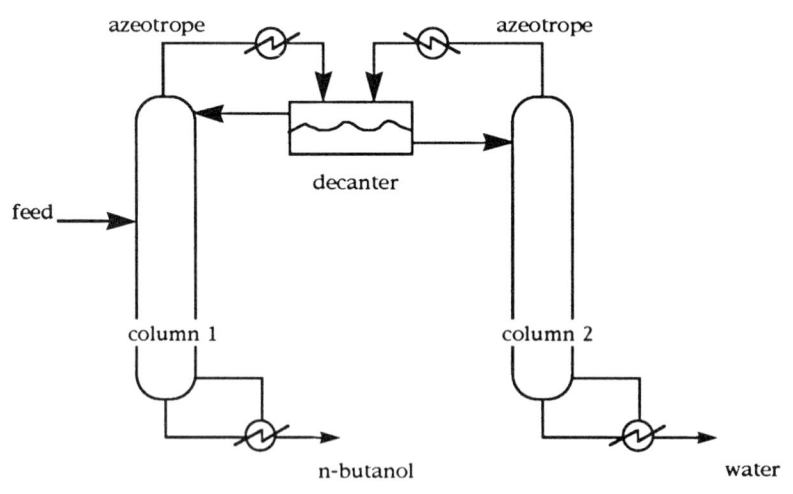

FIG. 11. Configuration presented in handbooks for separating n-butanol and water.

As will be true in virtually all synthesis problems, the secret to finding a solution is to find and use the right representation. For this problem, a convenient representation is shown in Fig. 12. We show a line that has the key compositions marked along it for the phase diagram for the system.

Starting with a feed at about 60% n-butanol, we see that we can distill it into two products: one close to the azeotropic composition as the distillate product and one, nearly pure n-butanol, as the bottoms product. The bottoms product is a desired product; however, we must still separate the top. We find that the top azeotrope lies in the region where it partitions into two liquid phases if we cool it. We therefore cool the azeotrope and put it into a decanter. Next, we decide to distill the water-rich phase (bottom phase in the decanter) into nearly pure water and azeotrope, again getting the azeotrope as a distillate product. We already know what to do with the azeotrope: cool it and feed it to the decanter. The only remaining stream is the upper decanter product, the one that is richer in n-butanol.

We could distill this upper product, getting n-butanol and azeotrope. However, the first column is already able to do this task, so we can choose to feed it into that column. We see that, with minor modifications, we have just invented the structure shown in the handbooks. In a similar manner, we should be able to invent two structures that can handle a feed to the water-rich side of the azeotrope. In one, the feed enters the decanter and in the other, the feed enters the second column (an alternative not mentioned in the handbooks).

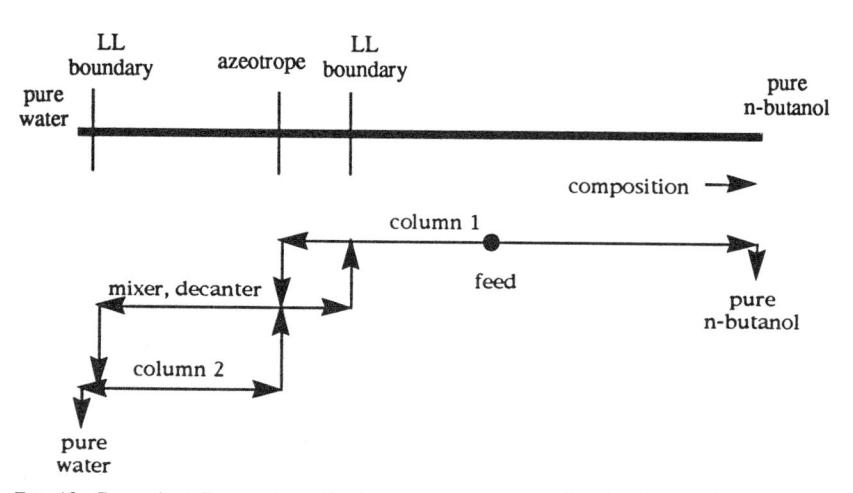

FIG. 12. Convenient diagram to synthesize a separation process for the n-butanol/water system.

VI. Synthesis Discussion

A. ANALYSIS-DRIVEN SYNTHESIS

When the species of interest display highly nonideal behavior, the synthesis of separation processes falls into the class of synthesis problems we might call *analysis-driven synthesis*. We use that term because the real work in this synthesis problem is to establish enough information (i.e., carry out complex *pre-synthesis analysis* computations) to begin the synthesis process. We have to find out about the behavior of the mixtures of the species (e.g., are there azeotropes) before we can even propose the types of equipment we should use to carry out a separation. Then, we must also worry about how these species will behave in any equipment we suggest. Finally, we must carry out a *post-synthesis analysis* in which we design the equipment to evaluate (e.g., determine costs for) the different alternatives. At a minimum, then, column design requires us to set the column pressure and compute the reflux ratio, the number of stages, and the column diameter. For nonideal behavior, these computations can be very difficult.

We thus see a pre-synthesis analysis that characterizes the behavior of the species and a post-synthesis analysis that requires us to design equipment. Both require that we compute equilibrium phase behavior (vapor/liquid, liquid/liquid, vapor/liquid/liquid, etc.)

B. IMPACT OF NUMBER OF SPECIES ON REPRESENTATION, ANALYSIS, AND SYNTHESIS METHODS

The synthesis of distillation-based separation systems is strongly supported for two- and three-species mixtures. There is much less support for mixtures of four species, and even less for five. There are several reasons for this, some of which are fundamental.

The pre-synthesis analysis allows us to understand the topology of the phase behavior of the species when mixed. With that understanding, we find that we can conjecture reasonable process alternatives. Unfortunately, we are limited to a three-dimensional world when it comes to human visualization. Thus we can appeal to visualization of complex topologies only when we can show them in two and, with difficulty, three dimensions. The triangular composition diagram is an excellent way to visualize complex behavior among species, but it is specific to three-species mixtures. When we go to four species, we stretch the ability of people to see what is going on; when we go to five, we are approaching impossibility.

Certainly, computers can work well at higher dimensions. Unfortunately, however, when we go to higher dimensions, what were lines often become planes, then volumes, and so forth. Whereas we can search for exotic behavior along a line with a "computational sledge hammer," it currently takes enormous effort to search a plane thoroughly for complex behavior. For volumes and higher dimensional spaces, then, we must use simpler and less complete searching at this time. This is the problem we face when analyzing systems with five or more species.

Another issue is the way we can use the degrees of freedom for a column. We argue that things change qualitatively when going from systems of three to four to five and higher species. In Fig. 13, we show what we might call the "natural" degrees of freedom for a column—the ones we generally pick when we wish to compute the performance of a column. These specifications lead to the easiest column calculation to converge. Using our intuition about columns, we argue that, if we specify

- the feed to a column (flowrate, composition, pressure, and temperature),
- the column operating pressure (P),
- the number of trays in the top (n_{top}),
- the number of trays in the bottom (n_{bot}),
- the reflux ratio (R), and
- the distillate product flow rate (D),

then the column will operate as expected. That is, a typical simulation of the column would tell us the top and bottom product composition, as well as the tray-by-tray temperatures, compositions, etc. Assuming our intuition is correct, we note that, once we specify completely the feed to a column, a column provides us with five more degrees of freedom. We shall now discuss how we might use these to analyze a column in other ways, for example, in designing the column.

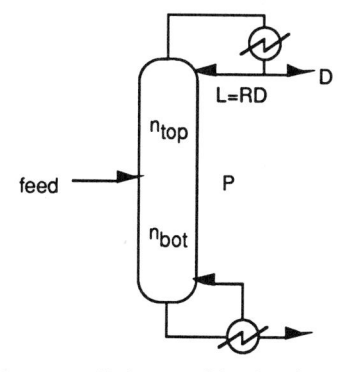

FIG. 13. "Natural" degrees of freedom for a column.

Of the five degrees of freedom, a designer typically fixes the column pressure to set the temperature of the condenser or the reboiler. Thus, we shall assume that P is not at our disposal. That leaves us with four degrees of freedom. Can we use these four degrees of freedom to fix the recoveries of four species leaving the column in the top distillate product? For example, can we design the column so that 99% of species A, 95% of species B, 1% of species C, and 0.5% of species D will leave in the distillate?

Let us carry out the following computational experiment involving the solving of a column model.

First we shall solve a column using the natural specifications—those listed above. Generally, such a model is easy to solve unless the species are highly nonideal in their behavior. We observe the recoveries for the two key species, finding that 85.1% of the light key leaves in the distillate while 74.3% of the heavy key leaves in the bottom product. We now change what we intend to compute and what we intend to specify as fixed for the model calculation. First, we decide to fix the recovery of the light key and ask the model to compute the required reflux ratio. We then alter the light key recovery to 85.0% (very slightly away from the value obtained above) and resolve the model, keeping all the other fixed variables (pressure, reflux ratio, number of trays top and bottom) at their current values. If the simulator only permits the natural specification to be made, then we can only change the reflux ratio iteratively until the recovery changes to the desired value. The reflux ratio should decrease slightly, allowing more of the heavy species to exit in the top product. Since the distillate flowrate is fixed, this will reduce the recovery of the light key.

Being more daring, we now change the recovery to 80% and then to 90%. In both cases there is no difficulty in converging to an answer. We discover that this trade is a practical one. We revert back to a recovery of 85.0%. Experience with many simulations suggests that, while it is a practical one, it is very difficult to converge with such a specification given at the start of the calculation.

Next we fix the recovery of the heavy key in the bottom product at 74.5% and ask the model to adjust the distillate top product flowrate accordingly. For this calculation we still want the light key recovery to stay at 85.0%, so the simulator adjusts the values for both the reflux ratio and the distillate top product flowrate, giving us our desired slightly altered key species recoveries. It has no difficulty in converging. We find that the larger changes also converge readily. So we can trade light and heavy key recovery specifications for distillate total flow and reflux ratio, provided we do it with a modicum of care.

We now ask if we can specify the recovery of a third species. The variable we trade is the number of trays in the top section of the column. The number of trays is a discrete quantity so, for a tray-by-tray model, we should not be able to fix the recovery exactly for the third species while holding the recoveries of the keys fixed at 85.0% and 74.5%. However, a collocation model (Cho and Joseph, 1983; Stewart *et al.,* 1984; Seferlis and Hrymak, 1994; Huss and Westerberg, 1994) allows us to model a column approximately such that the number of trays is a continuous variable. We assume the availability of such a model. With this model, we find that the changes we can accomplish in the recovery of the third species have to be very small. For example, we might ask that the amount of a species heavier than the heavy key have its recovery in the distillate top product decrease from about 1% to about 0.5% by adjusting the number of trays in the top section. Such a change will cause changes in the distillate flowrate, the reflux rate, and the number of trays in the top section. If we try to increase the recovery to 3%, the model fails. We return it to a 1% recovery. Further testing convinces us the trade is theoretically possible, but the range for changing the specification is quite small.

Finally, we ask if we can trade the number of trays in the bottom section for the recovery of a fourth species. Our intuition tells us we could be in numerical trouble here. In principle, we can make the trade; however, very small changes lead to very large changes in the number of stages and even more often, to computational failure. The fourth specification is theoretically possible but computationally nearly impossible.

What are the implications of this experiment? First, from a theoretical point of view, we can specify the recoveries of four species at most for a column. From a practical point of view, the first two are relatively easy (if we take care in doing it) to specify, adding a third specification is difficult, and a fourth virtually impossible. We cannot, as a result, specify completely the top product of a column for more than four species in theory and more than three in practice.

C. In Summary

Any method that we develop for three species starts to run into the practical limitations we described above when we try to extend them to four species. We have difficulties in presenting them to humans for visualization; moreover, difficult searches explode in size, no longer running along lines, but extending over planes, volumes, and higher dimensional spaces. Finally, we cannot specify product compositions for more than four species separations in theory, and we have real, practical difficulty with more than three.

For these good reasons, the literature explores methods to handle three species mixtures thoroughly and hesitates to extend them to mixtures with four and more species.

VII. Pre-analysis Methods

There are two classes of analysis methods that we shall explore. The first class allows us to understand the phase behavior of the species in the problem. The second allows us to analyze the behavior or design of a piece of separation equipment.

We need to understand the behavior of the species in our problem. As we noted earlier, it is not very useful to design a separation process assuming ideal behavior when such behavior does not exist. Such a design might not represent even a lower bound on the cost of the process we might actually develop. For example, toluene and water dislike each other so much that they readily separate into two liquid phases, each with little of the other species present. That is, they behave nonideally. If the compositions for these two phases meet our specifications, we can separate these two species by a relatively inexpensive decantation procedure, which is much more effective than designing a column predicated on ideal behavior.

Some of the analysis methods we are about to discuss rely on computing residue curves, so we first need to understand what such curves are and how to compute them.

A. EQUILIBRIUM-PHASE BEHAVIOR

Earlier in this paper we discussed a simple method to detect if an azeotrope will exist between two species, A and B. The method requires us to perform two bubble-point computations, one in which we have a trace of A in B and one in which we have a trace of B in A. We used the infinite-dilution K-values we computed in these bubble-point computations for the trace species to reveal where an azeotrope exists. If both infinite-dilution K-values are less than unity, there must be a maximum-boiling azeotrope between these species. (There could even be more than a single azeotrope between the species, such as two maximum-boiling azeotropes separated by a minimum-boiling azeotrope. We assume this is rare, but there must be at least one.) If the K-values are both greater than 1, by similar arguments there is at least one minimum-boiling azeotrope. Otherwise, we suggested, we could assume nonazeotropic behavior.

As we also discussed earlier, we can assess whether liquid–liquid behavior is likely by examining infinite-dilution activity coefficients.

In this section we wish to look in more detail at the nonideal behavior of mixtures of a given set of species. We shall start by examining residue and distillation trajectory plots that show the vapor/liquid behavior of mixtures when they are being distilled. Such plots are very informative in that they depict the VLE behavior in a manner that is useful for synthesizing separation processes for such mixtures. We shall then consider methods by which we might discover all the azeotropes for a given set of species. Finally we shall discuss methods for determining the multiple-liquid phase behavior for a set of species.

1. Residue and Distillation Curves

We look first at residue curves, which correspond to batch distillation, and then to distillation curves, which correspond to separation trajectories in columns operating under total reflux conditions.

a. Residue Curves. A residue curve (Hoffman, 1964; Doherty and Perkins, 1978) traces the composition of the liquid in a batch still in composition space versus time. Along this curve the temperature always increases, the composition of the heaviest species increases, and the composition of the most volatile species decreases.

Consider the batch still shown in Fig. 14, which contains a mixture that we boil away with time. What is the composition of the liquid in the still versus time? A dynamic material balance for the species i is

$$\frac{dx_i M}{dt} = x_i \frac{dM}{dt} + M \frac{dx_i}{dt} = -y_i V$$

We note that

$$\frac{dM}{dt} = -V$$

allowing us to write

$$M \frac{dx_i}{dt} = V(x_i - y_i)$$

Defining a dimensionless time as $\theta = t/(M/V)$, this equation becomes the fol-

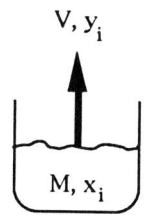

FIG. 14. Batch still.

lowing simple relationships:

$$\frac{dx_i}{d\theta} = x_i - y_i = x_i - K_i x_i$$

where setting $y_i = K_i x_i$ assumes that the vapor is in equilibrium with the liquid composition.

These equations can now be integrated versus dimensionless time. For a three-species mixture, we can plot the resulting compositions as a parametric function of θ on a triangular diagram. Note, the x_i's sum to unity so one of the equations is not independent. The last mole fraction can either be obtained by integrating the above equations or by integrating all but the last of these equations and computing the last mole fraction so the sum of mole fractions is unity.

b. Distillation curves. We can also compute a composition trajectory directly for a column by stepping from one tray to the next for a column operating at total reflux, as Fig. 15 illustrates. The material balance equations for this column are

$$V_{n+1} = L_n$$

and

$$y_{n+1,i} V_{n+1} = x_{n,i} L_n$$

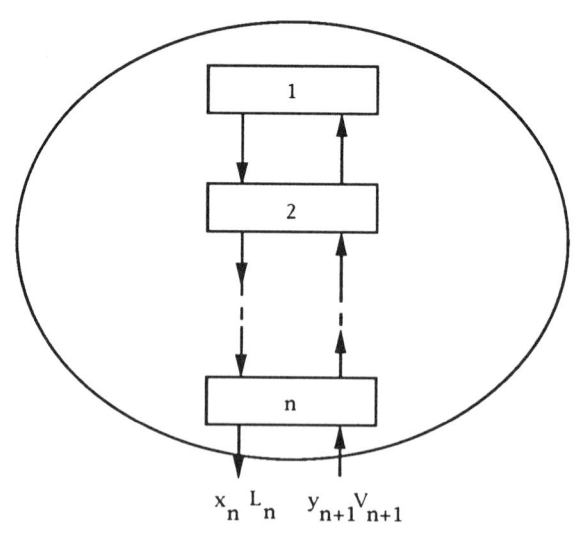

FIG. 15. Total reflux column (no feed or product).

therefore,
$$y_{n+1,i} = x_{n,i}$$

Equilibrium gives
$$y_{n+1,i} = K_{n+1,i} x_{n+1,i}$$

Propagation from one point to the next therefore combines these last two equations to give
$$x_{n,i} = K_{n+1,i} x_{n+1,i}$$

If we step down the column, we would know $x_{n,i}$ and would have to compute a dew point to find the composition $x_{n+1,i}$. Given that, we compute its dew point to step down to tray $n + 2$, etc. Stepping up the column requires that we compute a sequence of bubble points in a similar manner. Temperature increases as we step down a column, which is the same direction temperature moves when time increases for the residue curve computations.

2. Sketching Residue Curve Plots for Ternary Systems

Figure 16 shows a whole set of residue curve trajectories that might be computed for a stillpot containing initially a mixture of water, ethanol, and glycol.

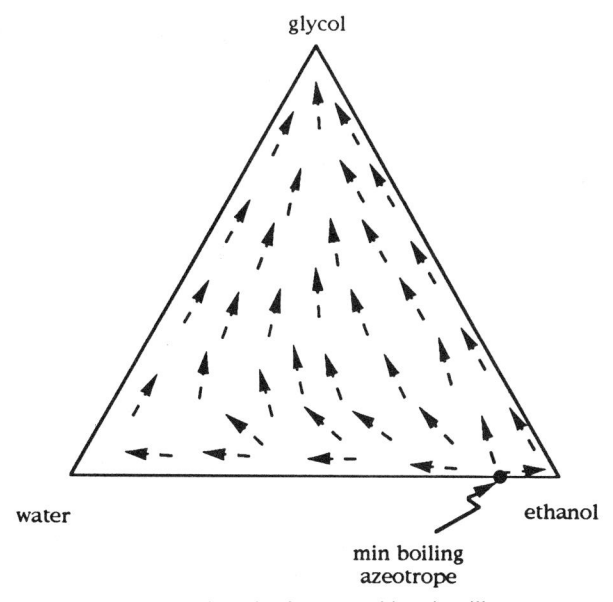

FIG. 16. Trajectories for compositions in stillpot.

(Alternatively, we could sketch distillation curve trajectories, which look very similar.) All of the residue curve trajectories move from the lowest temperature in the diagram—the ethanol and water azeotrope—and end at the highest temperature in the diagram—pure glycol. At any point in a residue plot, the equilibrium vapor composition is moving in the direction that is tangent to the curve, and in the opposite direction the trajectory is moving with time.

There are examples where the trajectories break the diagram into regions. Such a structure appears when there is more than one local minimum and/or maximum temperature in the diagram. We see such behavior in Fig. 17. The lower two species again have a minimum-boiling azeotrope between them; the third (top) species is the most volatile one. To guess the topological behavior for this diagram, we first place temperatures for the normal boiling point for the pure species and for the azeotropes onto the diagram. We then place arrows around the edges to indicate the directions for increasing temperatures, as shown. From just these few arrows, we observe that there are two local maximum temperatures in this diagram, one in the lower right and one in the lower left. Let us assume that there can be at most one ternary azeotrope and then attempt to sketch in trajectories within the diagram to expose its general structure.

Occasionally we can posit more than one structure possible, based on knowing only the temperatures for the pure species and the binary azeotropes. Since such structures can occur, care must be taken to see that all possible structures are discovered. Based on topological arguments, Zharov and Serafimov (1975; see also Serafimov, 1987) developed an equation among the number of "nodes" and "saddles" appearing in a residue curve map. Independently, Doherty and

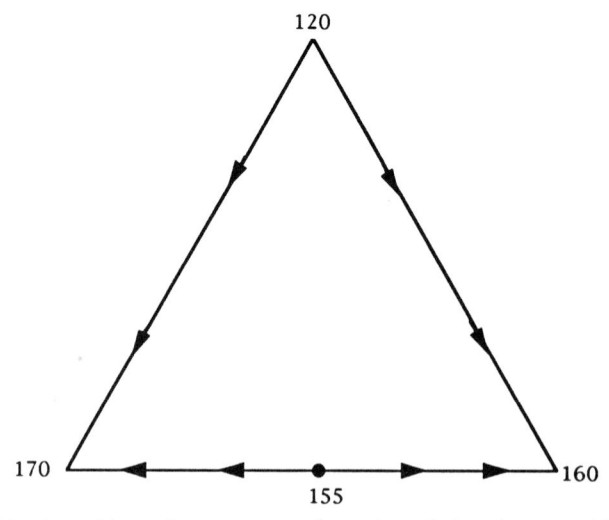

FIG. 17. Directions of increasing temperature along edges of triangular composition diagram.

Perkins (1979) developed a special version of this equation for three-species mixtures. Foucher *et al.* (1991) present detailed rules to sketch and to check the consistency of data on ternary diagrams. One version of this equation for three species mixtures is

$$4(N_3 - S_3) + 2(N_2 - S_2) + (N_1 - S_1) = 1$$

where N_i is the number of nodes and S_i is the number of saddles involving exactly i species on a ternary diagram.

A *node* is any point where all temperature trajectories enter or all leave, whereas a *saddle* is a point where some trajectories enter while others leave. The pure species in the lower left and right corners have all trajectories entering; the top pure species has all leaving. As each involves one species, each is a *single-species node*. We cannot yet classify the binary azeotrope along the lower edge. Its type will depend on whether trajectories enter it or leave it from the interior of the composition triangle. If they enter, then some trajectories will enter while those along the lower edge leave, making it a *two-species* or *binary saddle*. If they leave, all the trajectories leave, making it a *two-species node*. Points strictly inside the diagram are three-species points.

A recent review article by Fien and Liu (1994) describes how to apply this formula in some detail. Applying it for the ethanol–water–glycol example in Fig. 16, we see that $4(0 - 0) + 2(1 - 0) + (1 - 2) = 1$ satisfies this equation. If there can be only one ternary node or saddle, only $N_3 = S_3 = 0$ can satisfy this equation and the sketch must be unique.

For Fig. 17, we might ask if the azeotrope at 155° is a node or a saddle, or could it be either based on the information given. Substituting into the formula for both options, we get

If a node: $\quad 4(N_3 - S_3) + 2(1 - 0) + (3 - 0) = 1$

$$\text{or} \quad 4(N_3 - S_3) = -4$$

If a saddle: $\quad 4(N_3 - S_3) + 2(0 - 1) + (3 - 0) = 1$

$$\text{or} \quad 4(N_3 - S_3) = 0$$

The former would allow a ternary saddle (i.e., $N_3 = 0$, $S_3 = 1$) to exist while the latter permits only $N_3 = S_3 = 0$—i.e., no ternary node nor saddle—if our assumption is valid that there is at most one ternary node and/or saddle.

Figure 18 shows the diagram with the required ternary saddle where the binary azeotrope is a node. It has four *distillation regions* (labeled I to IV) separated by *distillation boundaries*, each of which has its own set of maximum and minimum temperatures within it. For example, region I has a maximum temperature of 170° and a minimum of 120°. It is this property of having its own unique minimum temperature (from which all trajectories emanate) and maximum temperature (at which all terminate) that characterizes a region.

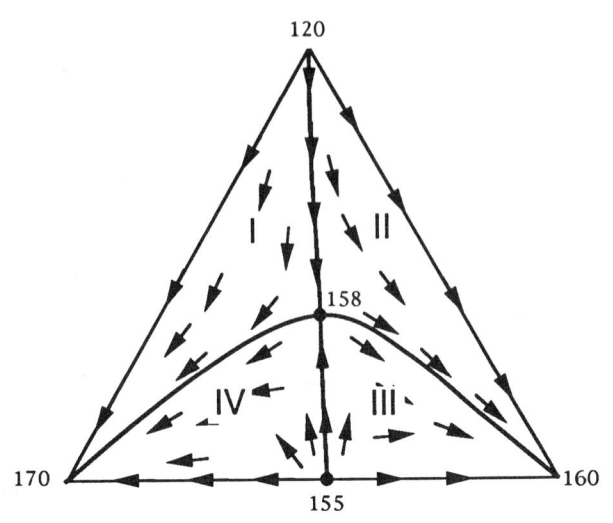

FIG. 18. Triangular composition diagram where binary azeotrope is a node.

Assuming the binary azeotrope is a saddle (as shown in Fig. 19), we note that there is a trajectory that starts at the upper vertex and passes through the minimum-boiling azeotrope on the lower edge. This trajectory is a distillation boundary that splits the diagram into two distinct distillation regions, labeled I and II. Each region has the same minimum temperature but a different maximum temperature within it.

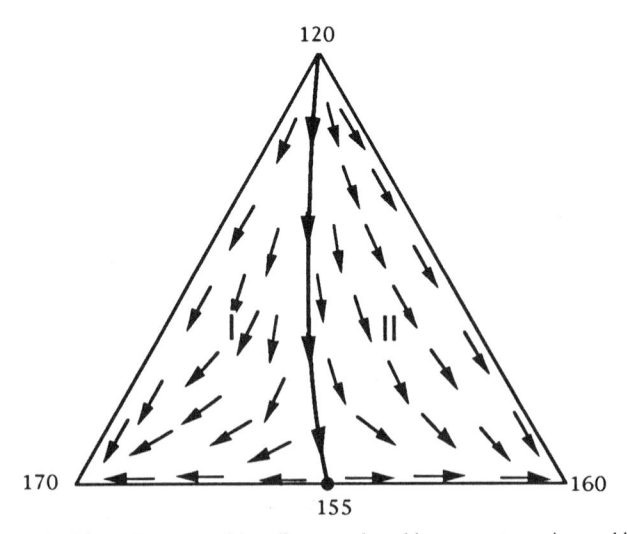

FIG. 19. Triangular composition diagram where binary azeotrope is a saddle.

The only way we can select between these two options is to obtain more information, either by computing the trajectories using a physical properties model we believe or by obtaining experimental data for the problem.

B. Distillation Column Behavior

1. Limiting Behavior

There are two limits at which we can examine the behavior of a distillation column. The first is at *total reflux* (i.e., with an infinite reflux ratio, which is often called *infinite reflux* conditions). The other extreme is to operate at *minimum reflux*. In this section we shall limit our discussion to the total reflux case; in later sections we shall look at operating columns at finite reflux (ratio) conditions. Intuitively, we tend to expect that a column will give its maximum separation when run at infinite reflux. While this is true for ideally behaving species, it does not have to be true when separating nonideally behaving species. Thus, we need to look carefully at running columns all the way from minimum to total reflux conditions.

2. Reachable Products for Total Reflux

One of the steps in developing alternative structures for a separation process is to discover the possible products for a proposed technology. There is no general method to accomplish this task for mixtures displaying complex behavior, even for distillation. For nearly ideally behaving mixtures, identifying possible products is a trivial task, sufficiently so that it is seldom recognized as a required step in posing solutions. For the special case of separating ternary mixtures using distillation columns that produce two products, Wahnschafft (1992) and Wahnschafft *et al.* (1992) show that all possible products for azeotropic mixtures can be determined with a relatively simple analysis which involves residue and distillation curves and pinch point trajectories.

To trace out these curves is to compute a sequence of flash calculations (a relatively easy task compared to computing distillation column performance). What is remarkable is the fact that we really need to use residue curves (those curves produced by solving ordinary differential equations) for a part of this analysis. Until now, we have assumed that these curves corresponded only to the time behavior of a batch still. Here, however, they become a necessary ingredient in the analysis of a staged distillation column behavior instead of a convenient approximation.

To understand the reachable-product problem, let us imagine separating an equimolar ideally behaving three-species mixture of species A, B, and C (A being most volatile, C the least) in a two product column. We illustrate these ideas in Fig. 20. Assume the column has a very large number of trays and that it will be run at very high reflux.

We first remove only one drop of distillate top product D, letting the rest of the feed exit from the bottoms product of the column. The distillate will be pure A, while the bottoms will be virtually all of the feed. We show these two where the distillate-to-feed ratio *D/F* is zero. We then draw off 1% of the feed in the top and 99% in the bottom. We will continue to remove pure A in the top and the rest of the feed in the bottom. Material balance dictates that the feed must lie on the straight line joining the distillate and bottoms compositions. The bottoms moves directly away from the feed toward the BC edge as we remove more and more A. When we reach the point where we are removing one-third of the feed in the top, we will remove essentially all the A in the top and all the B and C in the bottoms. If we remove 50% of the feed from the top, it will be all of the A and half of the B, with the remaining half of the B leaving in the bottoms together with all the C. When the distillate is two-thirds of the feed, we have all the A and B in the top and all the C in the bottoms. At 80%, we take all the A and B and some of the C out the top, with the bottoms being the rest of the C.

We can then backmix whatever is taken off in either product with the other to reach any compositions that lie between the product compositions. Thus, we can reach any product in the shaded bow-tie region on this figure.

At total reflux, the column itself (without the backmixing) can only reach products that lie both on the same *distillation curve* (not residue curve) and on a straight line passing through the feed so as to satisfy the overall column species

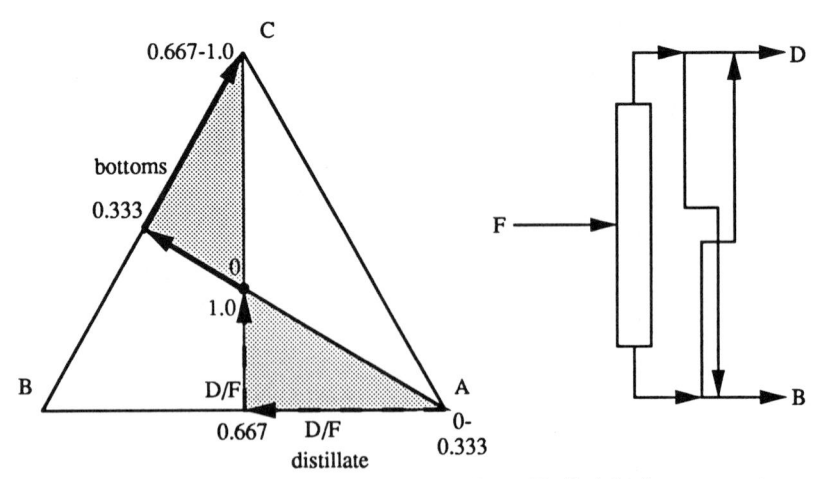

FIG. 20. Distillate and bottoms products versus *D/F* for an ideally behaving ternary mixture.

material balances. The limiting distillation curve for nearly ideal mixtures is that passing through the feed.

For this limit, think of removing one drop of top product, with the rest leaving in the bottoms while operating at near total reflux conditions. The top distillate product (one drop) will lie anywhere along the distillation curve passing through the feed, while the bottoms will be the feed. A similar argument holds for removing one drop in the bottoms at (near) total reflux conditions. Figure 21 shows the reachable region for such a column without the backmixing that we allowed before.

For nearly ideally behaving mixtures, the above completes the analysis needed to identify the reachable region at total reflux. For nonideally behaving mixtures, the shape of the distillation curves can lead to very interesting reachable-product regions. An S-shaped curve, for example, can lead to two disjoint reachable regions, a situation we shall examine later. The rule to remember for total reflux is that the column can produce any distillate product D and bottoms product B where

- the compositions for D and B lie on the same distillation curve, and
- the composition for the feed F lies on a straight line between the compositions for D and B.

VIII. Synthesis Method for Nonideal Mixtures

We now consider the separation of a nonideal mixture where the species do not display liquid/liquid behavior. Almost certainly, the technology of choice

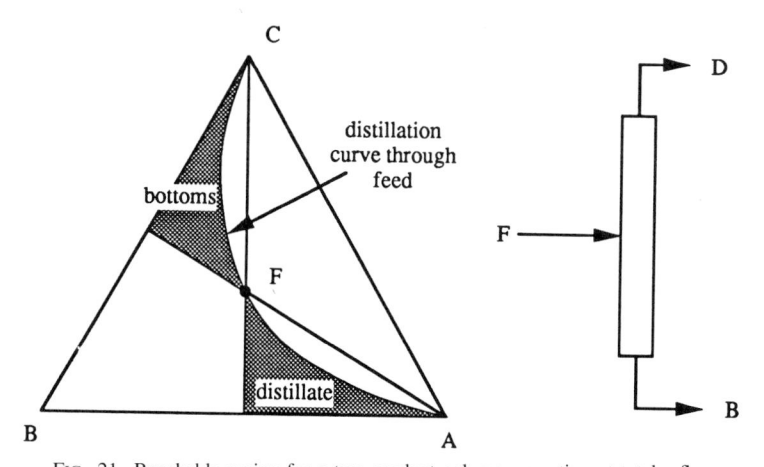

FIG. 21. Reachable region for a two-product column operating at total reflux.

108   ARTHUR W. WESTERBERG AND OLIVER WAHNSCHAFFT

will be distillation. In the following example, we examine the need to discover reachable interesting products when feeding a mixture into a distillation column. By selecting different interesting product sets, we generate design alternatives. As with the *n*-butanol/water separation example, we shall also see the need for recycle in completing the design. Here we explore two different reasons for using recycle: (1) to separate a mixture produced later in the process that is similar to an earlier one (i.e., discovering a *recursive solution*) and (2) to adjust the feeds to a column to allow it to produce two interesting products instead of one (i.e., *mixing to get better separation,* an apparent contradiction).

A. AZEOTROPIC SEPARATION: EXAMPLE 2

Consider the separation of the mixture of acetone, chloroform, and benzene shown in Fig. 22. How do we generate alternative structures systematically for solving this problem?

We start by sketching the "binary separation tasks" for this problem, as shown in Fig. 23. As we suggest separations, we can keep track of which portions of these tasks they accomplish. We use the arrows on Fig. 23 to show the minimum separation that must be accomplished between each of the pairs of species if the product specifications are to be met. The feed composition is also illustrated. To interpret this diagram, suppose the feed is 36% acetone and 24% chloroform, yielding a

$$\text{Benzene-free composition} = \frac{36}{36 + 24} = 60\% \text{ acetone}$$

We can see this on the sketch for the acetone/chloroform pair. The chloroform product has to be at least 99% chloroform so it can contain at most 1% acetone,

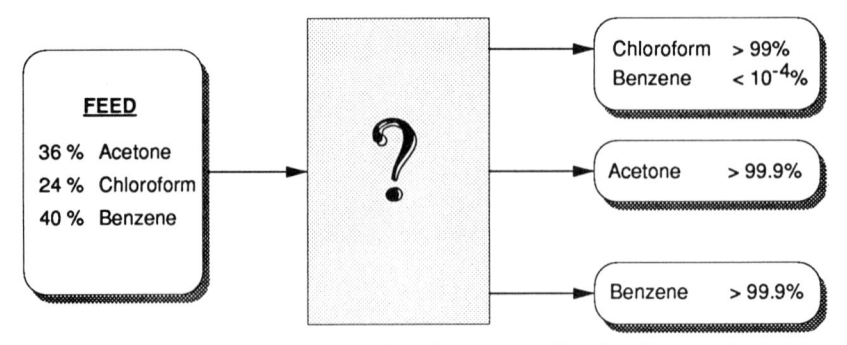

FIG. 22. Specifications for the separation of an acetone/chloroform/benzene mixture.

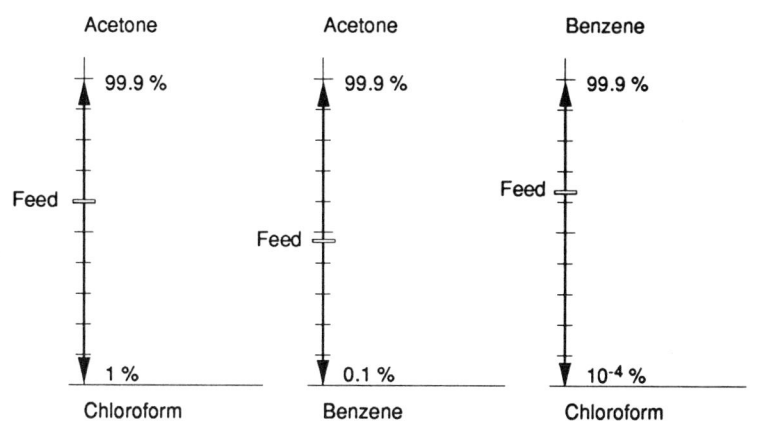

FIG. 23. Binary separation tasks and relative concentrations of the feed mixture.

while the acetone can have no more than 0.1% chloroform in it for it to be at least 99.9% pure.

We can map these tasks on the edges of our ternary composition diagram, as shown in Fig. 24. Note that a line projected from the pure acetone corner through the feed to the opposite edge has a constant ratio of benzene and chloroform all along it. Thus, the projected point on the benzene/chloroform edge

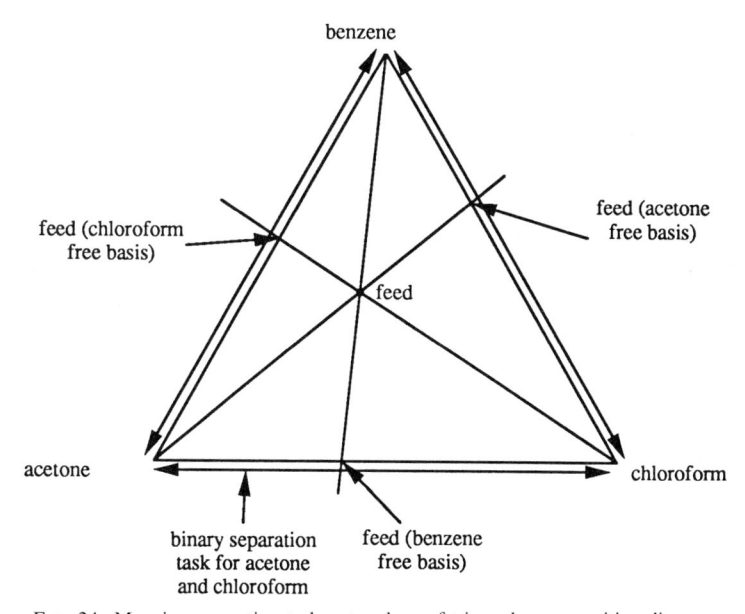

FIG. 24. Mapping separation task onto edges of triangular composition diagram.

is the composition of the feed on an acetone-free basis. Similar projections from the chloroform and benzene corners give the feed compositions on a chloroform-free and benzene-free (should be 60% acetone) basis, respectively. The mapping of the tasks is the result of similar projections. Note that the task along the bottom is to separate acetone from chloroform. Near the chloroform edge, we see a small gap, which says the chloroform product on that basis can be as much as 1% acetone.

We next need to know how these species will behave. First, their normal boiling points are 56.5°C, 61.2°C, and 80.1°C for the acetone, chloroform, and benzene, respectively. Thus, acetone is the most volatile, while benzene is the least. The lowest temperature, 56.5°C, is hot enough for cooling to be done using cooling water (i.e., above 25–30°C), so it makes sense to consider running the columns at a pressure of 1 atm.

We predicted their behavior earlier using infinite-dilution K-values, with the results at 1 atm shown in Table VIII. Only the acetone and chloroform appear to display azeotropic behavior. With this information and that for pure species boiling points at the pressure of interest, we can sketch the ternary diagram for this mixture. We can also use a computer code to generate it, which was done for Fig. 25. We see that there is one maximum-boiling azeotrope between acetone and chloroform.

Two features appear on the residue curve diagram in Fig. 25:

(1) A distillation boundary exits. We deduce this when attempting to explain the azeotropic behavior determined using infinite-dilution K-values.

(2) The boundary is curved. This, too, can be partially predicted by noting that the infinite-dilution K-values for acetone and chloroform in lots of benzene indicate that acetone is more volatile. Therefore, chloroform acts like an intermediate species in the benzene-rich end of the diagram. The residue curves start out aiming at chloroform from benzene.

What if we are dealing with a computer that does not draw or read these nice graphs, or what if we are dealing with more than three species? How might we proceed to find this behavior? We can carry out column simulations for a

TABLE VIII
INFINITE-DILUTION K-VALUES AND THEIR INTERPRETATION

		K^∞	
	Acetone	Chloroform	Benzene
In: Acetone	1.0	0.4 (max)	0.7 (normal)
Chloroform	0.6	1.0	0.4 (normal)
Benzene	3.0	1.5	1.0

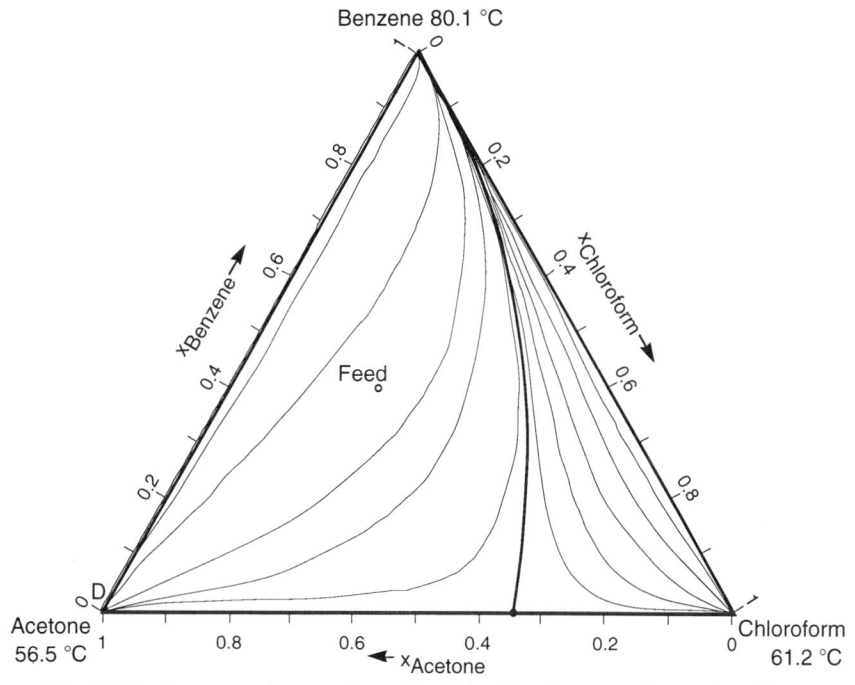

FIG. 25. Residue curve diagram for acetone/chloroform/benzene mixture. A residue curve boundary passes from the maximum-boiling azeotrope between acetone and chloroform to pure benzene.

column having lots of trays and a very high reflux ratio. The simulations can be carried out versus the amount of distillate product drawn off relative to the amount of feed, i.e., versus D/F. We shall assume for the moment that a column with lots of trays and a large reflux ratio will give us the maximum separations possible. That is true for ideally behaving mixtures, but (as we noted earlier and shall discuss later) it does not have to be true for azeotropic mixtures. Nonetheless, we proceed with that assumption.

Figure 26 is a sketch of the results obtained by carrying out these rigorous simulations. (It should be noted that some of these simulations can be very difficult to converge.) For small D/F, the top product of our column is pure acetone, the most volatile species in the region of the feed. At $D/F = 0$, the bottoms product is, in fact, the feed. As we take more and more overhead, the bottoms product moves on a trajectory away from the feed composition in the direction opposite the pure acetone vertex, continuing either until all the acetone is removed or until the bottoms product hits a distillation boundary. If we do not hit a distillation boundary, we should continue to get pure acetone until $D/F = 0.36$, the fraction of the feed that is acetone. However, at D/F values

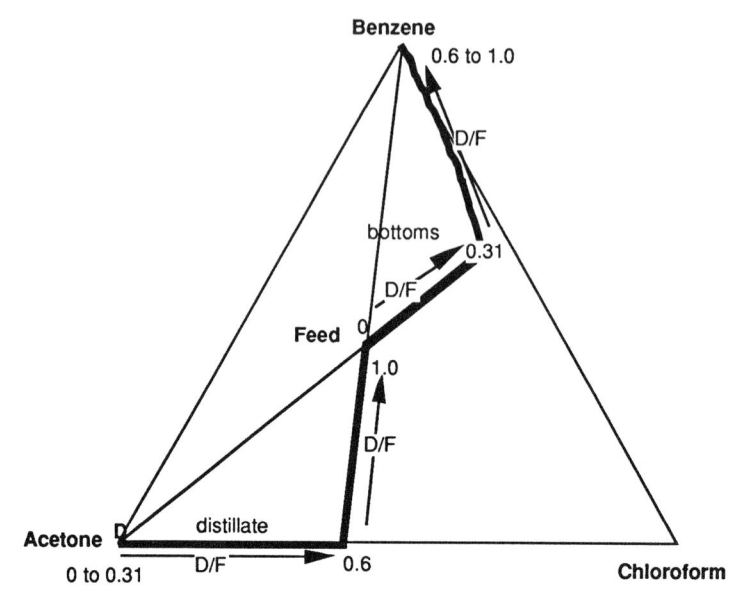

FIG. 26. Trajectories for distillate and bottoms products versus D/F for column.

above 0.31, the rigorously simulated column will no longer yield pure acetone in the top. We surmise we have hit a distillation boundary at the bottom of the column as the bottoms still contains acetone, and we now find chloroform showing up in the distillate.

Increasing D/F further causes more and more chloroform to appear in the tops while all of the benzene remains in the bottoms; it becomes purer and purer in benzene as we proceed. When D/F reaches 0.60, the bottoms is 40% of the feed, and it is essentially pure benzene. The feed was 40% benzene. Thus, the top must be essentially pure acetone and chloroform, which it is. Increasing D/F further forces some of the benzene to exit with the top product. At $D/F = 1$, the top product is the feed. The limiting composition for the bottoms is pure benzene, the point where the temperature is highest in the distillation region in which the column is operating.

We have, with this set of computations, discovered a portion of the residue curve boundary for this mixture. It is the portion that is most relevant for our feed mixture. This analysis also gives us a first estimate for the set of *reachable products* using distillation for this mixture. (The analysis is not complete, as we shall see later, because it does not discover what we might be able to reach with smaller reflux ratios.)

It is possible to show these results without using a triangular composition plot. In this form now to be proposed, we can imagine using a computer to detect the behavior we have just described.

First, we formally define the idea of a *binary separation range* for each of the binary pairs. This range is the *length* of the vectors on the plots we used earlier to indicate the part of a task that a column accomplishes. An equation to compute such a range for each value of *D/F* used in simulating the column is the following:

$$\text{separation range}_{ij} = \left| \left(\frac{f_i}{f_i + f_j} \right)_{\text{distillate}} - \left(\frac{f_i}{f_i + f_j} \right)_{\text{bottoms}} \right|$$

If both species go entirely to opposite products, the separation range is unity. If both are entirely in one product, there is no separation and the separation range has a limiting value of zero. We are looking for the maximum points in this plot, Fig. 27. (There are other useful representations we could plot, such as the molar flowrates for each of the species leaving in the distillate or bottoms or the splits on the species—e.g., 50% going into the distillate. Each of these gives a slightly different view, but all are aimed at indicating interesting products among the many possible.)

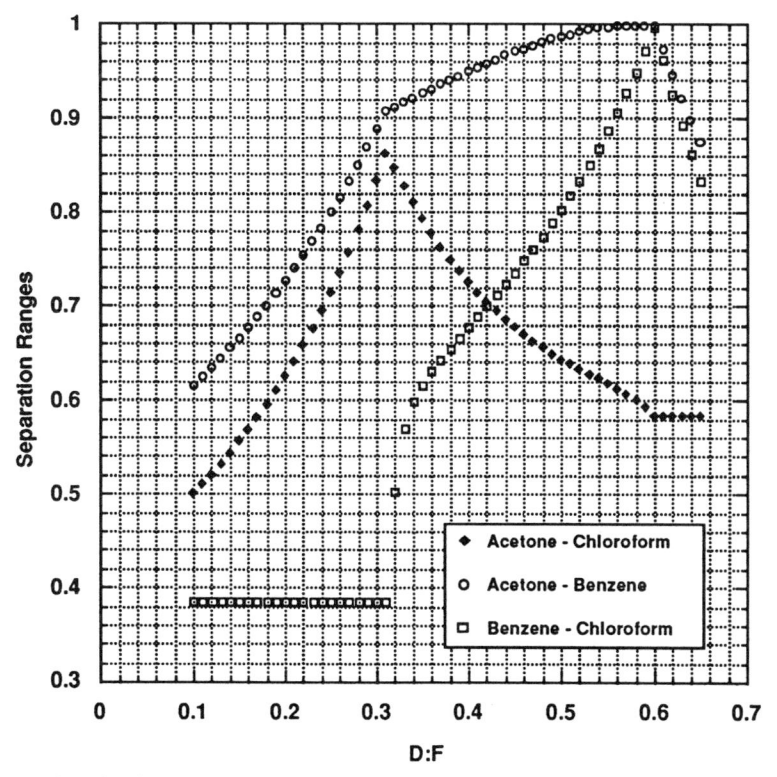

FIG. 27. Separation ranges obtained by rigorous distillation column simulation.

We would see the same interesting points we discovered before in these data:

- At **D/F of 0.31,** we *maximize* the separation of acetone to chloroform.
- At **D/F of 0.6,** both the acetone/benzene and benzene/chloroform ranges maximize at *unity*—implying that all the benzene is in one product and acetone and chloroform are in the other.
- At **just below D/F of 0.6,** we find a point where the amount of acetone in the bottoms product is so low that acetone can exit with the chloroform in a subsequent column while leaving the chloroform product contaminated with no more than 1% acetone, which is pure enough to meet specs.

We show these separations on a triangular composition plot in Fig. 28. The three interesting splits are as follows:

- The direct split: A top product of *pure acetone,* the bottoms being a mixture of all three species in significant amounts.
- The indirect split: A bottom product of *pure benzene,* the top being *only acetone and chloroform*—(two interesting products).
- The intermediate split: A bottom product which is a mixture of benzene and chloroform with a trace of acetone. The ratio of acetone to chloroform

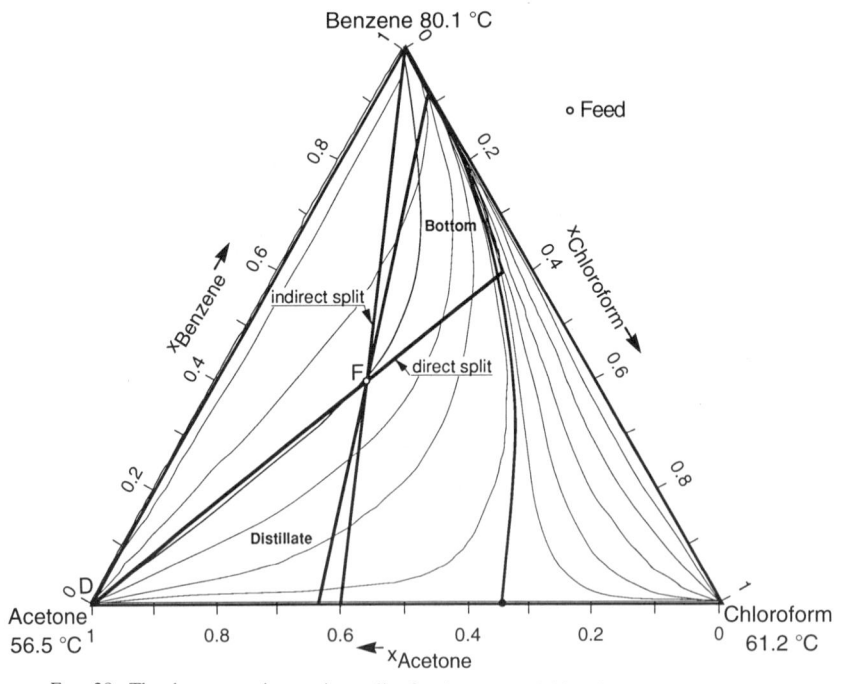

FIG. 28. The three most interesting splits for the acetone/chloroform/benzene problem.

in this bottom product is just that required for the chloroform product—1% acetone relative to the chloroform. The top is a mixture of acetone and chloroform.

1. Indirect Split: Alternative 1

The indirect split produces two interesting products in one column. However, we can quickly see that removing pure benzene as the bottom product in the first column is not a useful starting point. It leaves us with a binary mixture of acctonc and chloroform to separate. A second column will give us a top product of acetone and a bottom product which is the maximum-boiling azeotrope formed by acetone and chloroform. We will need some way to break this azeotrope, which, as we shall discover, can be done using benzene. Thus removing benzene first is, in fact, counterproductive. Anticipating better success by leaving benzene in, we shall rule out considering this option.

2. Mixing to Get Two Desired Products from One Column: Alternative 2

We might be able to use mixing to get two interesting products from a single separation. We note we can mix benzene with the original feed to move the material balance line so it permits acetone to be the top product and a fairly acetone-free benzene/chloroform mixture to be the bottom product. Doing so means we can get a solution to our problem that involves only two columns. The material balance lines for the mixing task and the two columns are shown in the triangular diagram in Fig. 29, along with the corresponding process flowsheet.

First, benzene is mixed with the feed, yielding the mix point M. This mixture feeds column 1, yielding a pure acetone top product and a bottoms of virtually all the benzene and chloroform plus a trace of acetone. This bottom product is fed to column 2, where it is separated into benzene product and chloroform product. Part of the benzene product is then recycled to mix with the original feed.

Recycling material to alter a column feed so a single separator can produce two desired products can often be a feature of these processes. In the next alternative solution, we shall again propose a recycle, but this time it allows us to separate an intermediate product.

3. Direct Split: Alternative 3

The next alternative flowsheet we shall consider starts with the direct split, where we take pure acetone off as product from the first column. The material

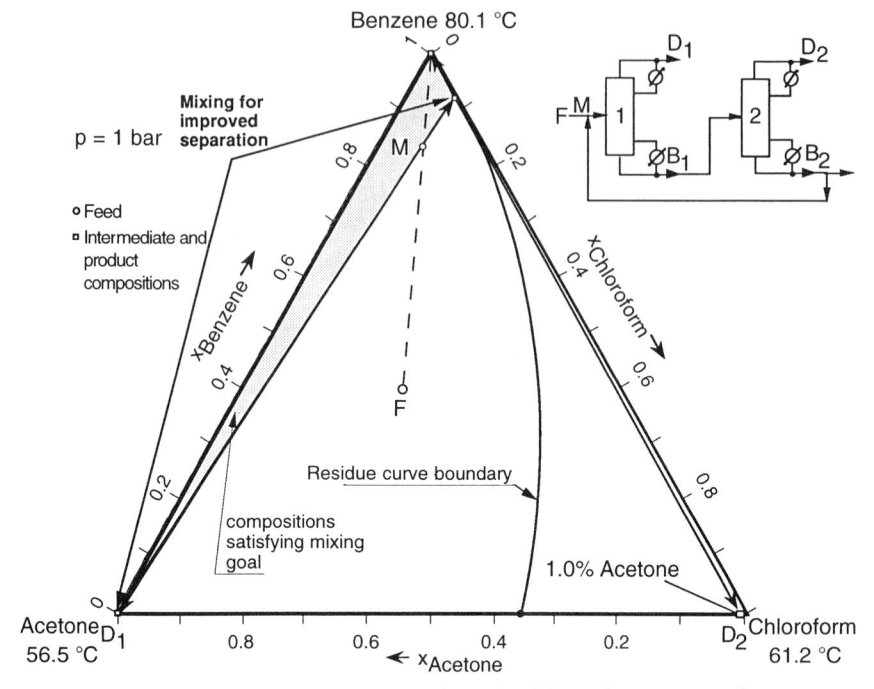

FIG. 29. Flowsheet resulting from use of recycle mixing to improve separations.

balance lines for the columns and mix points required are shown in the triangular plot in Fig. 30.

The first column produces nearly pure acetone from the top and a mixture of all three species in the bottom. This mixture is near the distillation boundary and occurs when D/F is about 0.31, as we saw earlier. We feed the bottom product to a second column, which separates benzene (bottom product) from the acetone and chloroform in its feed (top product). We find we can draw a material balance line through the mixture fed to column 2, which connects benzene to a mixture of acetone and chloroform in the other (right-hand side) region *because of the curvature of the distillation boundary.* This curvature is often important in devising separation schemes.

The top product from the second column, D_2, is separated into pure chloroform (top, D_3) and a mixture of acetone and chloroform very near the maximum-boiling azeotrope (bottom, B_3). We now seem to have an impasse, as we have an azeotrope to separate, just as we did for alternative 1. This time, however, the impasse is not "impassable."

We examine the separation ranges covered by the three columns already proposed. Figure 31 shows these ranges. To read this figure, examine the three ranges shown for column 1. The range covered by column 1 for the acetone/

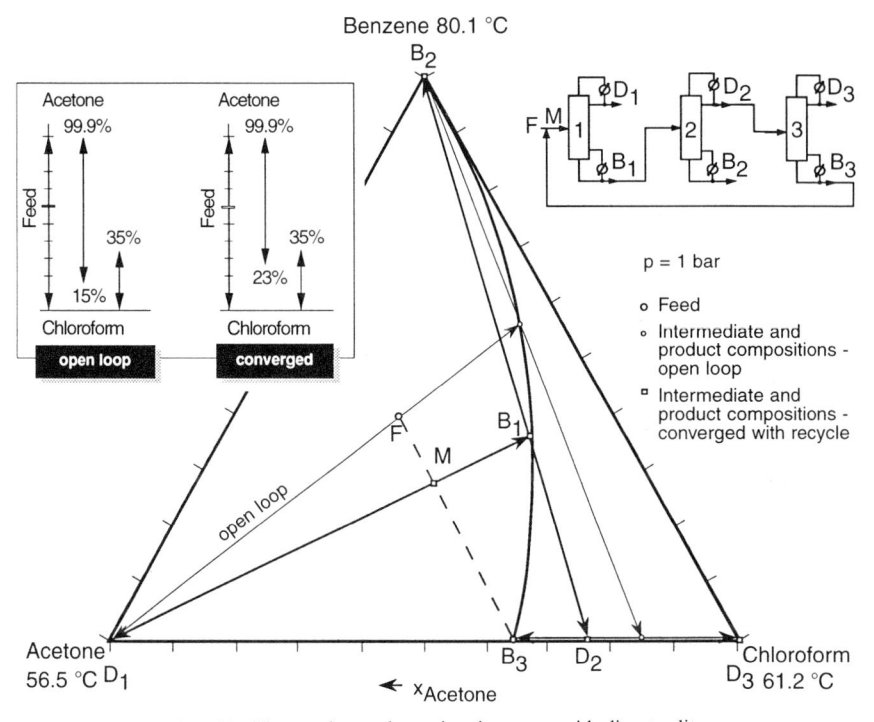

FIG. 30. Three-column alternative that starts with direct split.

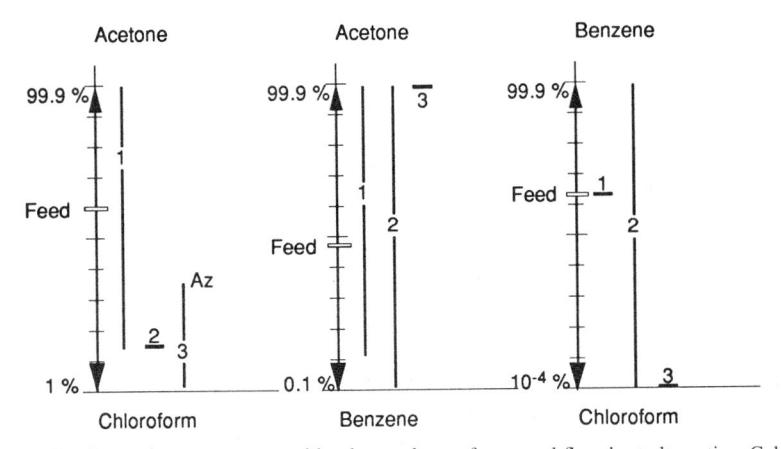

FIG. 31. Separation ranges covered by three columns for second flowsheet alternative. Column numbers are shown on their respective ranges.

chloroform binary mixture is from a top product of pure acetone to a bottom product that is about 15% acetone in chloroform—the mixture on the distillation boundary. We can discover the "15%" by projecting from pure benzene through the bottom product to the acetone/chloroform edge. This projection gives the composition of the mixture on a benzene-free basis.

Projecting from the chloroform vertex through the top and bottom products to the acetone/benzene edge gives the range covered for the acetone/benzene split by column 1. The binary range is from pure acetone (top) to a mixture of about 11% acetone in benzene (bottom).

Projecting from the acetone vertex to the benzene/chloroform edge shows that column 1 does not affect the ratio of benzene to chloroform. It is the same in all products as it is in the feed (for the pure acetone, this is a limiting condition). We show the range for column 1 for this binary mixture as being unchanged from the feed.

Figure 31 also shows the results for columns 2 and 3. Examining all these separations, we see that the entire ranges for all binary pairs are covered with the structure proposed so far. We have, in a sense, already solved the separation problem. We should therefore consider feeding the azeotropic mixture coming off the bottom of column 3 back into this process to separate it. We propose putting it back into column 1, as all three columns are needed to cover all the ranges that this feed requires for it to be separated. Figure 30 shows the flowsheet for this option with this recycle.

Feeding the azeotrope back shifts the composition of the total feed to column 1 to a point between F and B_3. We need to check that the resulting flowsheet will function as proposed. It does: the point M is the result of carrying out a rigorous simulation for the flowsheet, including the recycle. In the upper left of Fig. 30, we show the range covered by the acetone/chloroform split by the original structure (open loop—i.e., having no recycle) and by the final structure (with the recycle).

Finding that an intermediate column product can be recycled because a structure already exists to separate it is like discovering a recursive solution to our separation problem. This reason for recycling is different from the one for the previous alternative; it is a necessary part of many of these flowsheets.

4. Intermediate Split: Alternative 4

The last alternative we consider is to start with the intermediate split, where we remove a mixture of benzene, chloroform and a trace of acetone in the bottom product of the first column. (See Fig. 32.) Without a systematic approach, we would very likely miss this alternative. We separate the bottom product from the first column into benzene and chloroform in a second column. The top product from the first column is a mixture of acetone and chloroform,

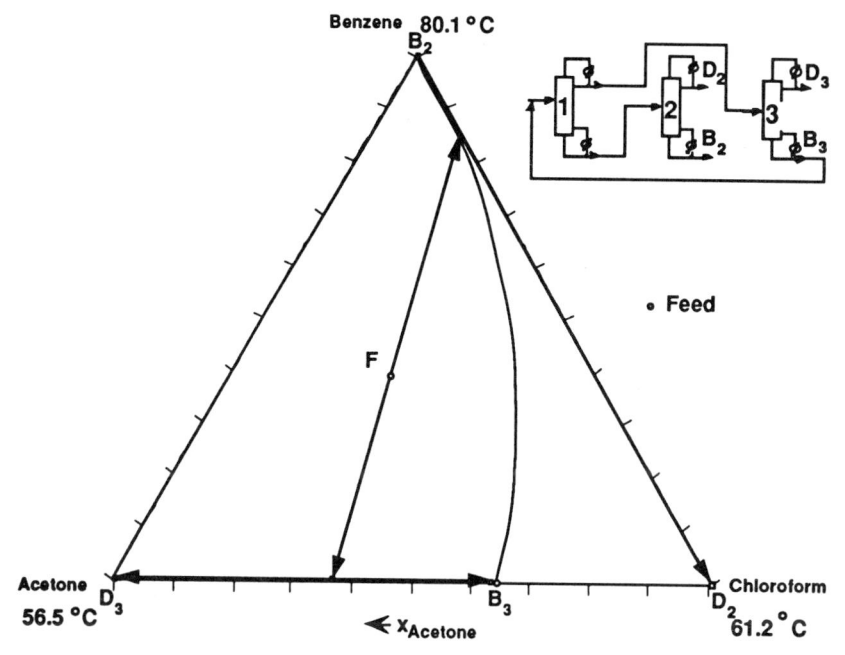

FIG. 32. Alternative three-column flowsheet. Again, note mixing of the azeotropic mixture, B_3, with the feed.

which we separate in a third column into acetone and the azeotrope between the acetone and chloroform. Using reasoning similar to that for the third alternative, we propose recycling this azeotrope back to the first column.

5. A General Approach to Creating Alternatives

Figure 33 summarizes the steps we take to generate alternatives for these complex separation processes. We can illustrate the approach by reviewing the steps we took for the current problem.

Starting with a feed mixture and product specifications, we analyze the feed and products to discover the binary splits required. We then place the feed onto a stack of streams to be processed. The stack is not empty, so we select the feed stream. It is not a product. We determine its binary concentrations and pass through the next two steps downward because no structure already exists that could serve as a partial structure to separate it. We select a separation method using our knowledge about the problem; here, we pick distillation. The "v" shown in this box means that several different options might be proposed here, with the different solutions branching along different paths. As we are considering ordinary distillation, we do not need to pick a separation agent.

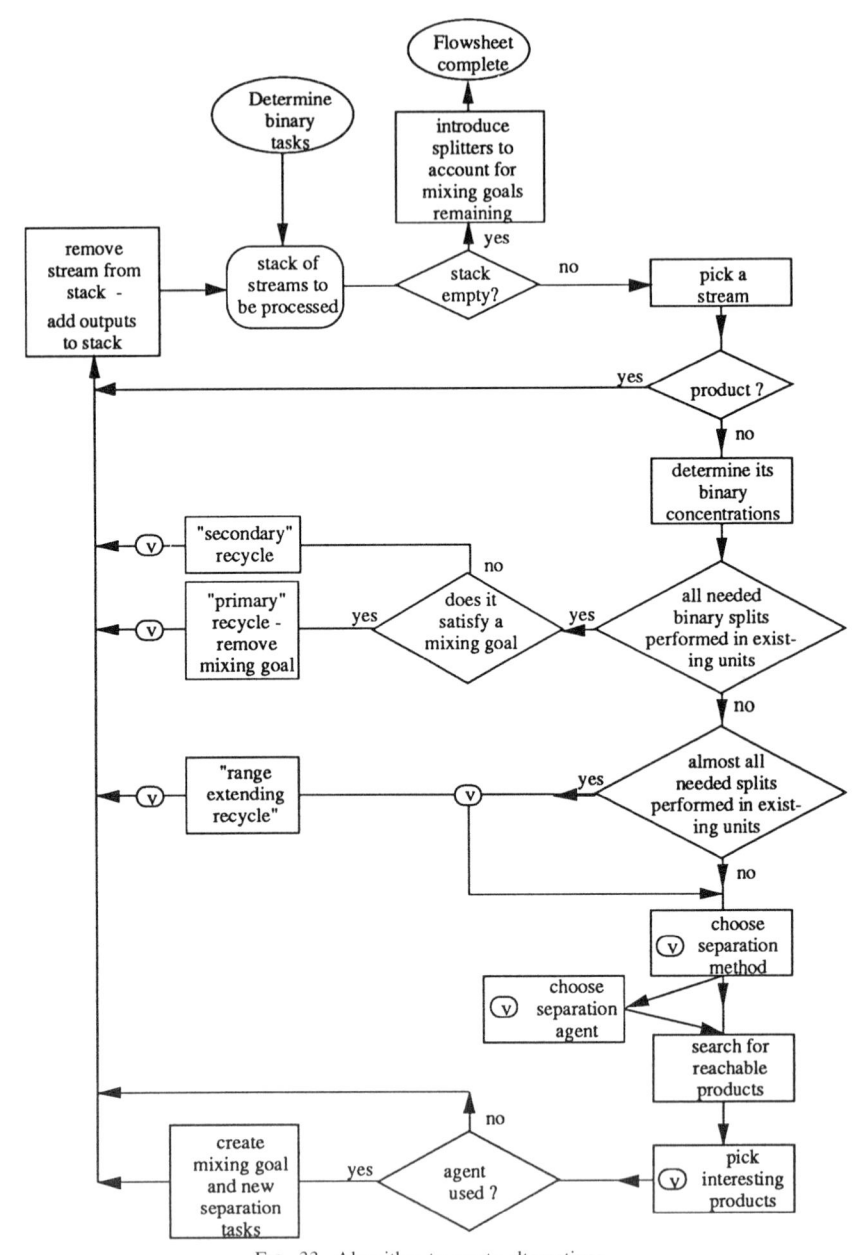

FIG. 33. Algorithm to create alternatives.

Simulations for differing D/F ratios sweep out the reachable separations. The interesting products among these are acetone, pure benzene, an acetone and chloroform mixture with no benzene, and a bottom mixture of benzene, chloroform, and a trace of acetone. We need to consider each of these as starting points for alternatives.

No separating agent was used, so we consider using mixing to create separation tasks that might allow two interesting products to occur in the same next step. We do see such a possibility—mixing something (which we selected to be benzene) with the feed to move it to a material balance line between acetone as a top product and the benzene/chloroform mixture with a trace of acetone as the bottom product. The need for something to mix with the feed is placed on the list as a new separation task. *Note:* Anything above the material balance line that can move the feed to that line is a legal stream for mixing. It just happens that benzene is a fairly obvious choice here.

We remove the feed from the stack and continue with the current option. We have a top product of acetone. Since acetone is a product, we cycle back from the step that tests if it is and remove it from the stack. We also have the benzene/chloroform mixture with a trace of acetone to process.

We make this mixture the stream to consider and proceed through the steps with it. The alternative flowsheets arise when we return to steps where we created alternatives, such as the one in the lower right where we have a number of interesting products not yet considered.

There are three recycle decision steps shown in the middle of the diagram. An instance of a primary recycle is the use of another stream (we used benzene) to move the feed to the material balance line between two interesting products so that both may be produced by one separator. Benzene, being a product, would be discovered in the step before last (introduce splitter to account for mixing goals remaining) rather than here, however. The recycling of the azeotropic mixture of chloroform and acetone back to the first column is an example of a secondary recycle. We have not seen the third case for recycle—a range extending recycle. There are times when a downstream separator does not have some of the species in its feed which, if they were there, it would separate to some extent. Thus we might still propose recycling a stream if the flowsheet proposed so far almost covers all the splits needed to process that stream.

We now consider a more complex example where liquid/liquid extraction and extractive distillation are among the processes to appear in the solution.

B. AZEOTROPIC SEPARATION: EXAMPLE 3

We wish to devise a separation process based on distillation, liquid/liquid extraction, and extractive distillation for the mixture of solvents shown in Fig.

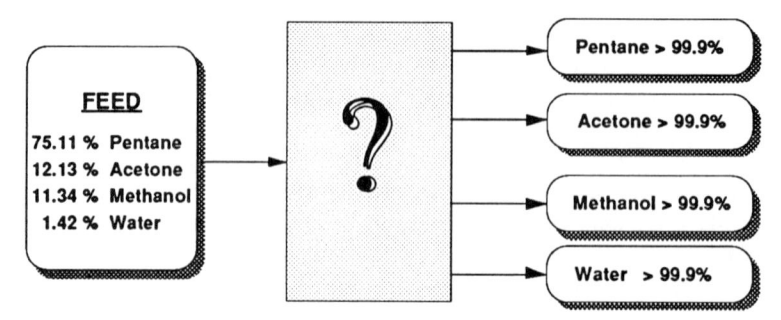

FIG. 34. Specifications for feed and products for Example 3.

34. Solvent recovery systems are often very complex systems to design, as they need to separate very different molecules.

We start by computing the infinite-dilution binary data shown in Table IX. The upper values are K-values at the boiling point of the more plentiful species. We will also need liquid activity coefficients if we wish to consider extraction processes; the lower values in each entry are infinite-dilution binary liquid activity coefficients at ambient conditions.

We now need to think of all the reasonable ways we might separate this very complex mixture. We shall use insights from the above data as well as any insights we have as chemical engineers. This step is a knowledge-intensive one.

Since pentane and water exhibit immiscibility, we might consider *decantation* as the first step. If it worked, it would be an inexpensive one to carry out. But a rigorous three-phase equilibrium calculation predicts that, in the presence of acetone and methanol, the small water fraction in the feed does not form a second liquid phase; so we reject this idea. The calculation also reveals that the feed mixture is almost at the azeotropic composition for the pentane/methanol binary pair.

TABLE IX
INFINITE-DILUTION PAIRWISE K-VALUES AND ACTIVITY COEFFICIENTS FOR EXAMPLE 3

$C_j \setminus C_i$	Pentane	Acetone	Methanol	Water
Pentane	1.0	3.0 (min)	5.9 (min)	71.4 (het)
	1.0	6.6	23.1	1537
Acetone	7.9	1.0	1.3 (min)	1.05 (min)
	4.7	1.0	2.0	7.4
Methanol	29.6	2.4	1.0	0.4 (ok)
	14.4	2.0	1.0	1.6
Water	8106	38.5	7.8	1.0
	3213	11.5	2.2	1.0

Next, we consider *distillation;* but, with the exception of water and methanol, all other pairs exhibit azeotropic behavior. Also, water is present in very a small amount here. We should be trying to get rid of the pentane first. Thus, distillation does not seem an appropriate first step.

If distillation is rejected, we might consider *extractive distillation.* The K-values of acetone and methanol at infinite-dilution (38.5 and 7.8, respectively) in water indicate that water could be used as an extractive distillation agent for the separation of these species. However, adding water would almost certainly introduce a second phase with the pentane.

If used, we would create two liquid phases in view of the calculations we did above to see if decantation is a good first step.

If water will force us to have a second phase, we might then consider using *liquid/liquid extraction.* We shall see that this is a good suggestion. Before evaluating it, however, we need to review some important ideas associated with liquid/liquid extraction.

Liquid/liquid extraction is typically used to remove small amounts of heavier species mixed with a large amount of a light species. Conventional distillation would require a large amount of the lighter species to be condensed overhead and would tend to be uneconomic. An example would be to remove a small amount of ethanol from a lot of diethylethyl ether. As a bulk separation method, liquid/liquid extraction is also suitable for isolating fractions of species with similar molecular structure from other species, as is the case when separating alkanes from aromatics. The extraction is done, for example, by finding a solvent such as acrylonitrile that likes the aromatics and forms a separate liquid phase with the alkanes. Extraction is usually performed at temperatures well below the boiling points of the species involved, which makes activity coefficients, γ, a better means of evaluation than vapor–liquid K-values.

The composition of species distributing between two liquid phases, I and II, is determined by equating their respective fugacities:

$$x_i^I \gamma_i^I f_i^{\circ I} = x_i^{II} \gamma_i^{II} f_i^{\circ II}$$

where x_i are mole fractions, γ_i are activity coefficients, and f_i° are standard-state fugacities. Assuming that the same standard-state fugacities are used for both phases (for example, pure liquid i at the temperature and pressure of the mixture), we see that the ratio of mole fractions is the inverse of the ratio of activity coefficients; i.e.,

$$\frac{x_i^I}{x_i^{II}} = \frac{\gamma_i^{II}}{\gamma_i^I}$$

To separate two species, we need to know how well the two phases differentiate between them in this ratio. Separability factors, $S_{i/k}^{I/II}$, are frequently used to indicate potential for separation:

$$S_{i/k}^{I/II} = \frac{x_i^I/x_i^{II}}{x_k^I/x_k^{II}} = \frac{\gamma_i^{II}/\gamma_i^I}{\gamma_k^{II}/\gamma_k^I} = \frac{(\gamma_i/\gamma_k)^{II}}{(\gamma_i/\gamma_k)^I}$$

Returning to our example solvent-separation problem, let us consider using liquid/liquid extraction to remove pentane from methanol and/or acetone. A suitable solvent is one that is immiscible with the bulk species, pentane. If at all possible, we would like the solvent to be present in the mixture so we do not have to introduce any other species into our separations problem. For our example problem, water is present. Noting the infinite-dilution K-values for water and pentane, we see that water will be highly immiscible with pentane.

Aside from being immiscible with the pentane, the solvent, water, has to effect a reasonably different distribution of the species to be separated, methanol and/or acetone and pentane.

We consider methanol first. For a quick estimate of the separation factors that water can produce, we consider the limiting selectivity that would be obtained if the methanol and pentane were infinitely dilute in both the extract phase (water-rich) and the raffinate (pentane-rich) phases if we were to use liquid/liquid extraction:

$$S_{\text{methanol/pentane}}^{\text{water-rich/pentane-rich}\infty} = \frac{(\gamma_{\text{methanol}}^\infty/\gamma_{\text{pentane}}^\infty)^{\text{pentane-rich}}}{(\gamma_{\text{methanol}}^\infty/\gamma_{\text{pentane}}^\infty)^{\text{water-rich}}} = \frac{(23.1/1.0)}{(2.2/3213)} = 33{,}740$$

Selectivity is excellent.

Although selectivity between acetone and pentane (using water and pentane as the extract and raffinate, respectively) is not quite as high (≈ 1850), it still suggests that we could separate these species, too.

We use rigorous simulation to determine feasible separations using water as a solvent. For a theoretical ten-stage liquid/liquid extraction process, we find that rather little water is needed to recover virtually all methanol from the pentane. At higher solvent flowrates the water-rich extract contains more and more acetone, but it cannot produce a complete separation of acetone and pentane. Thus, we select the solvent flow at which the methanol–pentane separation is sufficiently sharp. Figure 35 gives the separation selected.

As shown, the raffinate stream, F_{11}, leaving such an extractor essentially contains pentane and acetone plus a trace of water. This pentane-rich mixture can be separated in a distillation column, producing pentane as bottoms and a distillate which is limited to the minimum-boiling azeotrope between pentane and acetone. (The infinite-dilution K-values of 7.9 and 3.0 indicate that such an azeotrope exists.) Figure 36 gives the flowrates and relative concentrations between the major species of the streams entering and leaving the distillation column (labeled DI-2) that carries out this separation for the raffinate (pentane-rich) stream.

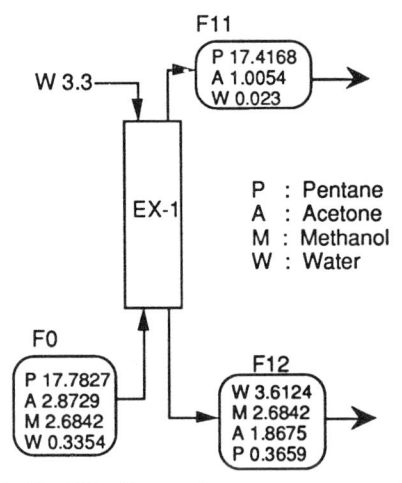

SYNTHESIS OF DISTILLATION-BASED SEPARATION SYSTEMS

FIG. 35. Liquid/liquid extraction step to recover methanol.

It is evident that we have produced one of our desired products here, nearly pure pentane. We now have to separate the near-azeotropic stream, F_{21}. An ideal situation occurs if we can recycle it back to the liquid/liquid extractor. If its composition is close to that of the original feed to the process, we could simply mix it with the feed. Our two separation devices then would process this feed, giving us a pure pentane product and, by material balance, a second stream which is the feed but with this same amount of pentane missing.

How might we decide if the stream F_{21} coming from the top of the second column is close enough in composition of the extractor feed to be recycled back and processed with it? Two characteristics of the stream F_{21}—its composition and its flowrate—must be important in this decision. If the compositions of the two streams are really close, we assume there should be no problem, no matter the flowrate. On the other hand, if there is very little flow in F_{21} relative to the feed (for example, only one drop), then we should almost always be able to

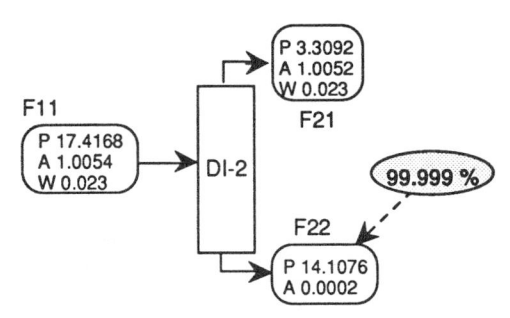

FIG. 36. Distillation column to remove acetone from pentane.

recycle it, no matter its composition. The total flowrate for the distillate stream F_{21} is 16% of that of the original total feed to the process. It is modest, but not negligible. It contains essentially only pentane and acetone, which are in the ratio of 3.3 to 1, while the feed has these same species in the ratio of 6.2 to 1.

We can perform approximate material balance calculations to see what happens to the process if we were to recycle the distillate and process it with the feed. We model each separator as a set of constant split factors for each species. For example, we see that 97.94% of the pentane, 35.00% of the acetone, and none of the methanol leave in the top stream from the extractor. Water enters as the extraction agent and also as a small part of the feed; 0.63% of the total water entering leaves with this top stream. We capture these results in the first three rows of numbers in Table X. We denote molar flow for species k in stream j leaving unit i by $\mu_{ij}[k]$, and the fraction of the flow of species k in the feed to unit i leaving in stream j by $\xi_{ij}[k]$.

Examining the distillation column, we find that 19.00% of the pentane entering it leaves in its distillate product. Similarly, 99.98% of the acetone and 10% of the water entering leave in the distillate. We assume these numbers do not change even if we recycle the distillate product from the second column. We capture these numbers in Table X, too.

The following equation gives the material balance for the total flow $\mu'_{0*}(k)$ of a species k entering into the extractor when we recycle stream F_{21}.

$$\mu'_{0*}(k) = \mu_{01}(k) + \mu_{02}(k) + \xi_{11}(k)\xi_{21}(k)\mu'_{0*}(k)$$

For example, the total flow of pentane into the extractor equals the pentane in the original feeds plus that which recycles, which is 0.9794 · 0.1900 times the total flow of total pentane into the extractor. Solving for the total flow of species k into the extractor, we get

$$\mu'_{0*}(k) = [\mu_{01}(k) + \mu_{02}(k)] \frac{1}{1 - \xi_{11}(k)\xi_{21}(k)}$$

TABLE X
APPROXIMATE MATERIAL BALANCES FOR FIRST TWO UNITS TO ESTIMATE IMPACT OF RECYCLING DISTILLATE PRODUCT BACK TO EXTRACTOR

	Pentane	Acetone	Methanol	Water
Total original feed, $\mu_{01} + \mu_{02}$	17.7827	2.8729	2.6842	3.654
Top product from extractor	17.4168	1.0054	0	0.0230
Fraction of feed in top product, ξ_{11}	0.979	0.350	0	0.0063
Distillate product from column	3.309	1.0052		0.0023
Fraction column feed in distillate, ξ_{21}	0.1900	0.9998		0.1000
$\dfrac{1}{1 - \xi_{11}(k)\xi_{21}(k)}$	1.229	1.538	1	1.0006
New feed (original feed times above factor)	21.9	4.42	2.68	3.64
Composition original feed	0.659	0.106	0.100	0.135
Composition of extractor input with recycle	0.670	0.136	0.082	0.112

which, when applied to pentane, gives

Total flow of pentane into the extractor = 17.7827/(1 − 0.1861) = 21.9

In a similar manner, we can compute the total flows for the other species, getting the results shown in the third line from the bottom of Table X. Finally, we compute the compositions for the original feed and for the feed after adding the recycle so we can compare them. We see that these compositions are surprisingly close to each other. The net result of recycling here is to increase the flow of the feed to the extractor unit by approximately 20% but with little impact on its composition. We therefore elect to recycle the distillate.

We now return to consider the extract stream F_{12}, from our first separation step. It still contains all species in nonnegligible fractions. We propose that this mixture be distilled. We check if distillation is reasonable by using rigorous simulation. While we do not show the details of these runs here, they were carried out by varying D/F ratios over a range of values as we did when we were investigating the acetone/chloroform/benzene example earlier. These runs indicate two interesting product sets. One allows for the complete removal of water. However, because water is a candidate separating agent for the acetone/methanol split, we put it aside and consider the other option—namely, to recover all the pentane.

Figure 37 shows the results from carrying out a rigorous simulation for this option. Because pentane forms azeotropes with acetone and methanol, these species appear in both products. Noting that although F_{31} is a much smaller stream than F_{21}, it has similar compositions, we decide to recycle it back to the extractor feed also. The bottom product of column DI-3 consists of acetone, methanol, and water–with *no pentane*.

Let us examine what we have accomplished with the liquid/liquid extraction step followed by the two distillation columns. The "upper" column, DI-2, produces essentially a pure pentane product, which we remove. The "lower" column, DI-3, produces a product with no pentane. We recycle all the other streams

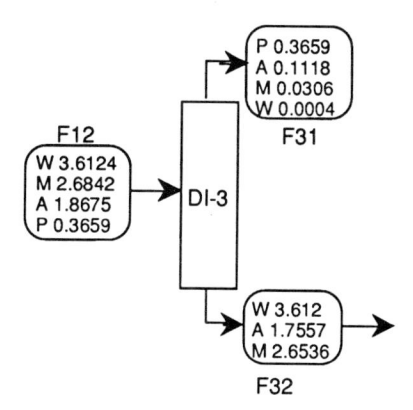

FIG. 37. Distillation column to remove pentane from methanol and water.

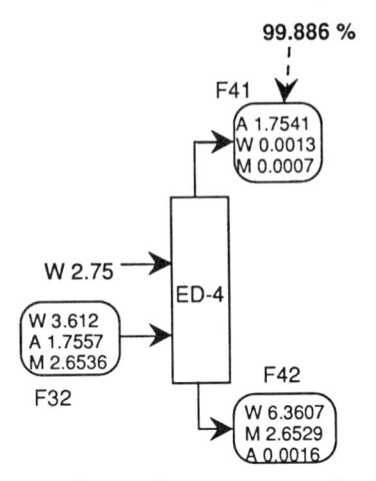

FIG. 38. Extractive distillation column to separate methanol from acetone.

so these three units, collectively, are there to remove pentane from the original feed.

So what do we do with the mixture coming off the bottom of the lower column, DI-3? Looking at our earlier data, we see that water boils at a higher temperature than the two remaining species and does not form an azeotrope with either of them. The K-values of acetone and methanol at infinite-dilution in water (38.5 and 7.8, respectively) suggest that water could be used as an extractive distillation agent for the separation of these species. In such a column, acetone, being decidedly more volatile, is recovered as a pure distillate product. It should be noted that, because of the tangent pinch between acetone and water, this column might best be operated below ambient pressure to exploit the improved vapor–liquid equilibrium near this tangent pinch. The results shown in Fig. 38 are from simulating such a column.

Finally, the bottom product of the extractive distillation column, ED-4, can be separated in a simple column since it contains the nonazeotropic species methanol and water only. This column is shown is Fig. 39, again based on a rigorous simulation.

The process produces all the desired pure species products. The water product from our last column, DI-5, also has to provide the water used as the extractive agent in the liquid/liquid extraction step and in the extractive distillation step. The complete set of steps is shown in the flowsheet in Fig. 40.

We now need to put in the proposed recycle flows to see if the structure discovered is possible when they are present. Rigorous simulation gives the results shown in Fig. 41. It is noteworthy that, despite the complexity of the nonideal mixture behavior, it has been possible to obtain all species as highly pure single-species products using only water which is already present as the separating agent.

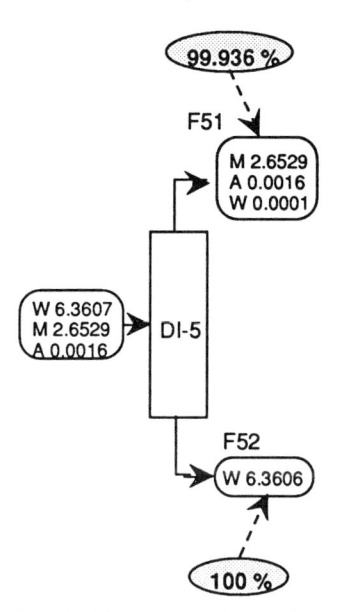

FIG. 39. Water/methanol separation.

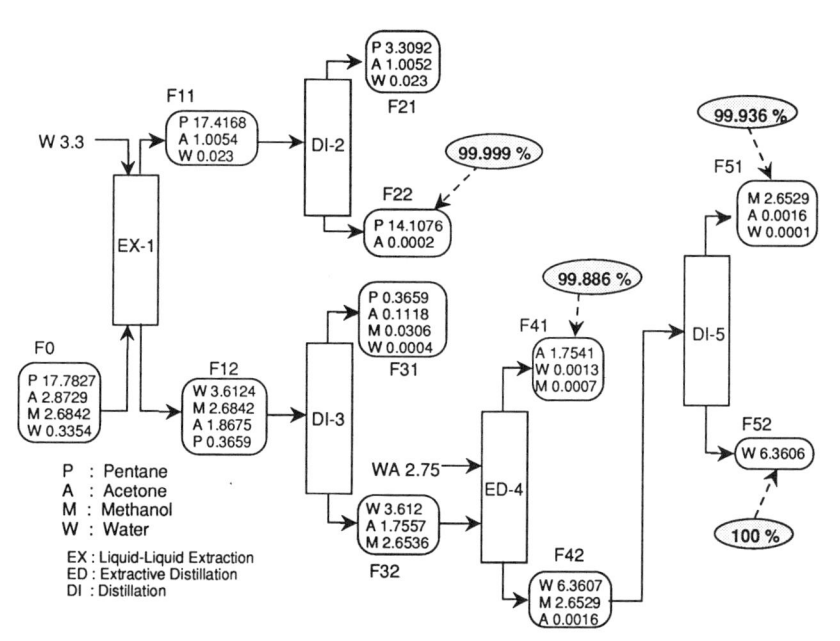

FIG. 40. Total process flowsheet before adding recycle streams.

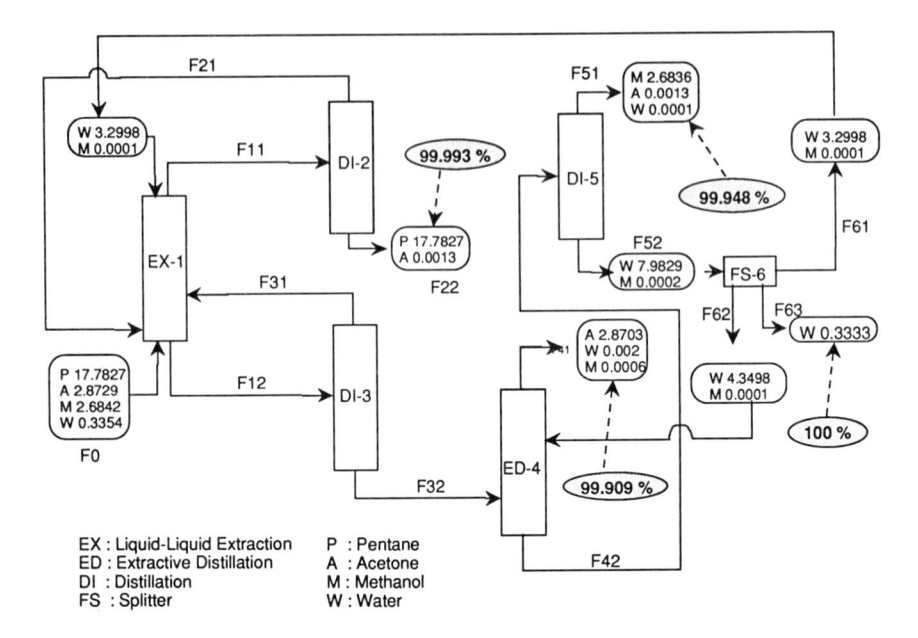

FIG. 41. Total flowsheet after including recycles.

We note that the liquid/liquid extraction step accomplishes separations across three relatively azeotropic compositions: namely, that between acetone and pentane, that between methanol and pentane, and that between water and pentane. It should also be noted that, since the extractor does not have to perform sharp separations, it requires an overall optimization to determine how many stages should be used. In principle, a single stage or, in other words, a simple decanter should suffice to make the process feasible.

The comparison between the simulation results for the original steps while they were being proposed and the final process with the recycles included shows that it is indeed possible to maintain the separation functionality of a sequence, and even to slightly improve the achievable purities—one of the essential premises of the sequential synthesis approach being advocated here.

Alternatives to this process are generated by returning to those steps where we made decisions and checking to see if alternatives decisions might exist. For example, we could look for a different extraction agent in the liquid/liquid extractor. We eliminated one obvious alternative when we discovered that adding more water to the liquid/liquid extractor did not allow us to remove all the acetone from the top product stream even though there is a good separation factor suggesting this alternative.

Sargent (1994) presents a related approach to the synthesis of distillation processes. His goal is to generate a superstructure of interconnected columns from a "state/task" network. The superstructure contains all the process alternatives as substructures. For ideally behaving species, he reproduces the superstructure proposed earlier by Sargent and Gaminibandara (1976).

For nonideal systems, Sargent first determines all the distillation regions. He then labels the pure species and all the azeotropes as pseudo-species in the system, giving the label A to the most volatile, B to the next, and so forth. For example, he would label the diagram in Fig. 25 for acetone, chloroform, and benzene with A for acetone, B for chloroform, C for the maximum-boiling azeotrope between acetone and chloroform, and D for benzene. The region on the left side of the distillation boundary has the species A, C and D, while the region to the right has the species B, C, and D. He then uses the "bow-tie" analysis we discussed earlier (and which we shall discuss in more detail in Section IX.B) to identify the reachable products for each of the regions. For example, in the left region, a first column can produce A, AC, CD, and D as products—where CD is a product along the distillation boundary. The product CD can be split in a second column. If the distillation boundary were straight, the distillate would be C (azeotrope) and the bottoms would be D (benzene). Here, however, the boundary is curved, indicating a mixture of B and C (azeotrope and chloroform) as the distillate product and D as the bottoms.

Sargent gives heuristic rules to suggest where to place recycle streams. Also, when two different columns give rise to the same product, he merges the products in the superstructure. By eliminating parts of the superstructure to form a substructure that produces all the products of interest, one generates a design alternative. Dropping different parts generates different alternatives.

To find the best substructure for a given feed, the idea is to optimize this superstructure to find the optimal substructure. Two potential difficulties arise: (1) It is difficult enough to solve the model equations for a single column displaying azeotropic behavior, much less optimize a superstructure containing several such columns with numerous recycles; (2) the superstructure almost certainly has local optima, requiring the use of much more costly approaches to find global optima. Although we shall overcome these restrictions in the next few years, they are currently serious difficulties.

IX. More Advanced Pre-analysis Methods

We re-examine how to assess species and equipment behavior when dealing with highly nonideal mixtures—this time in much more detail.

A. Species Behavior

In this section, we discuss advanced methods of finding azeotropes and determining if a mixture displays liquid/liquid behavior.

1. Finding Azeotropes

If a mixture forms an azeotropic composition, the vapor and liquid compositions at equilibrium are identical to each other. The following equations define this situation:

$$y_i = K_i x_i \quad \text{for } i = 1, 2, \ldots, n_C$$

$$K_i = K_i(T, P, \text{all } x_i, \text{all } y_i)$$

$$\sum_{i=1}^{n_C} y_i = 1$$

$$\sum_{i=1}^{n_C} x_i = 1$$

$$y_i = x_i \quad \text{for } i = 1, 2, \ldots, n_C - 1$$

There are $3n_C + 2$ variables (x_i, y_i, and K_i for all species, temperature T, and pressure P) in these $3n_C + 1$ equations. If we fix pressure, the model is completely fixed. Solving, we will determine the bubble-point temperature for the azeotrope as well as its composition. We note, therefore, that an azeotropic composition for a mixture is, as we already knew, pressure-dependent.

Suppose we have a mixture of water, pyridine, and toluene. We set the pressure to 1 atm and attempt to solve the above equations. Lacking any further insight, we set all the vapor and liquid compositions equal to 0.33333 and then attempt a solution using a Newton-based method. The problem with finding azeotropes becomes immediately evident. There are six solutions to these equations: the three binary azeotropes and the three pure species. There is no ternary azeotrope. To find all azeotropes for a mixture, we must find all solutions to the above equations. Finding multiple solutions to a set of highly nonlinear equations like these is usually a very difficult task.

We can attack this problem in two ways: (1) try to find a numerical procedure that will find all roots to a set of nonlinear equations, or (2) try to develop a method that uses physical insights to find all the roots.

We know of no numerical procedures that will guarantee finding all the solutions to an arbitrary set of nonlinear equations. "Continuation" methods are often capable of finding more than one solution if several exist. Fidkowski *et al.* (1993) propose using such a method along with discovering bifurcation points to compute all the azeotropic compositions for a mixture. Their homotopy function

$$\tilde{y} = \left[(1 - t) + t \frac{\gamma_i}{\phi_i} \right] \frac{P_i^{\text{sat}}}{P} x_i$$

causes the K-values for the species to move from those Raoult's law predicts to their nonideal values as the homotopy parameter t moves from 0 to 1. Except for very close-boiling species, there will be no azeotropes assuming Raoult's law. Thus, we know that only the pure species will satisfy these equations when the continuation parameter is 0. Their algorithm starts by solving these equations for $t = 0$ at each of the pure species. As the continuation parameter increases, these equations can become singular at one or more of the pure species' starting points. If they do, the singularity indicates the solution trajectory plotted versus the continuation parameter branches, one branch staying at the pure species (always a solution) and the other heading to the composition of a binary azeotrope along one of the adjacent binary edges of the composition diagram. An eigenvector analysis of the local linearized behavior of these equations indicates the direction for the bifurcating solution. The temperature along these branches increases for a maximum-boiling azeotrope and decreases for a minimum-boiling azeotrope. The solution along one of these branches may itself become singular, indicating a further bifurcation. Fidkowski and colleagues conjecture—and their computational experience suggests—that one will discover all azeotropes in the system by tracing all these solution trajectories starting from all pure species.

Wahnschafft (1994) has developed and tested extensively an interesting alternative method of finding all the azeotropes. The slight change that Wahnschafft makes has a significant impact on how to find azeotropes. He replaces the third and last of the equations above (i.e., $\sum_{i=1}^{n_C} y_i = 1$ and $y_i = x_i$ for $i = 1, 2, ..., n_C - 1$) with

$$K_i = 1 \quad \text{for } i = 1, 2, ..., n_C$$

This alternative set of equations is generally satisfied only at azeotropic points involving all n_C species. Assuming for the moment that each n_C species system has only one azeotrope—i.e., that a binary mixture has only one binary azeotrope, a ternary system has at most one ternary azeotrope, and so forth—the advantage of the above formulation is that there are no longer multiple solutions to the same set of equations, but rather a different set of equations to find each of the azeotropic points. In particular, the pure species generally do not satisfy the azeotrope condition for any set of $K_i = 1$ involving more than just one species. The latter could happen only if an azeotrope occurred exactly at a pure species point—we would be talking about the special situation of a tangent pinch.

In principle, we could use any root-finding method for solving problems involving the equations $K_i = 1$ for combinations of species representing different tuples, such as all binary combinations, all ternary combinations, quaternary, and so on, up to the one system of equation involving all n_C species. However, the root-finding methods may still suffer from convergence problems. Wahnschafft found that it is more robust to devise a continuation method that can lead up to the solution at the azeotropic points one wants to find.

Wahnschafft based his continuation method on a relaxation of the azeotropy condition for a given n_C species system. At an azeotropic point, all n_C K-values equal unity, which is a special situation of the condition that they are equal. In the surroundings of an azeotrope involving n_C species, the K-values are no longer equal to unity, but subsets of them will be equal to each other; in other words, the relative volatility between species pairs will still be unity at certain points close to an n_C species azeotrope.

Consider Fig. 42, where we plot three trajectories for a ternary system that involves three binary and one ternary azeotropes. At a binary azeotrope, the relative volatility between the two species involved, i.e., their K-value ratio, equals unity. We can seek out the trajectory emanating from this azeotrope into the three species region along which the relative volatility of these two species remains equal (i.e., along which their K-values remain equal). The K-value for the third species will generally be different. However, we find that, owing to the existence of the ternary azeotrope, there will be one specific composition along this trajectory where the K-value of the third species becomes the same as the other two; moreover that situation occurs as all three K-values become equal to unity. We can find the same point by tracing along another curve starting at either of the other binary azeotropes.

The trajectories along which two (or more) species have the same volatility have been called *isovolatility curves*. Based on the ability to trace out such curves in the composition space, we can outline a simple algorithm to robustly determine all azeotropic points that are predicted for a multispecies mixture:

(1) Fix the pressure at the value desired.

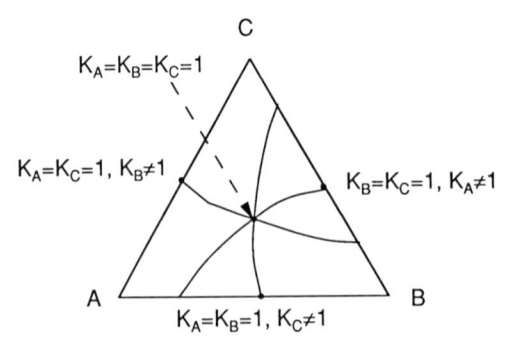

FIG. 42. Trajectories having subsets of all K-values equal to unity.

(2) Compute the infinite-dilution K-values for all pairs of species, as we did earlier to check for the existence of binary azeotropes. For this example, we would find that all three binary pairs display an azeotrope.
(3) Where a binary azeotrope exists (e.g., along the AB axis), use a Newton-based method to find the point along that axis where the two K-values are unity.
(4) At that azeotrope, compute the infinite-dilution K-value for one of the missing species.
(5) Increase the composition for that missing species—here, species C—monitoring its K-value. Solve the equations, keeping the K-values for the other two species equal to each other. Monitor the K-value of the newly introduced species (here, species C) to determine if its value gets close to unity (e.g., by checking for changing from values larger than 1 to values less than 1 or vice versa, and by checking the derivative of the K-value along the curve to see if it passes through a minimum or maximum). In either of these cases, we have approached a ternary azeotrope. Switch to solving the equations for the azeotrope (where all K-values are unity) using a Newton based method. If the trajectory hits the side of the composition space, stop searching along it.
(6) Given a ternary azeotrope and a fourth species in the mixture, compute its infinite-dilution K-value at this ternary azeotrope. Increase the amount of the fourth species, keeping the K-values for the other three equal to each other. Stop where the fourth K-value passes near 1 or shows a minimum or maximum along the trajectory being traced out. Solve directly for the four-species azeotrope. Stop searching this trajectory when it hits a side of the composition space.
(7) Repeat the second step above to find the remaining binary azeotropes. If more than a single binary, ternary, or higher-order azeotrope is suspected, march away from each of lower-order azeotropes toward it, as is done in the third and fourth steps above.
(8) And so forth.

The algorithm stops looking for azeotropes involving $n + 1$ species when there are no azeotropes involving n species. When applying this algorithm, we have to solve the equivalent of several flash unit calculations for each trajectory we trace. This algorithm very efficiently finds all azeotropes predicted by the physical property models being used for a multicomponent mixture. It does not require an eigenvalue/eigenvector analysis to spot bifurcations. Finally, it can be made to take advantage of the topological constraint that Zharov and Serafimov (1975) developed (see Section VII.A.1.c).

Doherty and Perkins (1979) examined the possibility of the existence of ternary azeotropes if there are no binary azeotropes. They show that the infinite-dilution K-value for one of other species must be exactly unity at a pure species node, which is equivalent to saying that the isovolatility trajectory starts at a point which is a pure species and a binary azeotrope at the same time. Obviously such situations will be extremely unlikely, but the proposed algorithm would not present any difficulty in moving to a ternary azeotrope for this case.

We are aware of no papers that prove this type of result for azeotropes involving more than three species; but, assuming such results do exist, the above algorithms should find all azeotropes.

a. Example: Ethanol and Water. We first detect the existence of the azeotrope by carrying out two bubble-point calculations, each at near-infinite dilution— the first with molar compositions of 0.9999 and 0.0001 and the second with 0.0001 and 0.9999 of ethanol in water. The desired K-value for each calculation is the ratio of the vapor composition, y, to the liquid composition, x, for the trace species; we get 1.29 (the K-value of water for a trace of water in ethanol) and 14.6 (the K-value of ethanol for a trace of ethanol in water), respectively. As discussed earlier, these numbers indicate a minimum-boiling azeotrope exists, very likely for an ethanol-rich composition.

We next search over the range of compositions from 0 to 1 for ethanol in water, computing a bubble point for each. At low ethanol composition, ethanol is more volatile. For a mixture that is nearly pure ethanol, the reverse is true. We are looking for the point where the K-value for ethanol changes from above unity to below. Near this point, we switch from a bubble-point computation to that for the equations above where we set the two K-values to unity. A Newton-based convergence algorithm converges easily to the desired azeotropic composition, $y_{\text{ethanol}} = 0.868$ (using Unifac) if we start the calculation close to the answer.

b. Example: Ethanol, Water, and Toluene. For a ternary mixture, we first discover one or more of the binary azeotropes, as we propose to use one of these azeotropes to start the search for a ternary azeotrope. We just discussed finding the ethanol/water azeotrope; we can also quickly discover the ethanol/toluene azeotrope.

In looking for a water/toluene azeotrope, we run into the problem that these two species generally form two liquid phases, a water-rich one and a toluene-rich one. We cannot ignore this two-liquid phase behavior, lest we get nonsensical answers. An azeotrope occurs when the composition for the combined liquid phases equals that for the vapor phase. Having K-values for a species set to unity, as we discussed above, means that the composition for that species for the combined liquid phases matches that for the vapor phase. Whether allowing for two liquid phases or not, water and toluene form a binary azeotrope for a

Ethanol 78.5 °C

p = 1 bar

78.2 C

76.7 C

liquid/liquid tie line

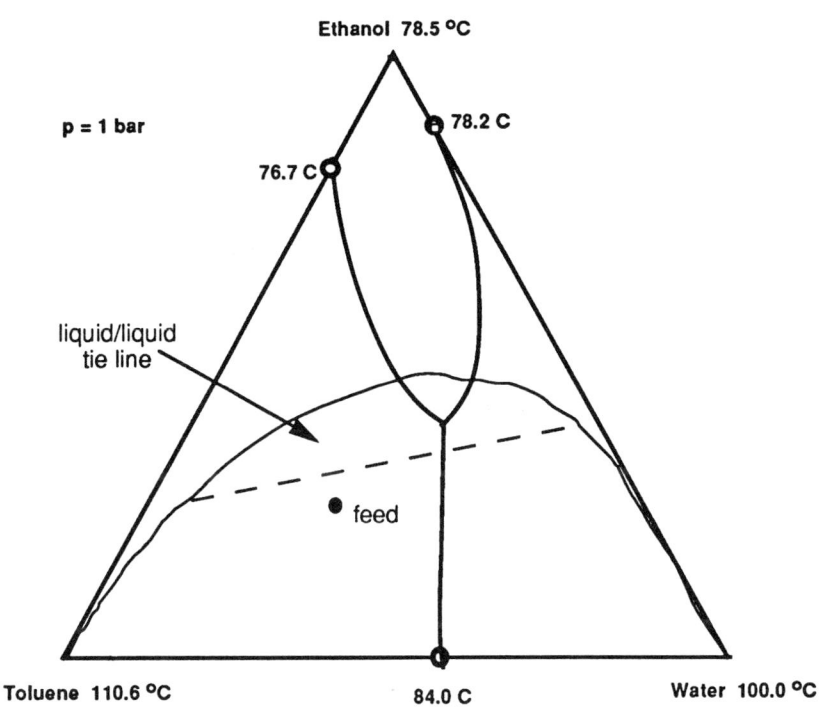

● feed

Toluene 110.6 °C   84.0 C   **Water 100.0 °C**

FIG. 43. Ternary diagram for water, ethanol, and toluene.

mixture that is somewhat more than 50% water. When allowing for two liquid phases, the temperature is, as shown below, at 84°C. If we do not permit two-liquid phase behavior, the temperature is much lower, namely, 65.6°C, a result that does not match at all with experimental data.

Figure 43 shows these azeotropes as well as a ternary one we now wish to find. Starting at the ethanol/water binary azeotrope, the infinite-dilution K-value for toluene is 2.78. Allowing for two liquid phases, the above algorithm locates the ternary azeotrope without difficulty. If we do not allow for two liquid phases, computations indicate there is no ternary azeotrope.

2. Discovering Liquid/Liquid(/Liquid) Behavior

Wasylkiewicz *et al.* (1993) present a method to find regions where mixtures partition in two or more liquid phases. They base it on the Gibbs tangent plane test (Michelsen, 1982, 1993) to decide if a current composition resides in a single- or a multiple-liquid phase region.

We shall first illustrate, using Fig. 44, why a mixture will split into two or more liquid phases by examining the shape of the Gibbs free energy for a binary

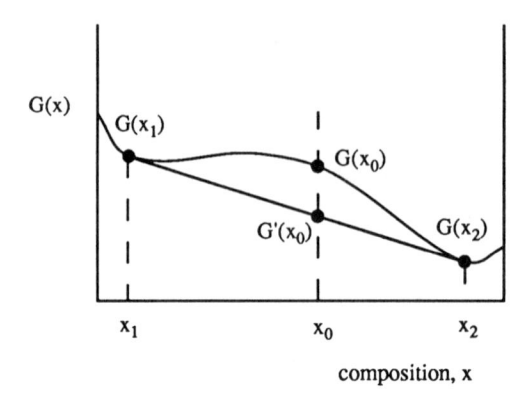

FIG. 44. Gibbs free energy for binary mixture that breaks into two liquid phases at any composition between x_1 and x_2.

mixture versus composition. We examine in particular the composition x_0. Its single-phase Gibbs free energy is above the tangent plane that supports the Gibbs free energy function at points x_1 and x_2. The single phase can reduce its Gibbs free energy by splitting into two phases having the two compositions x_1 and x_2. The total Gibbs free energy for the two phases is

$$G'(x_0) = \frac{x_2 - x_0}{x_2 - x_1} G(x_1) + \frac{x_0 - x_1}{x_2 - x_1} G(x_2)$$

a value that lies, as illustrated, on the tangent plane below $G(x_0)$.

Based on this insight, we place a plane that is tangent to $G(x)$ at the composition of interest. If that plane lies entirely below $G(x)$ for all x, then the given mixture will remain as a single-liquid phase at equilibrium; otherwise, it will split into multiple-liquid phases. Figure 45 illustrates for the point x_0. The tan-

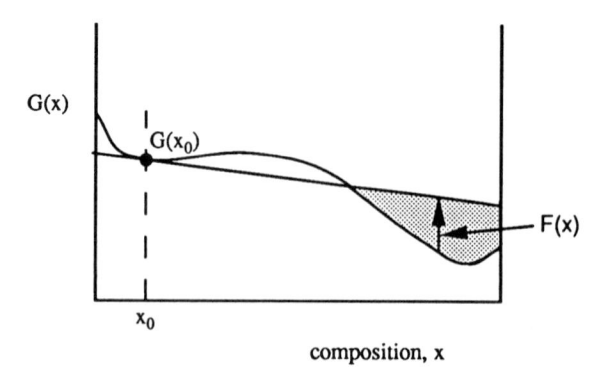

FIG. 45. Tangent plane supporting $G(x)$ at x_0. If $G(x)$ lies above this plane anywhere, the mixture at x_0 will break into multiple liquid phases.

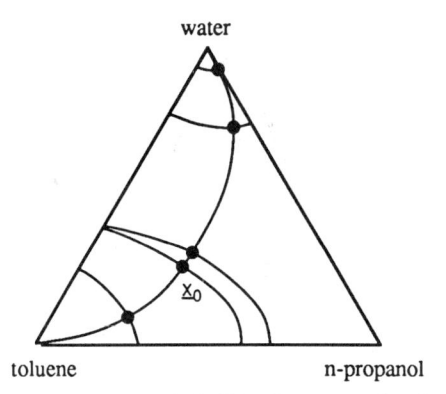

FIG. 46. Stationary points in the modified Gibbs free energy function over the composition space for multispecies mixtures. Diagram is approximate.

gent plane that supports $G(x)$ at x_0 lies partially above $G(x)$ on the right-hand side of the figure. This mixture lies between compositions x_1 and x_2 on the previous figure and will split into these two liquid phases. But how do we carry out this test? It would appear we have to cover the composition space with test points and hope that no regions are missed.

Wasylkiewicz *et al.* (1993) presented a method to carry out this test for general multispecies mixtures. They pick a point to test and create a tangent plane for it. The distance between the tangent plane and the Gibbs free energy surface defines a nonlinear surface above the composition triangle. The starting point is on a ridge in this surface. These authors propose tracing all ridges starting from this point to find all extreme point along these ridges in the surface (as opposed to searching the entire composition space). These paths can bifurcate. Figure 46 illustrates a typical path that their algorithm will trace. Only the extreme points need to be checked to discover if the tangent plane goes above the Gibbs free energy surface, thus indicating liquid/liquid behavior. The extreme points are good guesses for the phase compositions when they detect liquid/liquid behavior. Ridge-following and bifurcation detection involve evaluating eigenvectors and eigenvalues.

To test a single composition is a considerable amount of work. To develop a phase diagram for a mixture, we have to place compositions strategically over all of the composition space, detecting tie-regions (as lines, triangles, etc.). Fortunately, a tie-region covers all the compositions in it which no longer have to be explored. Figure 47, based on a figure in Wasylkiewicz *et al.* (1993), illustrates a completed phase diagram.

McDonald and Floudas (1994) and Michelsen (1994) also present methods to find the phase conditions for a given mixture. Both methods discover the number of phases and the compositions of these phases. McDonald and Floudas globally minimize the Gibbs free energy of the mixture using a computer pack-

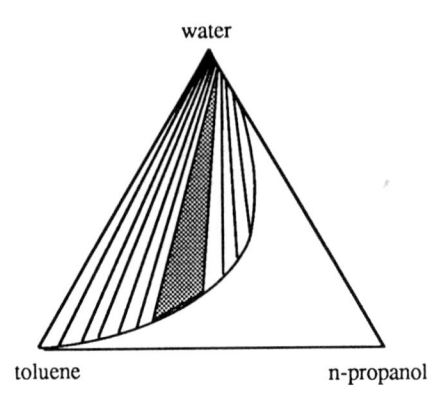

FIG. 47. Resulting liquid/liquid phase diagram. Diagram is approximate.

age they call GLOPEQ. Michelsen formulates the problem as a two-level problem. The outer level computes phase fugacity coefficients for all the possible phases. They show how to formulate the inner problem as a convex continuous-variable minimization problem.

B. LIMITING SIMPLE DISTILLATION COLUMN BEHAVIOR

We shall determine the limiting behavior for simple distillation columns by discovering all the products of such a column, regardless of the number of stages it has or the reflux ratio used in operating it. We present and extend here the ideas advanced in Wahnschafft (1992) and Wahnschafft *et al.* (1992).

1. Reachable Regions

To develop the alternative process configurations needed to separate a given feed mixture into a set of specified products, we need to know just what distillate and bottoms product compositions we can reach when using a conventional distillation column. We shall start by examining this problem for ideally behaving mixtures. We shall then look at the much harder problem in which the mixtures do not behave ideally.

a. The Material Balance Constraint. The following constraint holds for any number of species (see Treybal, 1968):

Material Balance Constraint: *For a conventional single-feed, two-product distillation column operating at steady state, the feed composition must lie on a straight line between the compositions for the distillate and bottoms products.*

b. Infinite Reflux. There are two convenient ways to imagine having infinite reflux conditions in a column. First we can consider having a real column where we turn off the feed and the product flows while maintaining a finite flow for the internal liquid and vapor, L and V. We get an infinite reflux ratio by having a zero denominator in the equation

$$R = \frac{L}{D}$$

The second extreme we can imagine is to maintain finite flows for the feed and products but increase the internal flows for L and V to infinite values. This second case cannot really occur, as we would need a column with an infinite diameter. It is a limiting case. Both ways to think of infinite reflux are useful. In the latter case the column is still thought of as producing its products.

At infinite reflux we can add a second constraint that must hold for the compositions of the two products:

Infinite Reflux Constraint: *The compositions for the distillate and bottoms products must both lie on the same distillation curve.*

This constraint follows from the definition of a distillation curve. Each point along a distillation curve represents both the vapor and the liquid compositions just above (or below) any tray, including those at the ends of the column where products would normally be withdrawn. This constraint, together with the material balance constraint, completely defines the reachable products for a column at total reflux.

Figure 48 illustrates some possible pairs of product compositions we might obtain for a given feed. It is for a relatively ideal three-species mixture, H/I/L.

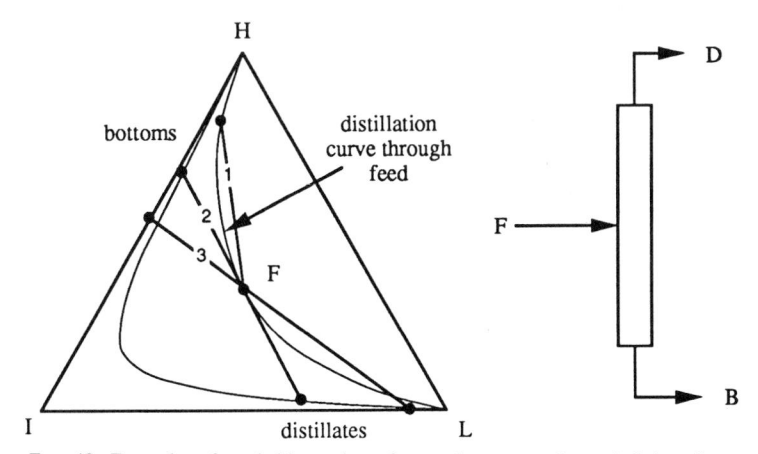

FIG. 48. Examples of reachable products for a column operating at infinite reflux.

The line marked with a 1 is one of two extreme situations we can see for this example. It has a bottoms composition on the distillation curve passing through the feed and a distillate composition which is exactly the feed. It is the limit of maintaining a finite feed while reducing the bottoms product flow to zero. If we take one drop of bottoms and the rest as distillate product, the distillate will have the same composition as the feed. The bottoms product is along the distillation curve at a point corresponding to the number of trays in the bottom of the column. Allowing for fractional trays, we can reach any point along this line.

Line 3 illustrates the other extreme—which is to connect a bottoms composition lying on the IH edge to one lying on the LI edge. There is a limiting distillation curve that passes from L to I to H on which these two points lie. Line 2 connects two points lying on a distillation curve that is between these other two extreme distillation curves for this example.

By plotting all such points, we create the shaded reachable region shown on Fig. 49. The straight edge from F to the LI edge occurs because any point along that edge can be a distillate if pure H is the bottoms product. All distillation curves passing through that edge will ultimately reach H, given an infinite number of trays.

If we bypass some of the feed and then mix what we bypass with either the distillate product or with the bottoms product, we can fill all the bow-tie area between the straight line passing from L through F to the IH edge and the line passing from H through F to the LI edge. The point marked a in Fig. 49 illus-

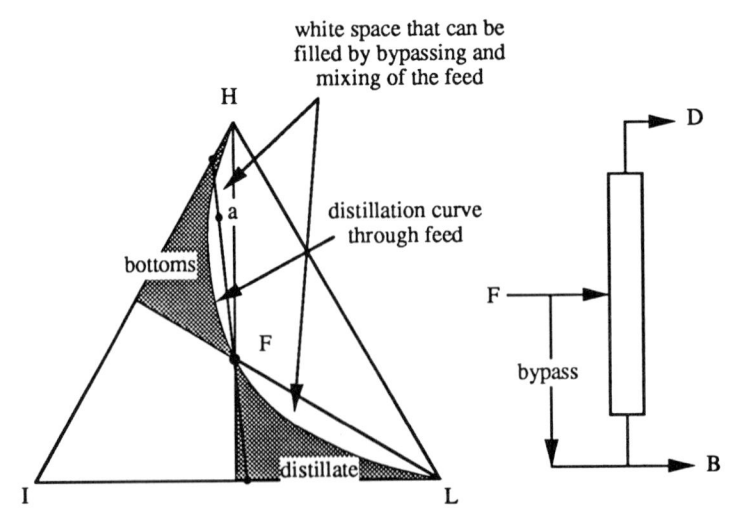

FIG. 49. All reachable products for a simple distillation column separating a nearly ideal mixture with the given feed composition F.

trates. This composition is reached by producing the two products at the extreme ends of the line as these are on the same distillation curve. Then by mixing the bypassed feed with the bottoms, we reach point a.

c. *Finite Reflux (Petlyuk, 1978; Petlyuk et al., 1981; Wahnschafft, 1992; Wahnschafft et al., 1992; Poellmann and Blass, 1994).* To understand what we can reach by reducing the reflux, we can write a material balance around the top of a column (as shown in Fig. 50):

$$V_{n+1} = L_n + D$$

This material balance says that the composition for V_{n+1} will lie between the compositions for L_n and D on the straight line connecting them.

We can imagine being on a composition diagram and stepping down the column starting at the top tray. Figure 51 illustrates the case when we have a total condenser and withdraw a liquid distillate product. (We leave it to the reader to discover that essentially the same analysis holds if we withdraw a vapor distillate product. If the distillate is a two-phase equilibrium mixture, the point D lies on the straight line joining the distillate's vapor and liquid compositions.) At the top of the column for a total condenser, the vapor leaving the top tray, V_1, condenses to form the distillate, D, and then refluxes back to the column, L_0. Therefore, they all have the same composition. We step along a distillation curve from that point to the composition for L_1, which is in equilibrium with that for V_1. It is exactly one stage along this curve. From the argument above, the composition for V_2 must lie on the straight line connecting L_1 to D. We step along the distillation curve, passing through V_2 one stage to L_2. We find V_3 along the straight line connecting L_2 to D.

We can keep on stepping along in this manner until we run into a situation where the compositions start to repeat—the characteristic of a pinch point for the column. Here, we see V_n leading to L_n. For a pinch situation, we find both V_n and V_{n+1} along the same straight line back to D from L_n, and these points coincide. Then L_{n+1} coincides with L_n, ad infinitum. Note that we are talking

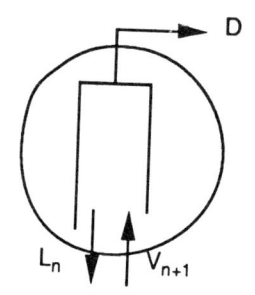

FIG. 50. Flow in the top of a column.

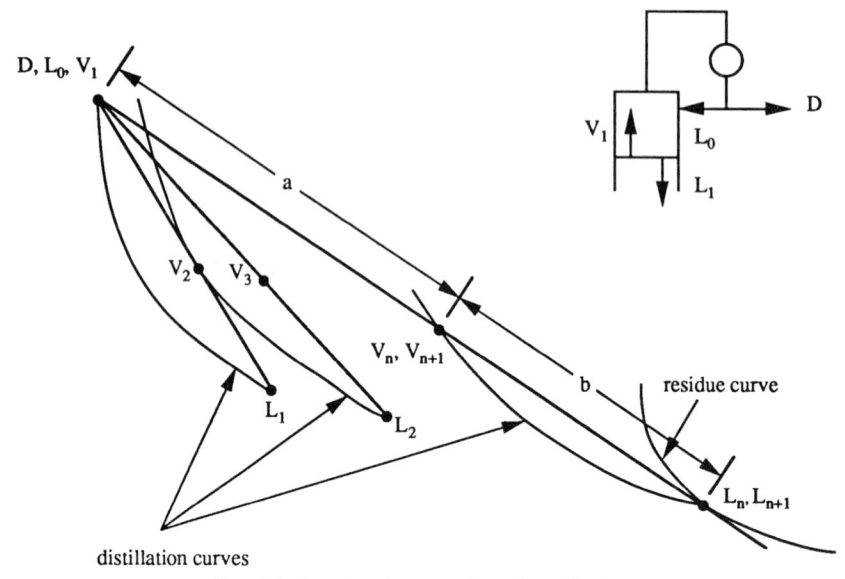

FIG. 51. Stepping down a column from the top.

about the same pinch points we discussed earlier. Thus we have already talked about how we might find such points by carrying out a computation similar to a flash computation.

There is a lot of interesting geometry in this diagram. First, we can look at where V_{n+1} is placed along the line relative to the positioning of L_n and D. From the level rule, we write

$$\frac{L_n}{D} = \frac{a}{b} = R_n$$

where a and b are the lengths shown in the diagram. This equation says that the reflux ratio defined in terms of the liquid flow leaving the nth stage divided by the distillate top product flowrate is the ratio of the line lengths a to b. If the reflux ratio does not change much, then this ratio does not change, and V_{k+1} is always about the same fraction of the distance along the line from L_k back to D. Figure 51 shows all the vapor compositions (V_2, V_3, and finally V_n) positioned in this manner.

Next we remember the equations that define a residue curve:

$$\frac{dx_i}{d\theta} = x_i - y_i$$

These equations have an interesting geometric interpretation. They say that the vapor composition V_n in equilibrium with the liquid composition L_n lies along a line that is tangent to the residue curve passing through the composition L_n. If that line also passes through D, then V_{n+1} is on the same line. It is possible to prove also that V_n and V_{n+1} must be coincident and therefore so are L_n and L_{n+1}. The proof follows from observing that if V_n and V_{n+1} are on the same line back to D, then L_{n-1} and L_n must also be coincident. If L_{n-1} and L_n are not coincident, then there must be at least two points where residue curves are tangent to that line. The proof takes some effort, but ultimately it demonstrates that these points are coincident.

Finally, from the results of these observations, we can prove that *a pinch point is any point where a line emanating from the distillate composition is tangent to a residue curve.* The vapor composition in equilibrium is on that same line. The reflux ratio to reach that pinch point is the ratio *a/b*.

Figure 52 illustrates stepping to a pinch point from a liquid distillate composition on a triangular diagram. This diagram is also rich in geometric interpretation. Starting at the distillate composition with a particular reflux ratio for the column, we step down the column as shown, finally ending at a pinch point. If we choose a slightly larger reflux ratio, we reach another pinch point farther

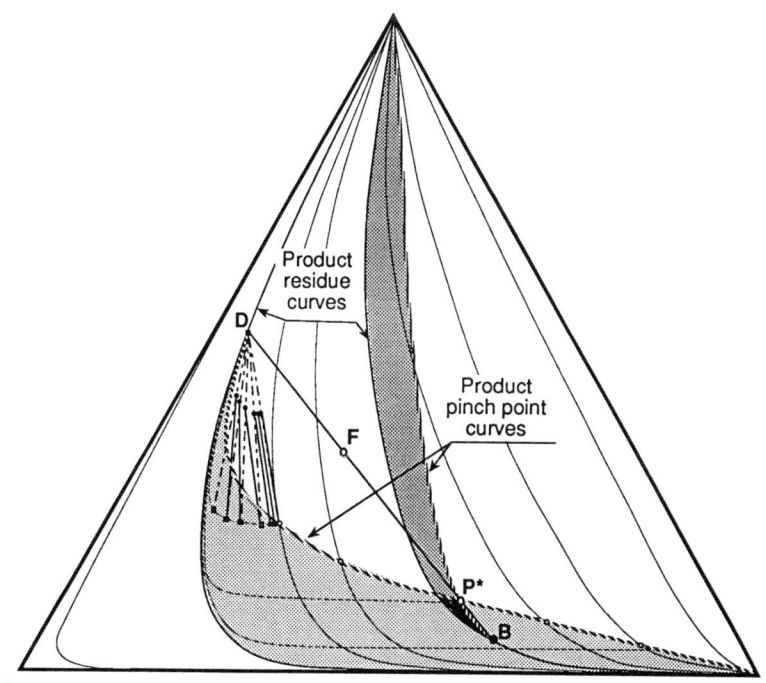

FIG. 52. All compositions that can be reached for both finite and infinite reflux conditions from the distillate D and bottoms B compositions by stepping away from them using a tray-by-tray calculation.

away from D. In this manner, we can construct a curve made up of pinch points. Which pinch point we will reach in a column depends on the reflux ratio; thus, this curve is parametric in the reflux ratio. We saw this parametric behavior earlier when we discussed pinch points just before developing insights into Underwood's method for estimating minimum reflux.

Any point between the distillation curve passing through the distillate and its corresponding pinch point curve is *reachable* from the distillate using tray-by-tray computations (allowing fractional trays). Here, for example, we could reach the bottoms product labeled B by choosing a reflux ratio that leads to a curve that traverses at first very close to the distillation curve through D until the tray-by-tray computations bend inward toward the pinch curve. We stop at B by choosing a finite number of trays (it takes an infinite number to reach the pinch curve) along that curve. We define the gray region as the *reachable region* for D. We can repeat a similar construction for the bottoms product, producing a pinch point curve and then a reachable region for B.

If the shaded regions do not overlap, no conventional column can produce these two products at the same time. We need to argue that the converse is true; namely, if the regions overlap, a column exists that can produce these two products. We would have no trouble making that statement for this particular column as B is reachable from D directly using tray-by-tray computations. We would feed the bottom tray in this column (very likely not the best way to run a column to get these two products).

There is a point P^* that resides on both pinch point curves in Fig. 52. We can make the following argument concerning P^*:

- By construction, P^* is simultaneously on both pinch point curves.
- As it is on the pinch point curve for D, the residue curve passing through P^* is tangent to the line from P^* to D.
- As it is on the pinch point curve for B, the residue curve passing through P^* is tangent to the line from P^* to B.
- Thus the line must be the same straight line (both pass through P^* and both have the same slope).
- By overall material balance for the column, the feed point F is on a straight line between D and B.
- Thus the point P^* must be on that line.
- Thus the point P^* must be a pinch point for F, too.

We therefore conclude that P^* is a pinch point for all three points: D, B and F.

For a reasonable topology for the residue map, we next can discover that the points between P^* and F may not be products from any column for which F is a feed. The construction in Fig. 53 illustrates. This figure corresponds either to a total condenser with a liquid top product or to a partial condenser with a vapor top product. If the top product is two-phase, the distillation curve passes through

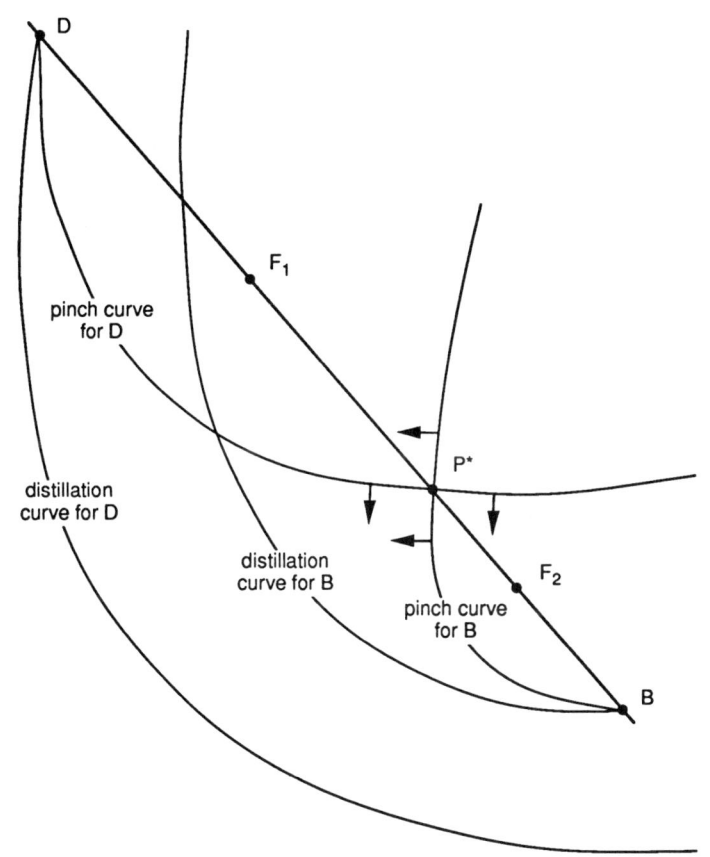

FIG. 53. Topology of the point P^*, the intersection point for a liquid distillate and bottom product pinch point curves.

both the vapor and liquid compositions for the distillate—they are in equilibrium with each other—while the distillate pinch point curve passes through D.

First, the column feed F must lie on the line between the products D and B. Both points F_1 and F_2 are possible feeds that could lead to the product D and B.

If the feed is F_1, a point lying between D and P^*, then the pinch curve through D lies entirely to one side of the point F_1 and curves back to cross at P^*. The corresponding distillation curve lies even farther away to the same side (the pinch point curve turns more sharply than the distillation curve). The reachable region for D excludes all points between D and P^*. Thus the points between F_1 and P^* are unreachable by the top part of the column. As D could move to be coincident with F_1, points beyond F_1 are not excluded.

If the feed is F_2, the pinch curve for B excludes all points between B and P^* by a similar argument. Again, the points between F_2 and P^* are unreachable. We have proved our assertion.

We now should understand the meaning of a reasonable topology. It is where the distillation curves turn in only one direction over the region of interest, a property they will have for ideally behaving species. We shall look in a moment at a case where the topology is more complex.

The region between F and P^* is a function only of F, the feed to the column, as we can find P^* by finding the pinch point curve for F from the arguments above. We have subtly turned our attention from D and B to F. Thus, we can map out this unreachable region by knowing only the column feed.

Excluding the option of bypassing and remixing any of the feed with either of the products, the diagram for an ideally behaving set of species in Fig. 54 maps out the entire set of reachable products for a given feed—for any reflux conditions from minimum to infinite. We can reach the lightly shaded region using total reflux. We can extend the region with the dark gray areas by using finite reflux.

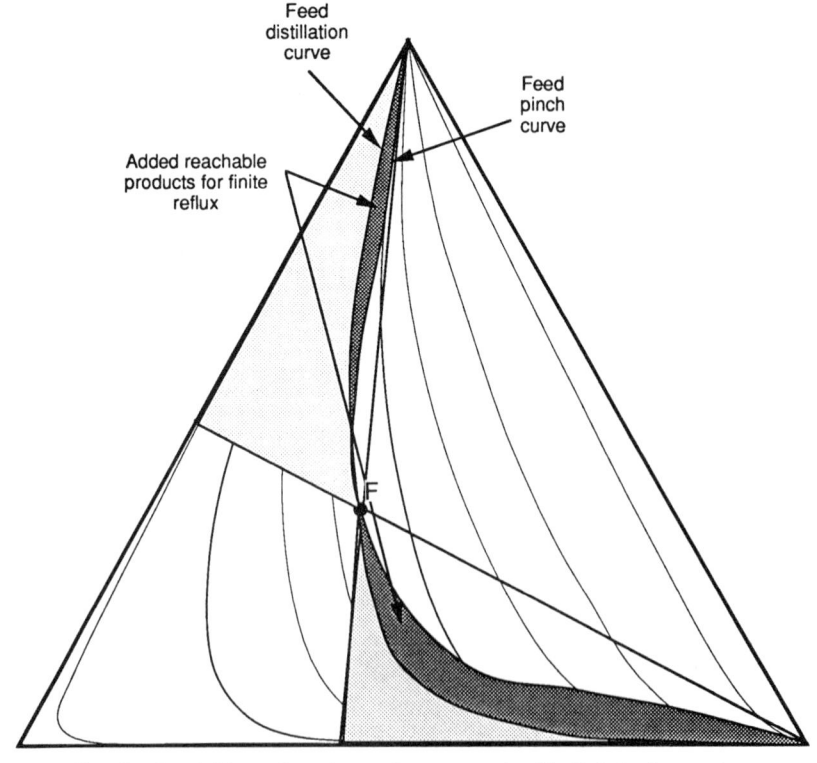

FIG. 54. Reachable products for a column separating ideally behaving species.

i. S-shaped residue and distillation curves. For nonideally behaving species, some of the residue and distillation curves take on a more complex shape. We shall examine the situation when some of them are S-shaped as in Fig. 55. S-Shaped residue curves appear to the left of the maximum-boiling azeotrope along the lower edge. They are also present in the residue diagram for acetone, chloroform, and benzene (Fig. 25), appearing just to the right of the maximum-boiling azeotrope along the lower edge. With little difficulty, we can demonstrate that this shape is quite common. We now examine its implications for finding reachable products for a column.

Figure 56 shows a region with S-shaped distillation curves, where the feed lies somewhere near the inflection point for one of these curves. We note some very interesting differences from what we have discussed so far. We know from earlier discussion that we can reach any products at total reflux when both lie at opposite ends of a straight line passing through F while also lying on the same distillation curve. B_1 and D_1 are such a pair. Interestingly, B_2 corresponds to two different distillate products, D_2' and D_2''. We see that the total reflux products are outside the region bounded by the lines passing to the minimum and maximum temperatures for the region (total reflux products were wholly inside this region before).

The shaded regions between the two S-shaped bounding curves are the reachable product regions for total reflux for the feed F. To see how we construct these two bounding curves, consider the two points marked a lying simultaneously on the same straight line passing through F and on the same distillation curve. The one to the left and below the feed lies where the straight line just brushes a distillation curve. If we rotate this straight line counterclockwise, it will no longer intersect this distillation curve. Thus, a column at total reflux cannot reach points to the left of the upper point a where this straight line intersects this same distillation curve. We create one segment of these bounding

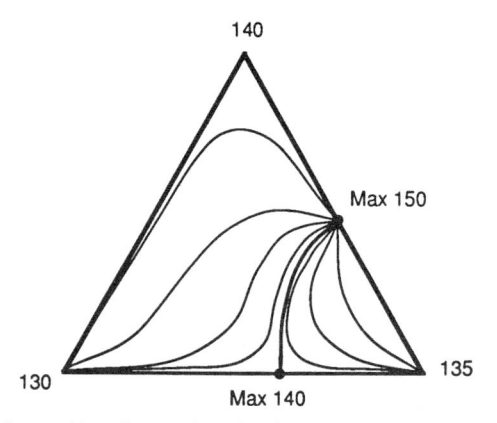

FIG. 55. Composition diagram featuring S-shaped distillation/residue curves.

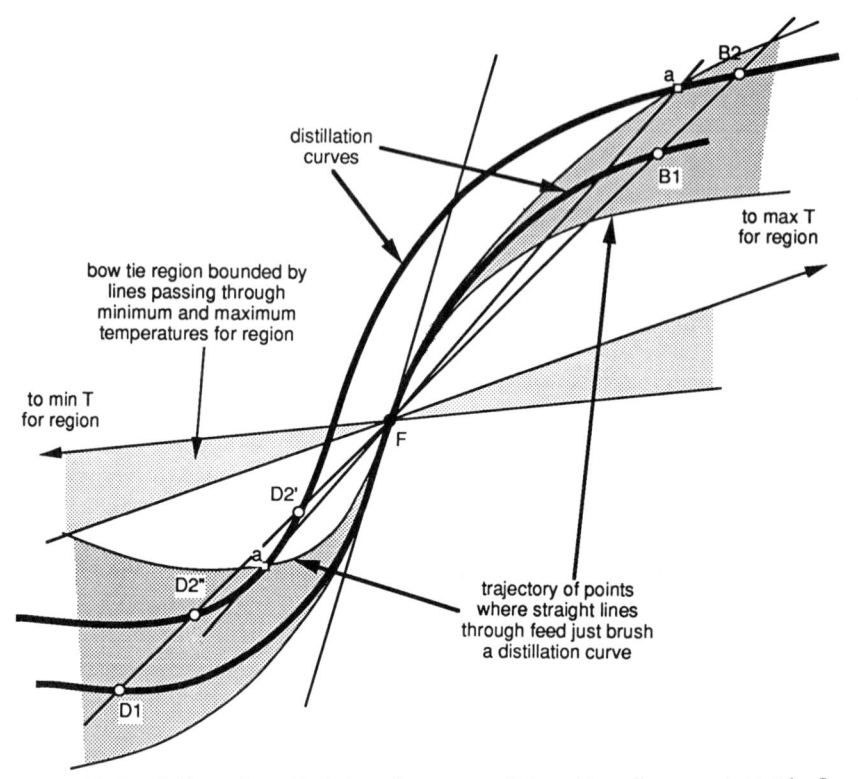

FIG. 56. Reachable products (shaded area between two S-shaped bounding curves) at total reflux for S-shaped distillation curves.

lines by finding the trajectory of points where a straight line passing through the feed just brushes a distillation curve and the other segment by locating the corresponding point on the other side of the feed where this line intersects the same distillation curve.

We find the same complexity for mapping out reachable products for finite reflux—see Fig. 57. We ask if we can distill the feed F into the distillate product D and bottoms product B. The shaded regions indicate the reachable regions for these two products. They overlap, suggesting we can reach them. We see there are two feed pinch points, P_1^* and P_2^*, along the straight line joining B, F, and D, whereas we saw only one before. In a manner similar to that used earlier, we can develop an argument that says that, if B lies between P_1^* and F, then D cannot lie between P_2^* and F. Similarly, if D lies between P_2^* and F, then B cannot lie between P_1^* and F. These segments between P_1^* and F and between P_2^* and F are reachable (in contrast to earlier findings), but they are not simultaneously reachable in the same column.

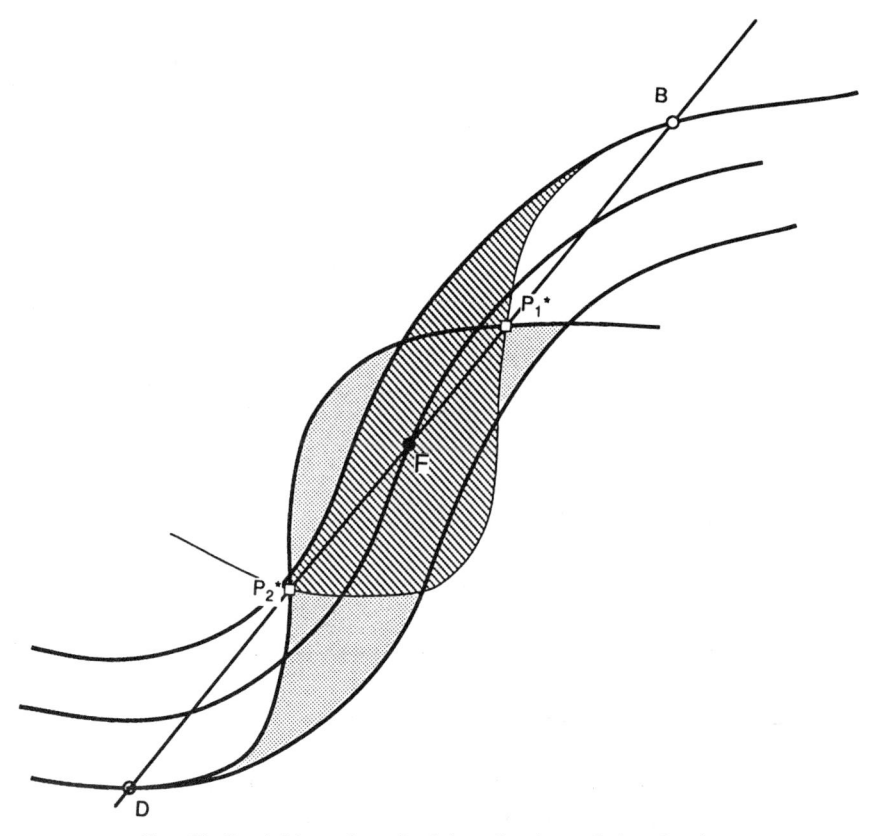

FIG. 57. Reachable products for finite reflux for an S-shaped region.

We see that the S-shaped region makes the process of discovering reachable regions more difficult (but not impossible).

ii. Crossing residue curve boundaries (Wahnschafft, 1992; Wahnschafft, *et al.*, 1992). Nikolaev *et al.* (1979) and Van Dongen (1983), among others, demonstrate by column simulations that one can cross residue curve boundaries when operating a column at finite reflux. Consider Fig. 58, which shows the distillation curves for acetone, chloroform, and benzene. We are going to be very particular and note that this plot features a *distillation* curve boundary as opposed to residue curve boundary. No column operating at total reflux can produce a distillate and bottoms product on opposite sides of this boundary because the two products must reside on the same distillation curve. However, this restriction does not hold for finite reflux. As we have shown above, we can step to a product by carrying out plate-by-plate computations from any com-

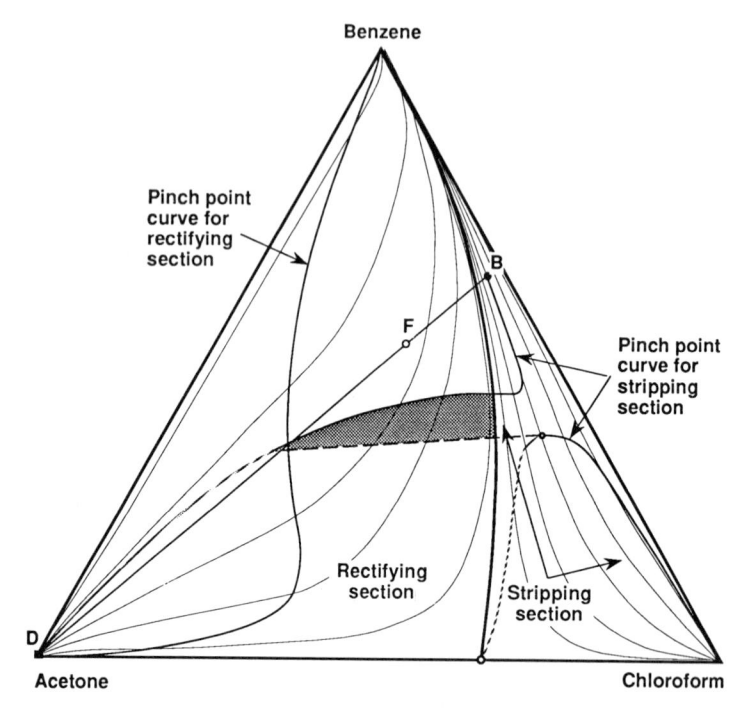

FIG. 58. Bifurcation of the pinch curve trajectory when a product is in another distillation region.

position residing between the distillation curve passing through a product and the pinch curve emanating from that product. We propose here a bottoms product in the right-hand-side region and a distillate product in the left-hand-side region, with the feed between as shown. Is this possible?

Examining Fig. 58, we see that there are two pinch point curves corresponding to B (i.e., to the stripping section of the column). Mentally treating the distillation curves shown as residue curves (they will be similar in shape), we see that a pinch point occurs where a straight line passing through the bottoms product composition is just tangent to a residue curve. The pinch curve passing through B, crossing into the left-hand-side region, and ending at the acetone node satisfies this requirement. However, so does a totally disjoint pinch curve starting at the maximum-boiling azeotrope, heading upward into the right-hand-side region, and ending at the chloroform node. For a small enough reboil ratio, we will quickly move from the distillation curve passing through B toward the pinch curve emanating from B. At total reflux, we will stay on the distillation curve, and, for an infinite number of stages, we will end up at the chloroform node which is on the disjoint pinch curve. Somewhere in between, we must jump from the one pinch curve to the other.

Figure 59 illustrates the process of stepping up the column away from the bottoms product end. This figure reminds us of the geometry involved. First, we note that L_n must lie on a straight line between B and V_{n+1}. Next, we remember that the residue curve passing through L_n points at the vapor composition in equilibrium with the composition corresponding to L_n. We consider three cases. (1) If the residue curve passing through the liquid composition points as shown, the composition in equilibrium with L_n will be down and to the right of L_n. In this case, the column compositions will move to the right as we step away from B, causing us to move toward the disjoint pinch trajectory. (2) If the tangent to the residue curve points straight at B, then L_n is a pinch point, and we would stop moving. (3) If it points to the left as we move away, the trajectory moves to the left. Thus we move to the disjoint pinch trajectory only if the composition of the liquid steps across the lower pinch curve, i.e., passes from case (3) through case (2) to case (1).

Associated with every pinch point is a vapor composition in equilibrium with it, a composition we already have from computing the pinch point itself. We are reminded by Fig. 59 of the geometry for computing the reboil ratio to reach a given pinch point. It is the ratio of the distance from B to L_n (b on the figure) divided by the distance from L_n to V_{n+1} (a on the figure). We can label each pinch point on both trajectories with a corresponding reboil ratio. The point B itself corresponds to a reboil ratio of zero. The reboil ratio increases to infinity along the pinch trajectory emanating from B as it passes to the pure acetone node. To see that it goes to infinity, we note that the distance b gets larger as

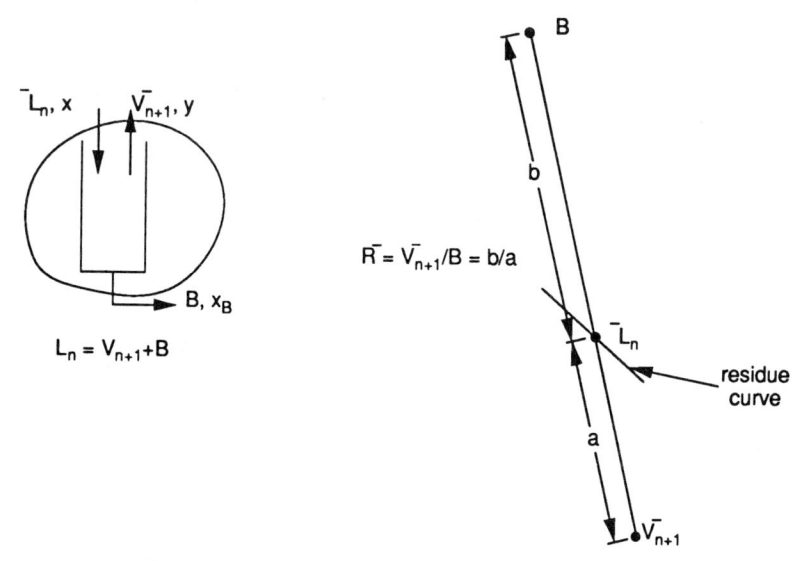

FIG. 59. Stepping up the column from the bottoms product.

we move away from B while the distance a must go to zero as we approach acetone.

Using a similar argument, we recognize that the reboil ratio for the disjoint pinch trajectory passes from infinity at the azeotrope to a finite but positive number and back to infinity as we approach the chloroform node. If we keep the reboil ratio less that the smallest reboil ratio along the disjoint trajectory, we cannot reach this trajectory. So, again, we see that we need to use small reboil ratios to cross a boundary.

iii. Bottoms compositions that permit crossing of a boundary. We are now in a position to map out those bottoms compositions that have a reachable products region that crosses the distillation boundary. Place a bottoms composition somewhere in the lower right of the composition diagram Fig. 58 for acetone, chloroform and benzene. If it is close enough to the chloroform node, two pinch point trajectories will again appear, but they will be qualitatively different from before. The one starting at the bottoms composition will end at the chloroform node. The second will move from the acetone node and end at the azeotrope. In this case, the reachable region for B will stay entirely in the right-hand-side region, lying between the distillation curve passing through B to chloroform and the pinch point curve emanating from B and passing to chloroform.

There must be one or more compositions somewhere between this one and the one shown Fig. 58, where the trajectories switch from the one shape to the other. We can imagine the two trajectories just touching and then trading branches. Let's explore where this will happen.

Examine Fig. 60. We show the residue curves that have the same S-shape as those in the lower left part of the right-hand-side region for acetone, chloroform, and benzene. Each curve is convex to the left at the top (curves toward the left) and convex to the right at the bottom, and each switches the direction it curves at its inflection point. Point a is such an inflection point on one of these residue curves. Draw a straight line through point a such that it has the same slope as the residue curve. Above the inflection point, this straight line is entirely to the right of the residue curve, and below it is entirely to the left.

Next, pick a bottoms product, B, that lies on this straight line above the inflection point. The pinch point curve emanating from B will move to the left initially because of the curvature of the local residue curves. It will move left until it encounters point a which, by construction, is a pinch point for B. It cannot cross this residue curve, however, because any residue curve an infinitesimal bit to its left can have no pinch point with B. On the other hand, a curve just to the right will have two pinch points with B, one just before a and one just after. The pinch point trajectory thus "reflects" off this residue curve. After encountering point a, it heads to the lower right and ultimately to the chloroform node.

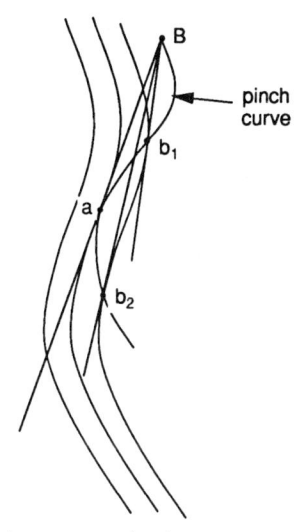

FIG. 60. The pinch point trajectory emanating from the bottoms product "reflects" off a pinch point that occurs at an inflection point of a residue curve.

Thus, any point B that lies on a line passing through an inflection point of a residue curve with the same slope as the residue curve cannot have a pinch point trajectory emanating from it that crosses the distillation boundary. We draw a family of lines through every such inflection point. To the left of all these lines are the bottoms products, B, that can be bottoms products with a top product in the left-hand-side region. Points on and to the right of these lines are not candidates.

Finally, we note with a similar set of arguments that no product in the left-hand-side region can have a pinch point curve that crosses the distillation boundary, thereby justifying the statement appearing in the literature that one cannot cross such boundaries from the "convex" side.

2. *Thoughts on the Geometry for Two and Four or More Species*

All the above insights are for three species. How do these insights extend to mixtures with two species and with four or more species. If we extrapolate the two-dimensional (planar) triangular diagram for three species down to two, we get a composition line. Each edge of a triangular diagram is such a line. For a four-species mixture, we need a three-dimensional tetrahedron such that each of its four sides is a three-species triangular diagram. Five species require four dimensions, one more than we can comfortably visualize, so we almost certainly must abandon visualization for five or more species. But we can attempt to describe the geometry of expressing equilibrium, distillation, and residue curves; operating "lines"; and, finally, pinch point trajectories.

Vapor–liquid equilibrium is a mapping from a liquid composition to a vapor composition. It can be done by including tie lines from one to the other for all compositions. On a line, such a mapping is very difficult to visualize, so we typically use a second dimension where we can plot vapor composition versus liquid composition as in a McCabe–Thiele plot or as in a temperature versus composition diagram. For three or four species, showing tie lines is fairly direct. The important point is that equilibrium is not a line (as we might think because of our familiarity with McCabe–Thiele plots) but a mapping.

Distillation and residue lines that we have plotted for three species on a planar triangular diagram remain as lines for all dimensional spaces. Each corresponds to a trajectory of composition points. A distillation curve corresponds to the curve we pass through the liquid compositions that occur on the trays of a distillation column operating at total reflux. A residue curve is the trajectory we get when we solve the differential equations

$$\frac{dx_i}{d\theta} = x_i - y_i$$

Each curve is "pinned" down by requiring it to pass through a particular composition.

For a fixed reflux ratio for a column, the operating line remains a line in higher-dimension composition space, as it too is a line passing through the sequence of liquid compositions appearing on the trays as we move up or down a column.

One line we have to think more carefully about is the pinch point line as we move to other dimensions. Each point on the line is the end point of a operating line for a distillation column operating with a fixed distillate (or bottoms) product, varying parametrically with the reflux ratio we use to define that operating line. Thus, a pinch point curve emanating from a fixed distillate composition is a sequence of points. It remains a line. It may bifurcate, as we have shown above, but it remains a line.

The compositions from which we can reach a distillate or bottoms product are those lying along a distillation curve if we operate at total reflux. If we operate with a finite reflux ratio, these trajectories move off this line and toward the pinch point trajectory emanating from that product, ending on it if we have an infinite number of stages. Thus, the reachable compositions should look like a ribbon moving through higher-dimension space. Almost certainly, this ribbon has bulges in it, but it should be a bounded two-dimensional surface between these two lines. However, this geometry is for reaching a particular product composition.

Thinking about all reachable products for a given feed is more difficult problem, particularly for four or more species. We still require the material balance constraint to hold. Thus, the points must line up on a straight line in whatever dimension space we examine. At total reflux, the points must also lie on the same distillation line. On a triangular diagram, all distillation lines lie in a planar surface, so we have little difficulty in seeing that these two constraints map out a two-dimensional portion of that diagram. But if four species are represented in a three-dimensional tetrahedron, do these same two constraints map out a three-dimensional portion or only a two-dimensional portion of that tetrahedron? Even if we answer this question, are we prepared to discover the shape of this space for each problem we face, and would we find it useful if we did? It was for this reason that we have proposed a different strategy to discover the reachable products for separating a mixture of four and more species: namely, the plotting of separation ranges as in Fig. 27. This representation does not provide the complete picture for three species, but it does give us one approach for such mixtures that neither grows combinatorially with the number of species nor defies our ability to visualize the results.

C. Extractive Distillation

Many researchers have contributed to the literature on extractive distillation, including Benedict and Rubin (1945), Hoffman (1964), Tanaka and Yamada (1965), Berg and Yeh (1985), Hunek *et al.* (1989), Pham *et al.* (1989), Ryan and Doherty (1989), Pham and Doherty (1990), Wahnschafft (1992), and Wahnschafft and Westerberg (1993). We shall base much of our discussion here on the last two references.

1. Difference Point

Suppose we would like to separate water (normal boiling point 100°C) from isopropanol (nbp 82°C). Isopropanol and water form a minimum-boiling azeotrope (nbp 80°C) at about 72% isopropanol and cannot be separated into pure products by ordinary distillation. We discover by computing infinite-dilution K-values that, in the presence of sufficient ethylene glycol (nbp 197°C), isopropanol is more volatile than water, and the two can be separated completely. We need to operate a distillation column such that, whenever isopropanol and water are together in the column, there is sufficient ethylene glycol present to make the isopropanol more volatile. Suppose we feed relatively pure ethylene glycol (the extractive solvent E) as a separate feed a few trays from the top of a column, as shown in Fig. 61. Being a high-boiling material, it will largely head straight down the column in the liquid phase. The feed (where A is the isopropanol and B is the water) enters several trays below the solvent. The section of the column between the two feeds washes out B. Just below the solvent feed, it must be essentially completely removed, or else it will end up in the distillate product.

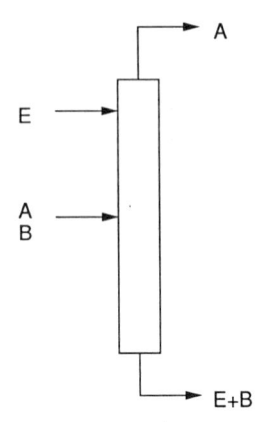

FIG. 61. Typical extractive distillation column.

The bottom section of the column is to remove species A. Species A heads down the column but is removed more and more until, just before the bottom tray, it is more or less completely depleted. The top few trays above the solvent feed are to separate A from E. As ethylene glycol is a very high-boiling material relative to isopropanol, two trays will remove it.

We can determine the reachable products for this type of column by extending the concepts developed for an ordinary column. We again develop our insights for a three-species mixture using a ternary composition diagram.

We start by writing a material balance for the section of the column between the two feeds, as shown in Fig. 62.

$$\hat{V}_{n+1} = \hat{L}_n + D - S$$

This equations looks just like a material balance for a normal column, except that we have replaced D by $D - S$.

Let us define Δ as a difference point (Hoffman, 1964)

$$\Delta = D - S$$

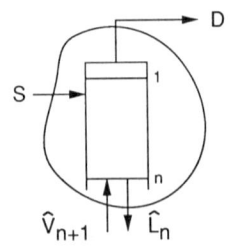

FIG. 62. Intermediate section of extractive distillation column.

and let it play the role of D in a normal column. Thus, Δ is the constant difference that we will see between the total (and species) vapor and liquid flows in this section of the column. For a normal column, the operating line must pass through the composition point for D. For the intermediate section of this column, the operating line must pass through the composition point for Δ, given by

$$x_{\Delta,i} = \frac{x_{D,i}D - x_{S,i}S}{D - S} = \frac{x_{D,i}D - x_{S,i}S}{\Delta}$$

Let us look again at the case of separating isopropanol from water using ethylene glycol. If we assume the solvent is essentially pure ethylene glycol and the top product essentially pure isopropanol, then

$$x_{\Delta,E} = \frac{0D - 1.0S}{D - S} = \frac{-S}{D - S}; \quad x_{\Delta,A} = \frac{1.0D - 0S}{D - S} = \frac{D}{D - S}; \quad x_{\Delta,B} = 0$$

For $D > S$, the first composition is negative and the second is greater than 1. The third (for water), being 0, says that this composition lies on the edge of the triangle opposite the node for pure water (B). A negative composition for the solvent E and a composition greater than unity for species A says the Δ point lies outside the normal composition triangle. Figure 63 shows where Δ will lie for the case of pure solvent feed and pure distillate top product. As noted, when $D > S$, $x_{\Delta,E}$ is negative which must be past the A/B edge. Composition $x_{\Delta,A}$ is greater than unity, which must be above the vertex for A. In

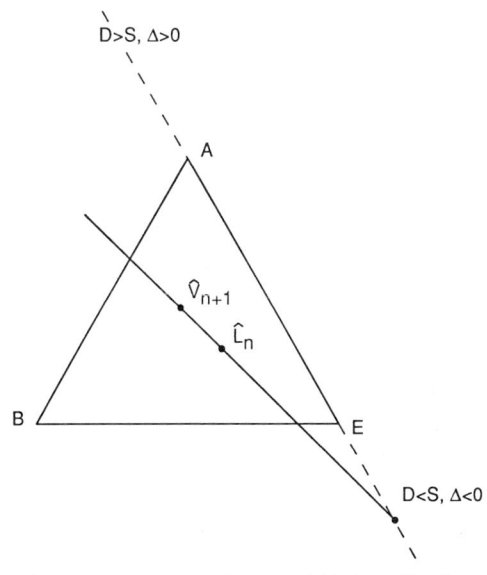

FIG. 63. Location of Δ point and corresponding material balance line for extractive distillation.

other words, the point must lie along the A/E edge extending above and to the left of the vertex for pure A.

If $D = S$, the Δ point is at infinity along this line. As S starts to exceed D, Δ jumps to negative infinity along this same edge. It moves toward the point E as S gets larger and larger relative to D. This makes sense. When $S = 0$, the Δ point should be at the distillate composition, here pure A. Adding solvent moves it upward along the line, extending the A/E edge. It should ultimately approach pure solvent as S becomes very large relative to D.

If $\Delta = D - S$ is positive, the material balance constraint says the vapor composition for $\hat{V}_{n+1}$ is a mixture of the liquid composition for $\hat{L}_n$ and the composition for Δ. Thus the point for $\hat{V}_{n+1}$ will lie between $\hat{L}_n$ and Δ. This puts $\hat{V}_{n+1}$ above and to the left of $\hat{L}_n$ in Fig. 63. If $\Delta = D - S$ is negative, then the same material balance equation says $\hat{L}_n$ is a mixture of the vapor composition for $\hat{V}_{n+1}$ and the composition for Δ, which now lies below and to the right (as we have sketched it in Fig. 63). In this case, $\hat{V}_{n+1}$ is again above that for $\hat{L}_n$ along the line passing through Δ. Therefore, while Δ jumps from plus infinity to minus infinity, the relative positions for $\hat{V}_{n+1}$ and $\hat{L}_n$ stay the same, and their positions move smoothly even as Δ jumps.

We stated above that isopropanol (A) is more volatile than water (B) in the presence of lots of ethylene glycol (E). Figure 64 is a sketch of how the residue curves appear for this mixture on a triangular composition diagram. Cover the upper left part of the diagram, leaving only the node for E showing—i.e., where we circle the node for E. This part of the diagram looks just like that for an ideal mixture where A is the most volatile species and B the intermediate. As

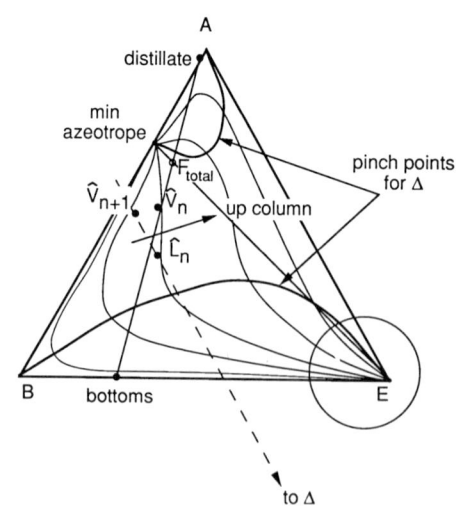

FIG. 64. Stepping up the middle section of an extractive distillation column.

one passes along a residue curve toward species A, the system becomes "aware" that A is not the lightest species around, but rather the minimum-boiling azeotrope is. The curves turn and head for it instead, terminating at A. They pick up a distinct S shape to them, as shown.

Let us assume we need to separate an azeotropic mixture of A and B. Since a conventional distillation column cannot carry out this separation, we plan to use solvent E as an extractive agent. We perform an analysis to see if it will work. The total feed, F_{total}, to the column is the sum of the azeotropic mixture and the solvent E. Its composition will lie along the line joining the compositions for the azeotrope and E. We want relatively pure A for the distillate (possibly contaminated with a small amount of B); so we sketch in about where we wish to find our desired distillate product composition. The bottoms product will be a mixture of species B and E. We place the products on our diagram, noting that both must lie on a straight line passing through the total feed composition.

If we sketch the compositions that can reach the distillate product in a conventional column, we find them limited to being near the A/E edge. To see this, follow the residue curve passing through the distillate product toward higher and higher temperatures. The pinch point trajectory emanating from the distillate product is a bit to the inside of this residue curve, but never too far from it here. All compositions that can reach the distillate are bounded by this pinch point trajectory and the distillation curves (close to the residue curve) passing through the distillate composition point. Similarly, we see that the compositions that can reach the bottoms are roughly those between the bottoms and the azeotrope. Not surprisingly, the two regions do not intersect, and thus we know we cannot separate the total feed mixture (azeotrope mixed with solvent E) in a conventional single-feed column.

Our intention is to use an extractive distillation column. The extractive agent is to be fed separately near the top of the column. We wish to see if this section has a composition trajectory that steps between a pair of compositions that the top and bottom sections can each reach from their respective product compositions.

Let us set the solvent feed initially equal to that of the distillate, which puts the Δ point at infinity. We choose this point initially only because it is easy to sketch the lines passing through it: they are parallel to the A/E edge. $\hat{V}_{n+1}$ and $\hat{L}_n$ for the column section between the two feeds will lie on such a line, as shown. Let us then locate $\hat{V}_n$, given $\hat{L}_n$. The slope of the residue curve passing through $\hat{L}_n$ points at $\hat{V}_n$. We see that $\hat{V}_n$ is thus to the right of $\hat{V}_{n+1}$. We sketch in a possible point. We note then that we move up the column when the tray number decreases from $n+1$ to n. Thus, $\hat{V}_n$ moves toward the A/E edge relative to $\hat{V}_{n+1}$. This is the exact direction in which we must move to connect the compositions reachable from the bottoms product to those reachable by the distillate. (If there were no extractive agent present, the Δ point would be located at pure A. For this case, $\hat{V}_{n+1}$ and $\hat{L}_n$ would be on a line passing through the node for A instead, and we would find ourselves moving toward the node for B rather than toward the A/E edge as we move up the column. That is incompatible with our intentions for this column.) With solvent fed as in an extractive column, we move from compositions we can reach from the bottoms product toward those along the A/E edge, from which we can step to the desired distillate product just a few trays above the solvent feed.

We moved toward the A/E edge because of the relative slopes of the material balance line passing through the Δ point and the residue curve passing through $\hat{L}_n$. Imagine that we have an $\hat{L}_n$ that lies much closer to the node for E, where the relative slopes are the reverse. Here, the column does not function as desired. The limit point for the desired behavior is where the slope of the line passing through Δ and the slope of the residue curve coincide. That is precisely *a pinch point* for Δ. With Δ at infinity, this occurs where a residue curve has a slope parallel to the A/E edge. We have plotted a trajectory of pinch points for the Δ point. It starts at the node for B and ends at the node for E here. Above this line, we move from left to right (the desired direction) in the section of the column between the feeds; below, we move in the other direction.

The S shape of the residue curves gives us another pinch point curve for Δ. It starts at the azeotrope and ends at the node for A. Above this curve, movement is again in the wrong direction. Thus, we have only the region between these two curves in which we can correctly operate the section of the column between the two feeds.

What does it mean to have two pinch point curves for Δ? To operate as we wish, we need to keep the composition of the liquid on the trays between these two curves for the section of the column between the feeds. We shall now show that one of the curves dictates a minimum reflux ratio and the other a maximum reflux ratio that must be used to operate the column for a fixed solvent ratio. We shall then show that there is a minimum solvent ratio we can use—one where the pinch point trajectories just touch. We will also show that larger ratios move the pinch curves apart, making it easier to effect the separation.

2. The Impact of Reflux Ratio for Fixed Solvent Ratio

The reflux ratio, R, for a column is the ratio of the liquid we reflux back to the column at the top stage relative to the distillate product flow; i.e., $R = L/D$, where L is the liquid flow in the top section of our column. Suppose we operate our column with a fixed solvent-to-distillate ratio; i.e., $R_s = S/D =$ constant. If R_s and D are fixed, the location of the Δ point is fixed because

$$\Delta = D - S = D - R_s D = (1 - R_s)D$$

With Δ fixed, we can find its pinch point trajectories, as we show in Fig. 65. When $\hat{L}_n$ sits exactly on such a trajectory, the compositions for $\hat{V}_{n+1}$ and $\hat{V}_n$

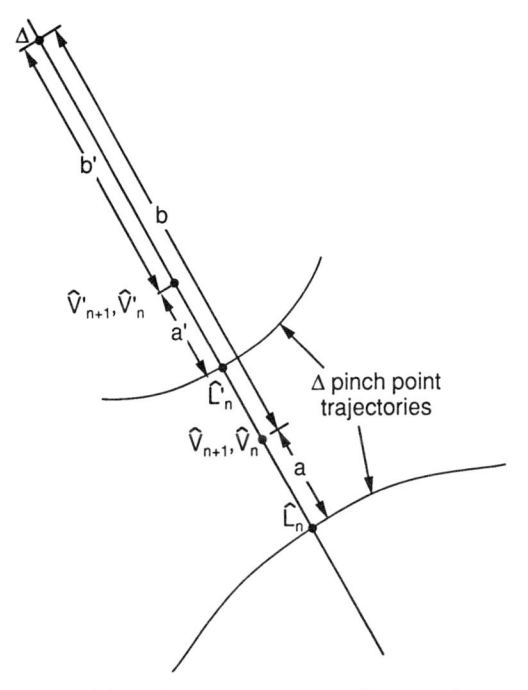

FIG. 65. Geometry to explain minimum and maximum reflux ratios for an extractive distillation column.

coincide because the line through the Δ point and the residue curve passing through $\hat{L}_n$ have the same slope at such a pinch point. The former points at $\hat{V}_{n+1}$, while the latter points at $\hat{V}_n$. We show the point $\hat{L}_n$ sitting on the lower curve and a second case in which $\hat{L}'_n$ sits precisely on the upper curve.

Let us first examine the point $\hat{L}_n$ sitting on the lower curve. We remember that a material balance for the section of the column between the feeds is

$$\hat{V}_{n+1} = \hat{L}_n + D - S = \hat{L}_n + \Delta$$

where we show the Δ point in Fig. 65 for the case in which Δ is positive ($R_S < 1$). Using the lever rule, we can write

$$\frac{b}{a} = \frac{\hat{L}}{\Delta} = \frac{L + q_S S}{D - S} = \frac{RD + q_S R_S D}{(1 - R_S)D} = \frac{R + q_S R_S}{(1 - R_S)}$$

where L is the liquid flow above the solvent feed and q_S characterizes the thermal condition of the solvent (typically the solvent will be subcooled with a q_S greater than unity). Thus the column reflux ratio is given by

$$R = \frac{b}{a}(1 - R_S) - q_S R_S$$

We can determine R for each pinch point by computing the distances a and b. $\hat{L}_n$ cannot actually sit on the pinch curve as shown because that requires a column with an infinite number of trays. We must place it between the two pinch curves at a point where b is smaller, which, by the above, requires R to decrease, assuming that the distance a does not change much with such a move. We argue, therefore, that the reflux ratio for a real column must be less than any R we see along this pinch point trajectory. This curve therefore defines a *maximum* allowed reflux ratio for operating the column.

From the geometry we see here, the maximum value of R along the pinch point trajectory is likely where the distance b is a maximum (but it may not be if a changes as we move along the trajectory).

The upper pinch curve produces the above equations, only this time in terms of the prime variables we show on Fig. 65. This time we must increase b, increasing R, to keep the liquid compositions inside the region between the two pinch curves. We must have R larger than the least value produced by all the points along this pinch curve.

We see that we have both an upper bound and a lower bound on the column reflux ratio. Does having an upper bound make intuitive sense? We have set the solvent flow proportional to the distillate product flow; i.e., $S = R_S D$. As we increase the reflux ratio R, the ratio of solvent feed flow, $R_S D$, to reflux flow, RD, decreases. This decreases the solvent concentration throughout the column, thus reducing its impact on the liquid activity coefficients that we are using to separate A from B. With an infinite reflux ratio, the solvent flow reduces to zero, and we have a normal column operating at total reflux which we know cannot separate A from B.

A similar set of arguments gives us the same results as above when the solvent rate exceeds the distillate rate, i.e., when the Δ point is to the lower right in Fig. 65. Again, the lower curve defines a maximum reflux ratio, and the upper a minimum.

There are other restrictions on the minimum value for the column reflux ratio. Both the bottom section (below the bottom feed) and top section above the solvent feed must be able to reach compositions that join with the trajectory traced by the middle section for a given reflux ratio. These must not restrict R to be outside the limits we just discovered, or else the column cannot operate.

3. The Impact of Solvent Ratio

In the previous section, we fixed the solvent ratio. Here we shall allow it to vary. We noted above that, as the solvent ratio increases, the Δ point moves away from the distillate composition toward infinity, jumps to negative infinity, and then moves toward the solvent composition, always along the line that passes through these two compositions. As we decrease the solvent ratio, the two pinch curves shown in Fig. 64 move closer together until they just coincide. Decreasing the solvent ratio below this value precludes the existence of a path for the composition trajectory for the column section between the feeds. Thus the column cannot function. Increasing the solvent ratio moves the two pinch point curves away from each other, giving an even larger region for the composition trajectory to pass for the section between the feeds.

We argue that the two pinch trajectories coincide at an inflection point for one (or more) of the residue curves. See Fig. 66. The line marked 1 in this figure has the same slope as the residue curve at its inflection point. For this line, Δ would be quite close to the pure A node (implying a small amount of solvent). If we move Δ away from A, there are two points, one to each side of the inflection point, which would point at the same Δ point (see the two lines marked 2). Only at the inflection point is there a single pinch point for the corresponding Δ point.

To find the minimum solvent rate, we draw all lines passing through all inflection points for the residue curves and project them to the line on which the Δ point must exist, i.e., the line passing through the composition points for the distillate and the solvent. The Δ point must be no closer to the distillate composition point than the most distant such intersection point to keep the Δ

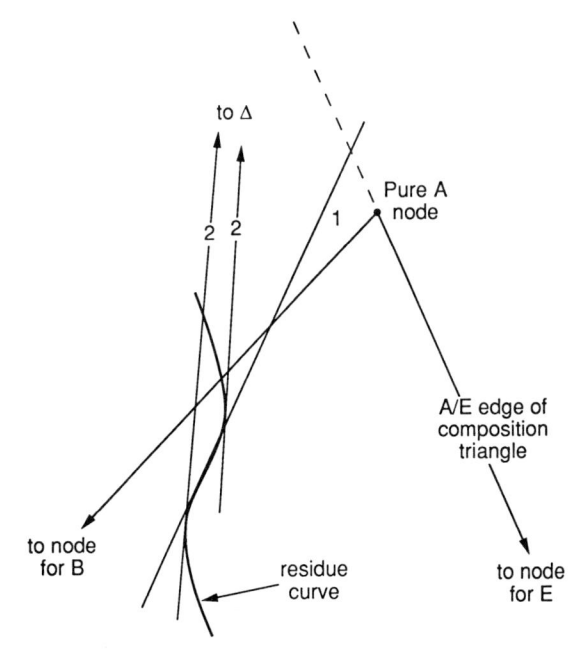

FIG. 66. Locating the Δ point that gives the minimum solvent flow.

pinch points for the corresponding residue curve from coinciding. This construction defines the minimum solvent ratio needed for the column.

X. Post-analysis Methods: Column Design Calculations

Column design represents the third major analysis activity in analysis-driven synthesis. However, our discussion here is short as column design is not the theme of this paper.

Once one has proposed alternative configurations for systems of separation devices to effect a desired separation, one must then design these devices so the various alternatives may be compared. For a distillation column, the first set of design decisions is to choose the number of trays, the feed tray location, and the reflux ratio at which to operate it. For a binary separation, the McCabe–Thiele diagram (or the concepts behind it) is an indispensable aid in making these decisions.

An approach to setting the reflux ratio often involves computing a minimum reflux ratio, a topic we have discussed several times in this article with respect to pinch points. Considerable work has been done on computing minimum reflux ratios for columns. Koehler (1991) reviews this work. As a rule of thumb, one sets the reflux ratio to be 20–100% or so above the minimum, a number that experience shows trades off the number of trays with column diameter in a near cost-optimal way.

With the reflux ratio fixed, one can step off the number of trays for a binary column using a McCabe–Thiele diagram or, if one wishes to account for heat effects, a Ponchon–Savarit diagram (Treybal, 1968), establishing the number of trays needed and where to feed the column.

For more than two species and a reflux ratio set to 1.2 times the minimum, a rule of thumb is to compute the total number of trays required for a total reflux column to produce the separation desired and then double this number as a first guess (Douglas, 1988). The next decision is select the tray on which to feed the column. For multispecies columns, the placement is not obvious. Typically, one must search by placing it on any one of a range of trays using a tray-by-tray simulation, thereby discovering which tray location requires the least reflux to effect the desired separation.

Using collocation models, which we mentioned earlier (Huss and Westerberg, 1994), one can formulate the design problem as a continuous-variable optimization problem with present worth as the objective function. The optimization problem will select the number of trays needed in both the top and bottom sections along with the reflux ratio—all of which are continuous variables in such a model.

Acknowledgments

The National Science Foundation, through its grant to the Engineering Design Research Center, and Eastman Chemical Co., through its support of the Computer Aided Process Design Consortium at Carnegie Mellon University, provided support for this work.

References

Andrecovich, M. J., and Westerberg, A. W. "A Simple Synthesis Method Based on Utility Bounding for Heat Integrated Distillation Sequences," *AIChE J.* **31**, p. 363 (1985).

Barnicki, S. D. and Fair, J. R. "Separation System Synthesis: A Knowledge-Based Approach. 1. Liquid Mixture Separations," *Ind. Eng. Chem. Res.* **29**, 421–432 (1990).

Bekiaris, N., Meski, G. A., Radu, C. M., and Morari, M. "Multiple Steady States in Homogeneous Azeotropic Distillation Columns," *Ind. Eng. Chem. Res.* **29**, 421–432 (1993).

Benedict, M., and Rubin, L. C. "Extractive and Azeotropic Distillation. 1. Theoretical Aspects." *Trans. Am. Inst. Chem. Eng.,* **41**, 353–370 (1945).

Berg, L. "Selecting the Agent for Distillation," *Chem. Eng. Prog.* **65**, (9), 52–57 (1969).

Berg, L., and Yeh, A. 1985. "The Unusual Behavior of Extractive Distillation—Reversing the Volatility of the Acetone—Isopropyl Ether System," *AIChE J.* **31**, (3), 504–506 (1985).

Bossen, B. S., Jorgensen, S. B., and Gani, R. "Simulation, Design, and Analysis of Azeotropic Distillation Operations," *Ind. Eng. Chem. Res.,* **32**, 620–633 (1993).

Cho, Y. S., and Joseph, B. "Reduced-Order Steady-State and Dynamic Models for Separation Processes," *AIChE J,* **29**, 261–269, 270–276 (1983).

Doherty, M. F. "The Presynthesis Problem for Homogeneous Azeotropic Distillations has a Unique Explicit Solution," *Chem. Eng. Sci.* **40**, 1885–1889 (1985).

Doherty, M. F., and Caldarola, G. A. "Design and Synthesis of Homogeneous Azeotropic Distillations. 3. The Sequencing of Columns for Azeotropic and Extractive Distillations," *Ind. Eng. Chem. Fundam.* **24**, 474–485 (1985).

Doherty, M. F., and Perkins, J. D. "On the Dynamics of Distillation Processes. I. The Simple Distillation of Multicomponent Non-Reacting Homogeneous Liquid Mixtures," *Chem. Eng. Sci.* **33**, 281–301 (1978).

Doherty, M. F., and Perkins, J.D. "On the Dynamics of Distillation Processes. III. The Topological Structure of Ternary Residue Curve Maps," *Chem. Eng. Sci.* **34**, 1401–1414 (1979).

Douglas, J. M. "Conceptual Design of Chemical Processes." McGraw-Hill, New York, 1988.

Ewell, R. H., and Welch, L. M. "Rectification in Ternary Systems Containing Binary Azeotropes," *Ind. Eng. Chem.* **37**, 1224–1231 (1945).

Fidkowski, Z. T., Malone, M. F., and Doherty, M. F. "Computing Azeotropes in Multicomponent Mixtures," *Comput. Chem. Eng.* **17**(12), 1141–1155 (1993).

Fien, G. A. F., and Liu, Y. A. "Heuristic Synthesis and Shortcut Design of Separation Processes Using Residue Curve Maps: A Review," *Ind. Eng. Chem. Res.* **33**, 2505–2522 (1994).

Foucher, E. R., Doherty, M. F., and Malone, M. F. "Automatic Screening of Entrainers for Homogeneous Azeotropic Distillation," *Ind. Eng. Chem. Res.* **30**, 760–772 (1991).

Hendry, J. E., Rudd, D. F., and Seader, J. D. "Synthesis in the Design of Chemical Processes," *AIChE J.* **19**(1), 1–15 (1973).

Henley, E. J., and Seader, J. D. "Equilibrium-State Separation Operations in Chemical Engineering," Wiley, New York, 1981.

Hlavacek, V. "Synthesis in the Design of Chemical Processes," *Comput. Chem. Eng.* **2**, 67–75 (1978).

Hoffman E. J. "Azeotropic and Extractive Distillation." Wiley (Interscience), New York, 1964.

Holland, C. D. "Fundamentals of Multicomponent Distillation." McGraw-Hill, New York, 1981.

Horsley, L. H. "Azeotropic Data III," Adv. Chem. Ser. No. 116. American Chemical Society, Washington, DC, 1973.

Hunek, J., Gal, S., Posel, F., and Gavic, P. "Separation of an Azeotropic Mixture by Reverse Extractive Distillation." *AIChE J.* **35**(7), 1207–1210 (1989).

Huss, R. S., and Westerberg, A. W. "Collocation Methods for Distillation Design," Paper 131c, Annual AIChE Meeting, San Francisco (1994).

Julka, V., and Doherty, M. F. "Geometric Behavior and Minimum Flows for Nonideal Multicomponent Distillation," *Chem. Eng. Sci.* **45**, 1801–1822 (1990).

King, C. J. "Separation Processes," 2nd ed. McGraw-Hill, New York, 1980.

Knight, J. R., and Doherty, M. F. "Optimal Design and Synthesis of Homogeneous Azeotropic Distillation Sequences," *Ind. Eng. Chem. Res.* **28**, 564–572 (1989).

Koehler, J. W., "Struktursynthese und minimaler Energiebedarf Nichtidealer Bektifikationen," Ph.D. Dissertation, Technical University of Munich (1991).

Laroche, L., Andersen, H. W., and Morari, M. "Homogeneous Azeotropic Distillation: Comparing Entrainers," *Can. J. Chem. Eng.* **69**, 1302–1319 (1991).

Levy, S. G., Van Dongen, D. B., and Doherty, M. F. "Design and Synthesis of Homogeneous Azeotropic Distillations. 2. Minimum Reflux Calculations for Nonideal and Azeotropic Columns," *Ind. Eng. Chem. Fundam.* **24**, 463–473 (1985).

Matsuyama, H. "Synthesis of Azeotropic Distillation Systems," Paper presented at the Japan-U.S. Joint Seminar, Kyoto, Japan (1975).

McCabe, W. L., and Smith, J. C. "Unit Operations of Chemical Engineering," 3rd ed. McGraw-Hill, New York, 1976.

McDonald, C.M., and Floudas, C. A. "Global Solutions for the Phase and Chemical Equilibrium Problem," Paper 220c, AIChE Annual Meeting, San Francisco (1994).

Michelsen, M. L. "The Isothermal Flash Problem. Part I: Stability," *Fluid Phase Equilib.* **9**, 1–19 (1982).

Michelsen, M. L. "Phase Equilibrium Calculations. What is Easy and What is Difficult," *Comput. Chem. Eng.* **17**(5/6), 431–439 (1993).

Michelsen, M. L. "Calculation of Multiphase Equilibrium," *Comput. Chem. Eng.* **18**(7), 545–550 (1994).

Modi, A. K. and Westerberg, A. W. "Distillation Column Sequencing Using Marginal Price," *Ind. Eng. Chem. Res.* **31**, 839–848 (1992).

Nikolaev, N. S., Kiva, V. N., Mozzhukhin, A. S., Serafimov, L. A., and Goloborodkin, S. I. "Utilization of Functional Operators for Determining the Regions of Continuous Rectification," *Theor. Found. Chem. Eng. (Engl. Transl.)* **13**, 418–423 (1979).

Nishida, N., Stephanopoulos, G., and Westerberg, A. W. "A Review of Process Synthesis," *AIChE J.* **27**, 321 (1981).

Perry, J. H. "Chemical Engineers' Handbook," 3rd ed. McGraw-Hill, New York, 1950.

Petlyuk, F. B. "Rectification of Zeotropic, Azeotropic, and Continuous Mixtures in Simple and Complex Infinite Columns with Finite Reflux," *Theor. Found. Chem. Eng. (Engl. Transl.)* **12**, 671–678 (1978).

Petlyuk, F. B., Serafimov, L. A., Avet'yan, V. S., and Vinogradova, E. I. "Trajectories of Reversible Rectification when One of the Components Completely Disappears in Each Section," *Theor. Found. Chem. Eng. (Engl. Transl.)* **15**, 185–192 (1981).

Pham, H. N., and Doherty, M. F. "Design and Synthesis of Heterogeneous Azeotropic Distillations. III. Column Sequences," *Chem. Eng. Sci.* **45**, 1844–1854 (1990).

Pham, H. N., Ryan, P. J., and Doherty, M. F. "Design and Minimum Reflux for Heterogeneous Azeotropic Distillation Columns," *AIChE J.* **35**,(10), 1585–1591 (1989).
Poellmann, P., and Blass, E. "Best Products of Homogeneous Azeotropic Distillations," *Gas Separ. Purif.* **8**(4), 194–228 (1994).
Rathore, R. N. S., VanWormer, K. A., and Powers, G. J. "Synthesis Strategies for Multicomponent Separation Systems with Energy Integration," *AIChE J.* **20**,491 (1974).
Reid, R. C., Prausnitz, J. M., and Poling, B. E. "The Properties of Gases and Liquids," 4th ed. McGraw-Hill, New York, 1987.
Ryan, P. J., and Doherty, M. F. "Design/Optimization of Ternary Heterogeneous Azeotropic Distillation Sequences," *AIChE J.* **35**,(10), 1592–1601 (1989).
Sargent, R. W. S. H. "A Functional Approach to Process Synthesis and its Application to Distillation Systems," Tech. Rep. Centre for Process Systems Engineering, Imperial College, London, 1994.
Sargent, R. W. S. H., and Gaminibandara, K. "Optimum Design of Plate Distillation Columns," in "Optimization in Action" (L. W. C. Dixon, ed.), 267–314. Academic Press, London, 1976.
Seferlis, P., and Hrymak, A. N. Optimization of Distillation Units Using Collocation Models, *AIChE J.*, **40**, 813–825 (1994).
Serafimov, L. A. "Thermodynamic Topological Analysis and the Separation of Multicomponent Polyazeotropic Mixtures," *Theor. Found. Chem. Eng. (Engl. Transl.)* **21**, 44–54 (1987) (translated from *Teor. Osn. Khim. Tekhnol.* **21**(1), 74–85, (1987)).
Smith, J. M., and Van Ness, H. C. "Introduction to Chemical Engineering Thermodynamics," 4th ed. McGraw-Hill, New York, 1987.
Stewart, W. E., Levien, K. L., and Morari, M. "Collocation Methods in Distillation," in "Foundations Computer-Aided Process Design (FOCAPD'83)" (A. W. Westerberg and H. H. Chien, eds.), pp. 539–569. Cache Corp., Ann Arbor, MI, 1984.
Stichlmair, J., Fair, J. R., and Bravo, J. L. "Separation of Azeotropic Mixtures via Enhanced Distillation," *Chem. Eng. Prog.* **85**,(1), 63–69 (1989).
Tanaka, S., and Yamada, J. "Graphical Solution of Operating Region in Extractive Distillation," *Kagaku Kogaku* (Abr. E. *Engl.*) **3**,(1), 40–43, (1965).
Terranova, B. E., and Westerberg, A. W. "Temperature-Heat Diagrams for Complex Columns. 1. Intercooled/Interheated Distillation Columns," *Ind. Eng. Chem. Res.* **28**, 1374–1379 (1989).
Thompson, R. W., and King, C. J. "Systematic Synthesis of Separation Schemes," *AIChE J.* **18**, 941–948 (1972).
Treybal, R.E. Mass-Transfer Operations, 2nd ed. p. 297. McGraw-Hill, New York, 1968.
Underwood, A. J. V. "Fractional Distillation of Multicomponent Mixtures—Calculation of Minimum Reflux Ratio," *J. Inst. Petrol.* **32**, 614 (1946).
Van Dongen, D. B. "Distillation of Azeotropic Mixtures. The Application of Simple-Distillation Theory to the Design of Continuous Processes," Ph.D. Dissertation, University of Massachusetts, Amherst (1983).
Van Dongen, D. B., and Doherty, M. F. "Design and Synthesis of Homogeneous Azeotropic Distillations. 1. Problem Formulation for a Single Column," *Ind. Eng. Chem. Fundam.* **24**, 454–463 (1985).
Wahnschafft, O. M. "Synthesis of Separation Systems for Azeotropic Mixtures with an Emphasis on Distillation-Based Methods," Ph.D. Dissertation, University of Munich, Munich, Germany (1992).
Wahnschafft, O. M. "A Simple and Robust Continuation Method for Determining All Azeotropes Predicted by a Multicomponent Vapor-Liquid Equilibrium Model," in preparation (1994).
Wahnschafft, O. M., and Westerberg, A. W. "The Product Composition Regions Azeotropic Distillation Columns. 2. Separability in Two-Feed Columns and Entrainer Selection." *Ind. Eng. Chem. Res.* **32**, 1108–1120 (1993).

Wahnschafft, O. M., Koehler, J., Blass, E., and Westerberg, A. W. "The Product Composition Regions of Single Feed Azeotropic Distillation Columns," *Ind. Eng. Chem. Res.* **21,** 2345–2362 (1992).

Wasylkiewicz, S., Sridher, L., Malone, M. F., and Doherty, M. F. "Gibbs Tangent Plane Anaylsis for Complex Liquid Mixtures: A Global Algorithm," Paper 151a, Annual Meeting AIChE, St. Louis, MO (1993).

Westerberg, A. W. "A Review of Process Synthesis, in Computer Applications to Chemical Engineering Process Design and Simulation," *ACS Symp. Ser.* **124,** 54–87 (1980).

Westerberg, A. W. "The Synthesis of Distillation-Based Separation Systems," *Comput. Chem. Eng.* **9,**(5), 421–429 (1985).

Zharov, W. T., and Serafimov, L. A. "Physicochemical Fundamentals of Distillations and Rectifications (in Russian)." Khimiya, Leningrad, 1975.

MIXED-INTEGER OPTIMIZATION TECHNIQUES FOR ALGORITHMIC PROCESS SYNTHESIS

Ignacio E. Grossmann

Department of Chemical Engineering
Carnegie Mellon University
Pittsburgh, Pennsylvania

I. Introduction	172
II. Overview of Previous Work	172
A. Process Flowsheets	172
B. Review of Subsystem Synthesis	175
III. Mathematical Programming Approach	178
IV. Representation of Alternatives	180
V. MINLP Modeling	187
A. Example of a MINLP Model	192
VI. MINLP Algorithms	197
A. Basic Algorithms	197
B. Extensions of MINLP Methods	202
C. Logic-Based Methods	207
D. Computational Example	209
VII. Solution Strategies for MINLP Synthesis Problems	213
A. Handling Zero Flows	213
B. Large Size of MINLP Problems	215
C. Handling Nonconvexities	219
VIII. Applications	224
IX. Concluding Remarks	237
Acknowledgments	239
References	239

This paper presents an overview of the mathematical programming approach for process synthesis. First, the methods for process synthesis are reviewed with an emphasis on algorithmic methods. The mathematical programming approach is covered next in a discussion of basic concepts on representations for synthesis, modeling of mixed-integer nonlinear programming (MINLP) problems, MINLP algorithms, and solution strategies. As is shown, these four components are basic elements in the algorithmic methods for process synthesis. Also, it is shown, both through the derivation of methods and their application to several examples, that MINLP optimization has reached a stage where it can solve practical problems of significant size. Finally, several future directions of research are also discussed.

I. Introduction

A major step in process synthesis has been its formalization through the mathematical programming approach (Grossmann, 1990a,b; Grossmann and Daichendt, 1994). Major motivations include the increased need for the automation of the synthesis process and the need to explore a larger number of alternatives at the preliminary stages of design in order to improve the economics and other design criteria. In this paper we provide a comprehensive overview of the algorithmic approach to process synthesis, particularly the one that relies on the use of MINLP optimization.

This paper is organized as follows. We first present an extensive overview of previous work in process synthesis, emphasizing algorithmic methods. Next, we outline the major steps that are involved in the mathematical programming approach: representation of alternatives, mathematical modeling, and solution of the corresponding mathematical programming problem, which in general involves a mixed-integer nonlinear programming (MINLP) problem. For the former step, we discuss several major alternative representations that can be used, and clarify the use of representations at different levels of abstraction, particularly superstructures and aggregated models. We then concentrate on the modeling, highlighting the usefulness of propositional logic for deriving discrete constraints. Next, we give a unified presentation of the major MINLP algorithms that have emerged, emphasizing their common features. Then we discuss the role of solution strategies to address difficulties with zero flows, large-scale problems, and nonconvexities. Finally, several example problems are presented to illustrate some of the major points in this paper. The paper concludes with a discussion of future research directions.

II. Overview of Previous Work

A. PROCESS FLOWSHEETS

Reviews on earlier developments in the area of process synthesis can be found in Hendry *et al.* (1973), Hlavacek (1978), and Nishida *et al.* (1981). In the late sixties, work began to develop a systematic approach to process synthesis based mostly on the use of decomposition and heuristic rules (Rudd and

Watson, 1968; Rudd, 1968; Masso and Rudd, 1969). Algorithmic methods for selecting the optimal configuration from a given superstructure also began to be developed through the use of direct search methods for continuous variables (Umeda *et al.*, 1972; Ichikawa and Fan, 1973) as well as branch and bound search methods (Lee *et al.*, 1970).

In terms of process flowsheets, the first computer-aided process synthesizer for generating initial structures, AIDES [Adaptive Initial DEsign Synthesizer], was developed by Rudd and his students (Siirola *et al.*, 1971; Siirola and Rudd, 1971; Powers, 1972); using a high level representation of tasks, it relied on the use of heuristics and linear programming, which were coordinated through a means–ends-analysis search. The second computer-aided process synthesizer to be developed was BALTAZAR (Mahalec and Motard, 1977a,b). It also relied on heuristics and linear programming, and used a tree search within the framework of theorem proving. Neither AIDES nor BALTAZAR incorporated equipment costs directly, but they employed heuristics as indicators of economic performance.

The current state of flowsheet synthesis is represented by two different approaches: (1) *hierarchical decomposition* (Douglas, 1985, 1988, 1990) and its computer implementation PIP [Process Invention Procedure], (Kirkwood *et al.*, 1988) and (2) *mathematical programming* (Grossmann, 1985, 1990a,b) and its initial computer implementation in PROSYN [PROcess SYNthesizer], (Kravanja and Grossmann, 1990). It has been pointed out (Rippin, 1990) that these two approaches are concerned with different aspects of design and can be regarded as complementary.

The hierarchical decomposition technique breaks the synthesis procedure into five discrete decision levels: (1) batch versus continuous, (2) input/output structure of flowsheet, (3) recycle structure and reactor considerations, (4) separation systems, and (5) heat exchanger network. At each decision level beyond the first, the economic potential of the project is evaluated and a decision is made whether or not further work on the project is justified. This method utilizes heuristics, short-cut design procedures, and physical insight to develop an initial base-case design. This approach is motivated by Douglas' claim that only 1% of all designs are implemented in practice; thus this screening procedure avoids detailed evaluation of most alternatives. Relying on heuristics, this approach cannot *rigorously* produce an optimal design; and although heuristics often lead to good designs, they are *fallible* (e.g., see Fonyó and Mizsey, 1990). Furthermore, owing to the *sequential* nature of the flowsheet synthesis, interactions among the design variables at the various decision levels may not be properly accounted for, as it is necessary to solve for them *simultaneously*. For instance, Duran and Grossmann (1986a) and later Lang *et al.* (1988) have shown that simultaneous optimization and heat integration of flowsheets generally produces significant improvements in the profit compared to the sequential approach. Despite these shortcomings, hierarchical decomposition provides a useful approach for generating an initial flowsheet and alternatives (i.e., a base-case design and superstructure). And, when coupled with the concept of simultaneous synthesis of the complete flowsheet, it also provides a framework for decomposing the synthesis problem into a hierarchy of detailed and aggregated models; it is therefore simpler to solve than the entire flowsheet, while still reflecting the presence of downstream tasks.

The mathematical programming approach utilizes optimization techniques to select the configuration and parameters of the processing system (Grossmann, 1985, 1990a,b). A superstructure containing alternative processing units and interconnections is modeled as discrete, binary variables (0–1) to depict the existence (1) or nonexistence (0) of that unit. Initially, the synthesis of total process systems (chemical plant, heat exchanger network, utility plant) was formulated as a mixed-integer linear program (MILP) (Papoulias and Grossmann, 1983a,b,c). To explicitly handle the nonlinearities in process models, the outer-approximation algorithm for MINLP has been developed and successively refined (Duran and Grossmann, 1986b; Kocis and Grossmann, 1987, Viswanathan and Grossmann, 1990). It is now widely available in the program DICOPT++ within the modeling system GAMS (Brooke *et al.,* 1988). This algorithm partitions the problem into two parts: (1) an NLP subproblem, where the continuous variables for a single flowsheet configuration are initially optimized and then the remaining alternative substructures are suboptimized for the given flows; (2) linearization of the nonlinear equations to obtain an MILP master problem, which then determines a new optimal flowsheet configuration (i.e., new set of binary variables) for the next NLP subproblem.

For convex MINLP problems (Duran and Grossmann, 1986b), the NLP optimization yields an upper bound for a cost-minimization problem while the MILP yields a lower bound. Thus, when the values of the objective functions for both problems are identical (or they cross), the global optimal solution is obtained. For the more general nonconvex MINLP problem (Viswanathan and Grossmann, 1990), the algorithm terminates when no improvement in the NLP objective function is obtained. In this case one cannot ascertain whether a local or the global optimal solution is obtained. This is because the linearizations of nonconvex equations may cut off portions of the feasible region, including the global optimal solution. Although the algorithm provides a reasonable degree of reliability for finding the global optimum, it can get trapped in a poor local solution (Daichendt and Grossmann, 1993a,b). The algorithmic approach presupposes that a superstructure is available or can be generated. At present, depending on the problem formulation, the number of binary variables needed to model the superstructure places an upper bound to the size of the problem that can be solved. It should also be noted that in order to address the MINLP optimization of process flowsheets more effectively, Kocis and Grossmann

(1989b) developed the modeling/decomposition (M/D) strategy in which the basic objective is to solve NLP subproblems pertaining only to the existing part of the superstructure. This strategy not only avoids the solution of NLP problems of larger dimensionality, but also reduces such numerical difficulties as singularities that arise in the case of nonexisting units with zero flows. The M/D strategy has been implemented in PROSYN and extended in various ways by Kravanja and Grossmann (1994).

B. Review of Subsystem Synthesis

Most of the research work in process synthesis has concentrated on the study of subsystems in which a major emphasis has been the management of energy. Pinch analysis has been widely used for setting design targets (see Linnhoff, 1993, for a recent review). However, recent efforts have been directed to the better understanding of synthesis issues in nonideal separation systems (Van Dongen and Doherty, 1985; Doherty and Cardarola, 1985) and reaction systems (Glasser *et al.*, 1987).

1. Heat-Exchanger Network Synthesis

Heat-exchanger network synthesis (HENS) is by far the most developed technique, and many methods and software packages are available for it. An extensive review can be found in Gundersen and Naess (1988). The discovery of the heat-recovery pinch (Umeda *et al.*, 1979; Linnhoff *et al.*, 1982), which is derived through thermodynamic analysis, provided the basis for the advancements in developing synthesis techniques for HENS. The most widely used method, commonly known as "pinch technology" (Linnhoff and Hindmarsh, 1983), relies on the use of targets (energy, number units, area) and is based on a user-driven approach. SUPERTARGET and ADVENT are two major pieces of software implementing this approach. As for algorithmic methods, there has been a gradual evolution from LP/MILP/NLP methods, which are based on targets (Cerda and Westerberg, 1983; Papoulias and Grossmann, 1983b; Floudas *et al.*, 1986; Colberg and Morari, 1990; Gundersen and Grossmann, 1990), to simultaneous MINLP models in which networks are automatically synthesized and energy, area, and number of units are optimized simultaneously (see Yee and Grossmann, 1990; Ciric and Floudas, 1991). Examples of software include MAGNETS for the target-based methods and SYNHEAT for the simultaneous MINLP. Also, the utility target for the HENS problem has the advantage that it can be effectively represented in an aggregated form (Duran and Grossmann, 1986a), and thus can be embedded within other synthesis models to perform simultaneous optimization. It should also be noted that the ideas of pinch analysis are being expanded beyond heat-exchanger networks to include total sites and assessment of environmental problems (see Linnhoff, 1993, for a review). Also, an important related problem is that of synthesis of utility systems (Papoulias and Grossmann, 1983a; Colmenares and Seider, 1987) and synthesis of integrated refrigeration systems (Shelton and Grossmann, 1986a,b).

2. Distillation Sequencing

A general review on distillation synthesis can be found in Westerberg (1985) and Floquet *et al.*, (1988). The synthesis of simple separation sequences based on heuristics is fairly well developed. Enumeration search methods (dynamic programming, branch-and-bound) have been proposed by Hendry and Hughes (1972) and Gomez-Muñoz and Seader (1985), while evolutionary search procedures have been described by Stephanopoulos and Westerberg (1976), Seader and Westerberg (1977) and Nath and Motard (1981).

Thermodynamic analysis of the effect of composition on the work of separation has been used to suggest nonconventional ways (i.e., nonsharp splits) of separating mixtures by distillation (Petlyuk *et al.*, 1965). Further work in this area provides sequencing heuristics (Gomez-Muñoz and Seader, 1985). Graphical techniques (Carlberg and Westerberg, 1989a,b) have been developed that provide insight into the heat flows of these complex columns. An evolutionary synthesis strategy has been proposed (Koehler *et al.*, 1992) that decomposes the problem into a selection of sequences based on decision factors starting from an initial superstructure. Then the sequence is further refined by considering the introduction of side-stream strippers and enrichers.

Complete thermodynamic analysis, based on reversible distillation, takes into account the effects of finite temperature and composition driving forces as well as nonuniform heat distribution and hydraulic resistance (Fonyó, 1974a,b). The effect of nonuniform heat distribution (i.e., adiabatic distillation) can be mitigated by the introduction of intercoolers/interheaters (Terranova and Westerberg, 1989; Dhole and Linnhoff, 1992).

Algorithmic formulations for heat-integrated distillation sequence synthesis involving simple columns (Rathore *et al.*, 1974a,b; Andrecovich and Westerberg, 1985a,b; Kakhu and Flower, 1988; Floudas and Paules, 1988) have been developed as well as a graphical approach using the pinch design method (Smith and Linnhoff, 1988). A more accurate thermodynamic targeting model has recently been reported by Wahnschafft *et al.* (1993). In addition, work on superstructure optimization has been performed for synthesizing complex columns (Sargent and Gaminibandara, 1976; Eliceche and Sargent, 1986; Rigg, 1991). The synthesis of systems involving multiple feeds and mixed products has also been considered (Floudas, 1987; Wehe and Westerberg, 1987; Quesada and Grossmann, 1995a).

As for the synthesis of azeotropic columns, most of the developments have been based on geometric representations with residue curves (Doherty and Cardarola, 1985; Knight and Doherty, 1989; Van Dongen and Doherty, 1985). Guidelines have also been developed by Stichlmair et al. (1989), some of which Laroche et al. (1992) have questioned as general synthesis rules. Finally, Wahnschafft et al. (1992) have developed the program SPLIT, which is based on a ruled-based system and evolutionary search method.

3. Mass-Exchange Networks

Motivated by applications in waste recovery systems, El-Halwagi and Manousiouthakis (1989a,b, 1990) have considered the problem of synthesizing mass-exchange networks. For the simpler case in which concentration targets are specified for single components, interesting analogies can be drawn with the heat-exchanger network problem, both in terms of the pinch point (El-Halwagi and Manousiouthakis, 1989a) and in terms of the MILP transshipment model (El-Halwagi and Manousiouthakis, 1990). For the case of concentration specifications for multiple components, the synthesis problem becomes more complex and must be formulated as an MINLP problem (El-Halwagi and Manousiouthakis, 1989b; Papalexandri et al., 1993). Bagajewicz and Manousiouthakis (1992) have extended the representation of mass-exchange networks for synthesizing heat-integrated distillation columns. These authors use a "state-space" approach for avoiding a simultaneous superstructure optimization, and they incorporate the constraints by Duran and Grossmann (1986a) for the pinch location method to represent the heat integration. A simultaneous MINLP model has been proposed recently by Papalexandri et al. (1993).

4. Reactor Network Synthesis

Synthesis of reactor networks poses a more difficult modeling problem as these must usually be described by differential-algebraic equations. Compared to HENS and distillation systems, however, the combinatorial part in reactor networks tends to be smaller. Heuristic-based approaches are generally limited to simple reactions. Reactor networks have been synthesized using a superstructure representation of serial recycle reactors without bypass for isothermal and nonisothermal reactors (Chitra and Govind, 1985a,b). A geometric approach, based on reactor trajectories, describes the attainable region of composition space generated by reaction and mixing for isothermal, constant-density systems (Glasser et al., 1987), and was later extended to include adiabatic, variable-density systems (Hildebrandt et al., 1990). An algorithmic procedure using a constant-dispersion model has been discussed (Achenie and Biegler, 1986). Subsequent work converted the synthesis problem into an optimal control problem

(Achenie and Biegler, 1988) and to an MINLP optimization problem (Kokossis and Floudas, 1991). Recent work has relied on targeting models (Balakrishna and Biegler, 1992a) in which a reactor network representation is based on a segregated mixing zone and multiple zones of maximum mixedness (multiple-compartment model), that taking advantage of the methods developed by Glasser *et al.* while overcoming the dimensional constraints by using two-dimensional projections of the multidimensional composition space. The simple segregated model is proposed as an initial solution, which can be posed as an LP and checked for global optimality. If it is not optimal, an iterative scheme is invoked that increases the complexity of the model by adding the first maximal-mixedness compartment, then checks to see if this more complex model can extend the attainable region. This method has also been extended to consider simultaneous heat integration (Balakrishna and Biegler, 1992b) using the aggregated model by Duran and Grossmann (1986a). Glavic *et al.* (1988) have developed thermodynamic criteria for the appropriate placement of reactors in a process flowsheet. Also, Omtveit and Lien (1993) have extended the representation of the attainable region so as to account for recycles in flowsheets.

III. Mathematical Programming Approach

From the review given in the previous section, it is clear that a significant number of synthesis models have been developed that are based on mathematical programming. It follows that the process synthesis problem can generally be stated as follows: Given specifications of input and of output streams—which may correspond to raw materials and desired products in flowsheets or simply to process streams in subsystems (e.g., heat-exchanger networks)—the problem consists in integrating a process system that will convert the inputs into desired outputs so as to meet desired specifications while optimizing a given objective or goal function.

The synthesis of such a system involves the selection of a configuration or topology, as well as its design parameters. One has to determine which units should integrate the system and how they should be interconnected, as well as the sizes and operating conditions of the units. The former clearly imply making discrete decisions, while the latter imply making a choice within a continuous space. Thus, from a conceptual standpoint the synthesis problem corresponds to a nonlinear discrete/continuous optimization problem which mathematically gives rise to an MINLP problem.

In general, the major steps involved in approaching this problem are as follows:

Step 1. Postulate a representation of alternatives whose embedded designs are candidates for a feasible and optimal system.

Step 2. Model representation in step 1 as the MINLP problem

$$\min Z = f(x,y)$$
$$\text{s.t.} \quad h(x,y) = 0 \quad \text{(MINLP)}$$
$$g(x,y) \leq 0$$
$$x \in X, \quad y \in Y$$

Here y represents a vector of 0–1 variables that denote the potential existence of units (0-not included, 1-included), while x represents a vector of continuous variables which correspond, for instance, to material/heat flows, pressures, temperatures, and sizes of equipment.

Step 3. Obtain the optimal design embedded in the representation by solving the corresponding MINLP problem.

It is important to note that the representation of alternatives can be specified at various levels of detail. The two extreme cases are superstructures and aggregate models. Superstructures are relatively detailed representations of a process in which the potential existence of units and streams of a flowsheet are explicitly considered. Aggregate models are representations at higher levels of abstraction in which the synthesis problem is simplified by the use design targets (e.g. minimum utility, maximum yield). This may often simplify the mathematical programming problem, reducing the MINLP to a nonlinear programming (NLP), mixed-integer linear programming (MILP), or linear programming (LP) problem. Of course, several levels of abstraction are often possible for a given synthesis problem.

Umeda *et al.* (1972) were probably among the first authors to advocate the optimization of superstructures for process synthesis. At that time, however, the problem was formulated as a nonlinear programming problem involving only continuous variables, and solved with direct-search techniques. Next, Papoulias and Grossmann (1983c) formulated the simultaneous synthesis of total systems as a mixed-integer linear programming problem in order to explicitly handle 0–1 variables, and resorted to standard branch-and-bound computer codes. It is not until very recently, however, that algorithmic developments have allowed synthesis problems to be explicitly formulated as MINLP problems (Duran and Grossmann, 1986c; Kocis and Grossmann, 1989b; Viswanathan and Grossmann, 1988). As for the targeting approach, Cerda and Westerberg (1983) and Papoulias and Grossmann (1983b) were among the first to develop aggregated models for heat integration (LP transportation and LP transshipment models) with the objective of minimizing utility cost. Other targeting models include the ones proposed by El-Halwagi and Manousiouthakis (1989a) for mass-exchange networks, and the one by Balakrishna and Biegler (1992a) for reactor networks.

The two crucial steps in the approach just described are step 1, for generating the representation of alternatives, and step 3, for solving the MINLP problem. As it turns out, however, step 2 is also extremely important because the way in which one models MINLP problems can have a great impact on the performance of the algorithms. In the next sections, we will discuss the three major steps in the optimization methodology.

IV. Representation of Alternatives

In order to formulate the synthesis problem as an optimization problem, one has to develop a representation of alternatives that will systematically consider the candidates for the optimal solution. Developing an appropriate representation is clearly of paramount importance, as the optimal solution can be only as good as the representation being used.

We consider superstructure representations first. In general, these can be explicit or implicit. The former generally give rise to networks while the latter give rise to trees. As an example, consider the separation of a single feed of four components, A,B,C, and D, into pure products. As is well known (Hendry and Hughes, 1972), the alternative separation sequences consisting of sharp splitters can be represented through the tree shown in Fig. 1. This tree representation lends itself to decomposition where the alternatives can be enumerated implicitly through a branch-and-bound search. However, in using this representation, the MINLP problem must be converted into the separable discrete optimization problem over the discrete space Y_{D_i}, $i = 1, ..., m$:

$$Z = \min_{y_i} \sum_{i=1}^{m} C_i(y_i)$$

$$\text{s.t.} \sum_{i=1}^{m} g_{ij}(y_i) \le a_j, \quad j = 1, ..., t \qquad (1)$$

$$y_i \in Y_{D_i}, \quad i = 1, ..., m$$

This requires that continuous variables be selected independently at each node of the tree, often leading to suboptimal solutions even if the branch-and-bound search is performed rigorously. Also, although fewer nodes may be examined in the tree with the use of heuristics, this increases the likelihood of obtaining suboptimal solutions.

On the other hand, consider the network representation shown in Fig. 2 that is implied by the MILP model of Andrecovich and Westerberg (1983). Here, in contrast to the tree, every node corresponds to a distinct separator. Furthermore, alternative sequences can be represented by using the same subset of nodes [e.g.,

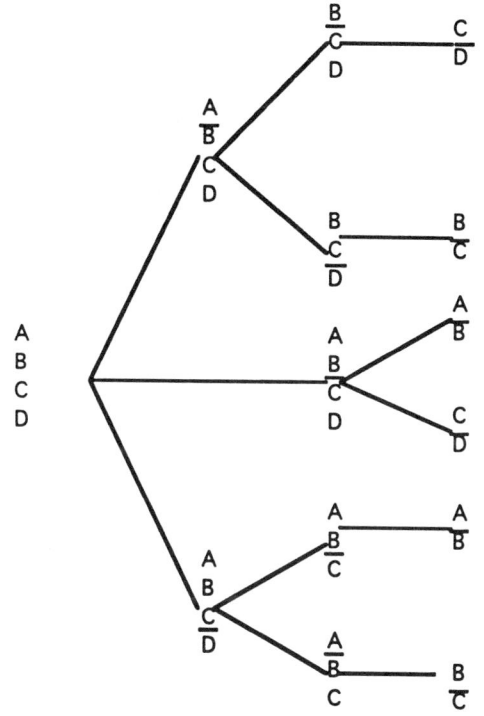

FIG. 1. Tree representation of separation sequence.

sequence 1 (A/BCD, B/CD, C/D) and sequence 2 (AB/CD, A/B, C/D) share the node C/D]. Using this network representation is more compact for modeling the problem explicitly as an MINLP problem. The advantage here is that the optimization can be performed rigorously because the continuous variables can be optimized simultaneously with the selection of the configuration. The disadvantage is that one loses the capability of performing the straightforward decomposition that is possible with the tree. It will be shown later in the paper, however, that one can still resort to more sophisticated decomposition schemes to rigorously solve the MINLP problem for the network.

Thus, from the above discussion, it follows that network representations that are explicitly modeled as MINLP problems provide a more general and rigorous framework for the optimization.

The next question we address is how to actually postulate or derive the superstructures. As experience—our own and that of other researchers—has shown, this is an easier task for homogeneous systems than for heterogeneous systems. An example is the network in Fig. 2, although it is restricted to sharp separations. We can, however, extend this representation to separation problems involving mixed products, with single feed columns that allow the possibility of bypasses and nonsharp splits as shown in Fig. 3 (see Floudas and Anastasiadis, 1988; Quesada and Grossmann, 1995a).

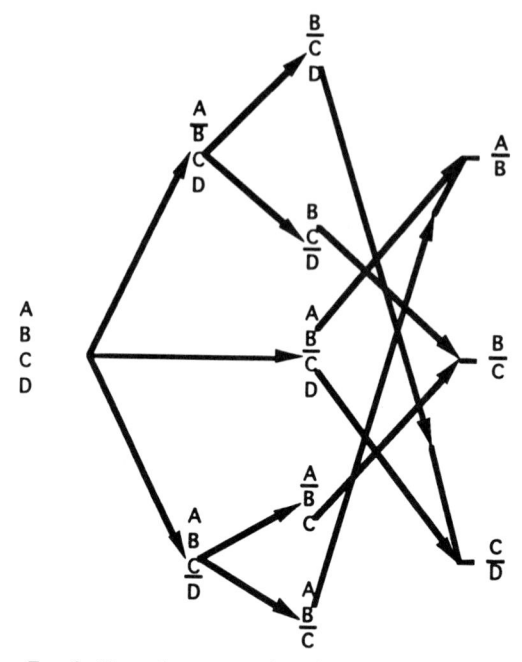

Fig. 2. Network representation of separation sequence.

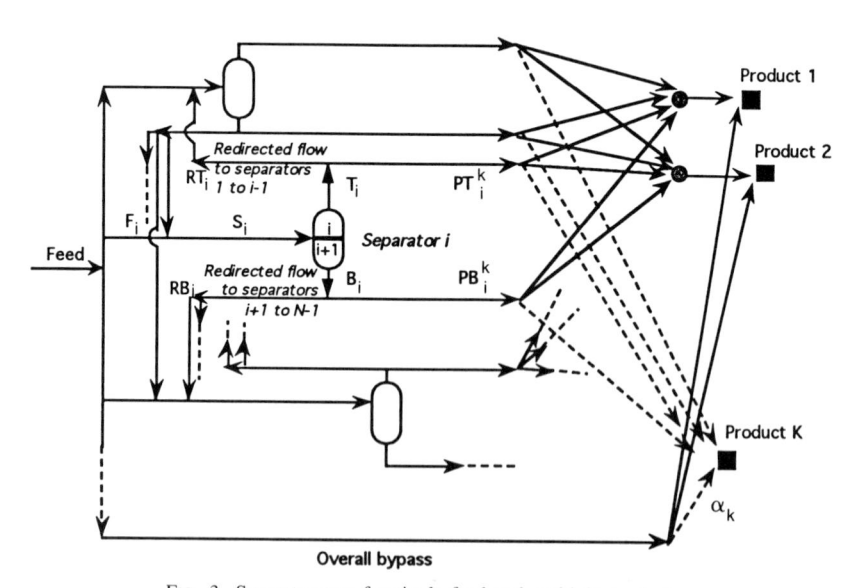

Fig. 3. Superstructure for single feed and multiple products.

As an additional example of an homogeneous system, consider the synthesis of heat-exchanger networks. Here, one possible representation that allows for systematic stream splitting and mixing is shown in Fig. 4. In this superstructure each exchanger unit corresponds to a potential match of pairs of streams. By activating the appropriate exchangers and setting flows to zero, all network configurations can be generated (see Floudas *et al.*, 1986; Ciric and Floudas, 1988). A different representation is given in Fig. 5. Here the superstructure for the network is represented by a sequence of stages where all potential stream matches can take place in each stage (Yee *et al.*, 1990a). Although this superstructure is more restricted, it has the advantage that, under assumptions of isothermal mixing, no flow variables are required in the MINLP and all the constraints become linear. Also, in this representation one can easily control the complexity of the network by imposing constraints to disallow the splitting of streams.

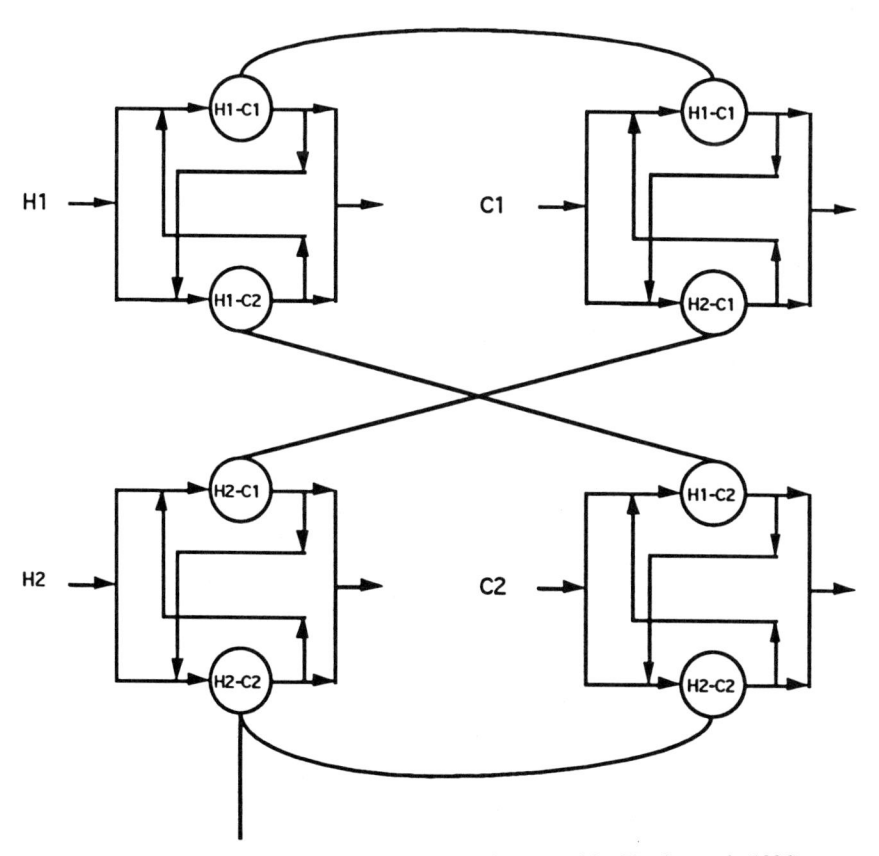

FIG. 4. Superstructure for heat-exchanger network proposed by Floudas *et al.* (1986).

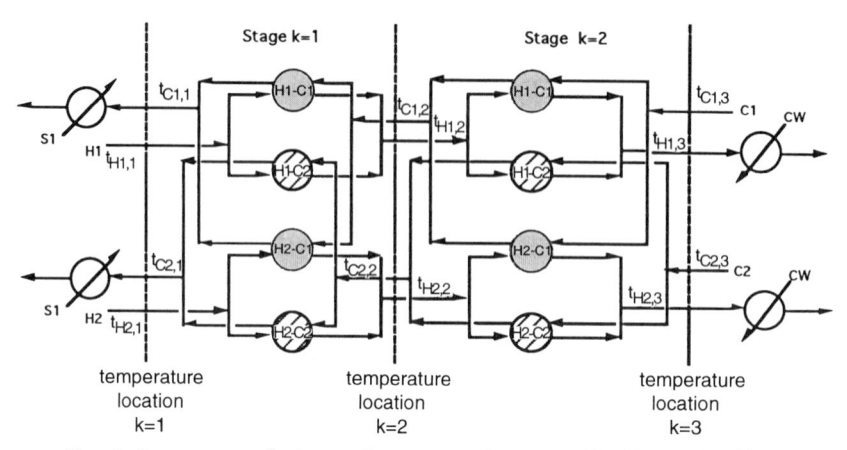

FIG. 5. Superstructure for heat exchanger networks proposed by Yee et al. (1990a).

In the previous examples, a one-to-one correspondence exists between the units and the tasks (e.g., in Fig. 2 each node performs a particular separation task). It is possible, however, to develop more general superstructure representations in which a one-to-many relationship exists between the units and the tasks. An example of a one-to-many relationship is the superstructure for separation shown in Fig. 6; proposed by Sargent and Gaminibandara (1976), this superstructure accommodates sharp splits and has the Petlyuk column embedded as an alternative design. Note, for instance, that column 1 does not have a prespecified separation task. From this example it is clear that superstructures that have one-to-many relationships between units and tasks tend to be richer in terms of embedded alternatives. On the other hand, the more restricted one-to-one superstructures tend to require simpler MINLP models that are quicker to solve.

Aggregated models, or higher-level representations of superstructures, can be developed to deal with simplified design objectives or targets. Here again, the example *par excellence* is the heat-exchanger network problem, where the targets for minimum utility cost and minimum number of units can be respectively modeled as LP and MILP transportation (Cerda and Westerberg, 1983) or transshipment problems (Papoulias and Grossmann, 1983b). Figure 7 shows the heat cascade representation used in the transshipment model for predicting the minimum utility cost. This representation, which is simply given in terms of heat flows across temperature intervals, has been shown by Daichendt and Grossmann (1993a) to be an aggregation of the superstructure of Fig. 5 by Yee et al. (1990a).

When the stream data are variable (flowrates and temperatures), the utility target problem can be modeled as a system of inequalities (see Duran and Grossmann, 1986a); and when only the flowrates are variables, it can be represented

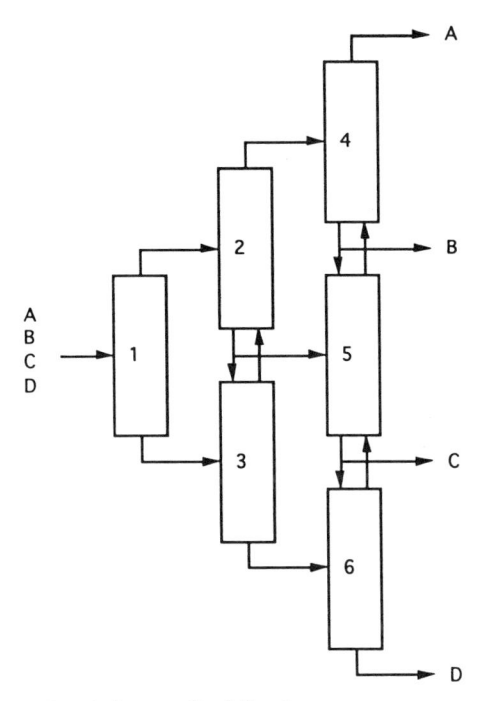

FIG. 6. Sargent–Gaminibandara superstructure.

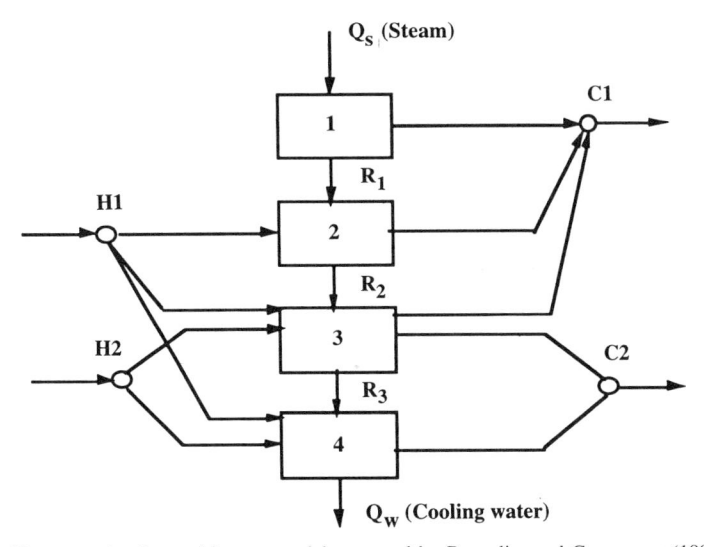

FIG. 7. Heat cascade of transshipment model proposed by Papoulias and Grossmann (1983a,b,c).

by the transportation or transhipment equations (see Papoulias and Grossmann, 1983c; Andrecovich and Westerberg, 1985b). But while these representations clearly simplify the synthesis problem, they do not provide all the information to automatically synthesize a network.

The systematic development of superstructures for heterogeneous systems is, in principle, a more difficult task. Consider, for instance a process flowsheet that is composed of reaction, separation, and heat integration subsystems. In theory, one could develop a superstructure by combining the superstructures for each subsystem. However, this approach could lead to a very large MINLP optimization problem.

Therefore, one approach is to assume that some preliminary screening has been performed (e.g., through heuristics) in order to postulate a smaller number of alternatives in the superstructure (Kocis and Grossmann, 1989b). While this approach seems restrictive, it does provide a systematic framework for analyzing specific alternatives at the level of tasks. As an example, consider the synthesis of an ammonia plant. A preliminary screening would indicate that the following major options: for the reactor, multibed quench or tubular; for separation of product, flash condensation or absorption/distillation; for recovery of hydrogen in purge, membrane separation or cryogenic separation. Figure 8 displays the superstructure for these alternatives. This superstructure has embedded at least eight different configurations.

As this example indicates, generating superstructures for heterogeneous systems based on specific alternatives at the level of tasks is actually not a very

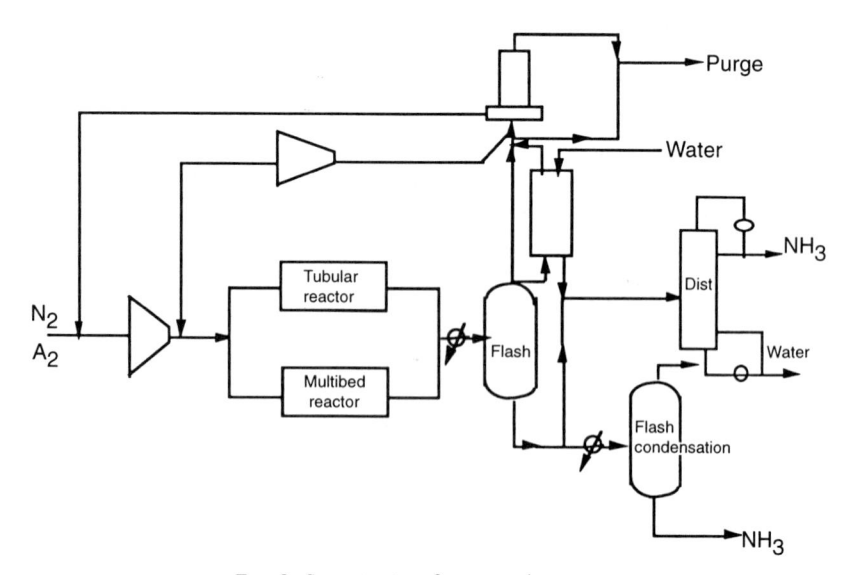

FIG. 8. Superstructure for ammonia process.

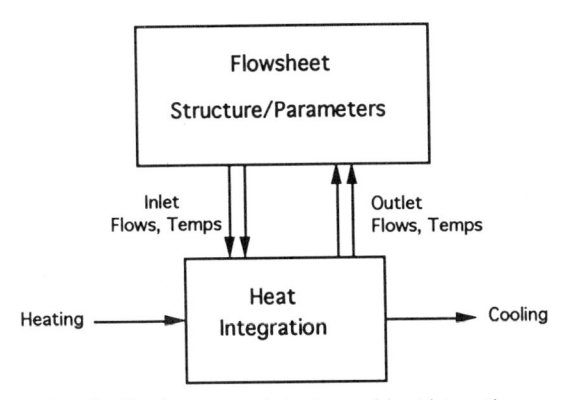

FIG. 9. Simultaneous optimization and heat integration.

difficult task. Another approach to developing synthesis models for heterogeneous systems is to use an aggregated representation for one of the subsystems in order to facilitate the simultaneous optimization of the entire system. For instance, this approach has been applied in the simultaneous optimization and heat integration of flowsheets in which utility loads, rather than a network structure, are predicted for the latter (see Fig. 9). In the work by Papoulias and Grossmann (1983c), the heat integration was represented through the transshipment model with variable flows, while in the work by Duran and Grossmann (1986a), Lang et al. (1988), and Balakrishna and Biegler (1992b), the heat integration is represented through a set of inequalities that are a function of variable flows and temperatures. Interestingly, the largest benefit of this approach is the fact that the simultaneous strategy yields designs exhibiting greater profit and overall conversion than those observed in sequential strategies. Sequential strategies tend to produce suboptimal solutions since interactions and trade-offs cannot be accounted for properly by designing each subsystem one after another.

Finally, it should be noted that one of the other approaches being examined for heterogeneous systems relies on physically based representations for reaction and mass and heat transfer, which are commonly treated as aggregated models. Also, Friedler et al. (1991, 1992, 1993) have provided graph-theoretic algorithms to systematically derive superstructures for process networks, given a list of unit operations and process streams.

V. MINLP Modeling

Once a representation for the candidate designs has been developed, the next step involves the modeling of the MINLP optimization problem. The major feature in such models is the modeling of discrete decisions, typically with

0–1 variables. For most applications it suffices to assign these variables to each potential unit in the superstructure as the interconnecting streams are activated or deactivated according to the selection of units. There are, however, cases when it is also necessary to assign 0–1 variables to the streams.

The handling of 0–1 variables allows the specification of constraints that are extremely relevant for synthesizing practical flowsheet structures. Typical examples include the following:

(a) Multiple choice constraints for selecting among a subset of units I:

Select only one unit:
$$\sum_{i \in I} y_i = 1 \tag{2}$$

Select at most one unit:
$$\sum_{i \in I} y_i \leq 1 \tag{3}$$

Select at least one unit:
$$\sum_{i \in I} y_i \geq 1 \tag{4}$$

(b) If–then conditions:

If unit k is selected, then unit i must be selected:
$$y_k - y_i \leq 0 \tag{5}$$

In addition, 0–1 variables can be related to activate or deactivate continuous variables, inequalities, or equations. As an example, consider the condition for the continuous variable x:

$$\text{If } y = 1 \rightarrow L \leq x \leq U, \quad \text{if } y = 0 \rightarrow x = 0$$

which can be modeled through the constraint
$$Ly \leq x \leq Uy \tag{6}$$

This constraint is often used in conjunction with linear cost models with fixed cost charges, which again requires the use of 0–1 variables:
$$C = \alpha^T y + \beta^T x \tag{7}$$
$$Ly \leq x \leq Uy$$

While deriving integer constraints like those outlined above is not too difficult a task, deriving others can be more complex. The modeling of integer and mixed-integer constraints can, however, be facilitated by the use of propositional logic. It has been shown by a number of authors (e.g., Williams, 1988; Cavalier and Soyster, 1987) that virtually any propositional logic statement can be translated into a set of 0–1 inequalities by the systematic application of rules. For this, one must consider how basic logical operators can each be transformed into an equivalent representation in the form of an equation or inequality. These transformations are then used to convert general logical expressions into an equivalent mathematical representation (Cavalier and Soyster, 1987; Cavalier et al., 1990).

Consider logic propositions that are composed of literals P_i which represent a given selection or action. To represent the truth of that literal, a binary variable y_i is assigned. The negation or complement of P_i ($\neg P_i$) is given by $1 - y_i$. The logical value of true corresponds to the binary value of 1 and false corresponds to the binary value of 0. The basic operators used in propositional logic and the representation of their relationships are shown in Table I. From this table, it is easy to verify, for instance, that the logical implication $P_1 \lor P_2$ reduces to the inequality in (4).

With the basic equivalent relations given in Table I (e.g., see Williams, 1988), one can systematically model an arbitrary propositional logic expression that is given in terms of OR, AND, IMPLICATION operators, as a set of linear equality and inequality constraints. One approach is to systematically convert the logical expression into its equivalent *conjunctive normal form* representation, which involves the application of pure logical operations (Raman and Grossmann, 1991). The conjunctive normal form is a conjunction of clauses, $Q_1 \land Q_2 \land \cdots \land Q_s$. Hence, for the conjunctive normal form to be true, each clause Q_i must be true independent of the others. Also, since a clause Q_i is just a disjunction of literals, $P_1 \lor P_2 \lor \cdots \lor P_r$, it can be expressed in the linear mathematical form as the inequality,

$$y_1 + y_2 + \cdots + y_r \geq 1 \qquad (8)$$

The procedure to convert a logical expression into its corresponding conjunctive normal form was formalized by Clocksin and Mellish (1981). The systematic procedure consists of applying the following three steps to each logical proposition:

(1) Replace the implication by its equivalent disjunction,

$$P_1 \Rightarrow P_2 \Leftrightarrow \neg P_1 \lor P_2 \qquad (9)$$

(2) Move the negation inward by applying DeMorgan's theorem:

$$\neg(P_1 \land P_2) \Leftrightarrow \neg P_1 \lor \neg P_2 \qquad (10)$$

$$\neg(P_1 \lor P_2) \Leftrightarrow \neg P_1 \land \neg P_2 \qquad (11)$$

(3) Recursively distribute the OR over the AND by using the following equivalence:

$$(P_1 \land P_2) \lor P_3 \Leftrightarrow (P_1 \lor P_3) \land (P_2 \lor P_3) \qquad (12)$$

TABLE I
Constraint Representation of Logic Propositions and Operators

Logical relation	Comments	Boolean expression	Representation as linear inequalities
Logical OR		$P_1 \vee P_2 \vee \cdots \vee P_r$	$y_1 + y_2 + \cdots + y_r \geq 1$
Logical AND		$P_1 \wedge P_2 \wedge \cdots \wedge P_r$	$y_1 \geq 1$ $y_2 \geq 1$ $\vdots$ $y_r \geq 1$
Implication	$P_1 \Rightarrow P_2$	$\neg P_1 \vee P_2$	$1 - y_1 + y_2 \geq 1$
Equivalence	P_1 if and only if P_2 $(P_1 \Rightarrow P_2) \wedge (P_2 \Rightarrow P_1)$	$(\neg P_1 \vee P_2) \wedge (\neg P_2 \vee P_1)$	$y_1 = y_2$
Exclusive OR	Exactly one of the propositions is true	$P_1 \underline{\vee} P_2 \underline{\vee} \cdots \underline{\vee} P_r$	$y_1 + y_2 + \cdots + y_r = 1$

Once each logical proposition has been converted into its conjunctive normal form representation, $Q_1 \wedge Q_2 \wedge \cdots \wedge Q_s$, it can be easily expressed as a set of linear equality and inequality constraints.

As an example, consider the logic condition "if the absorber is selected or the membrane separator is selected, then do not use cryogenic separation." Assigning the Boolean literals to each action P_A = select absorber, P_M = select membrane separator, P_{CS} = select cryogenic separation, the logic expression is given by

$$P_A \vee P_M \Rightarrow \neg P_{CS} \tag{13}$$

Removing the implication as in (9) yields

$$\neg(P_A \vee P_M) \vee \neg P_{CS} \tag{14}$$

Applying De Morgan's Theorem as in (10) leads to

$$(\neg P_A \wedge \neg P_M) \vee \neg P_{CS} \tag{15}$$

Distributing the AND over the OR gives

$$(\neg P_A \vee \neg P_{CS}) \wedge (\neg P_M \vee \neg P_{CS}) \tag{16}$$

Assigning the corresponding 0–1 variables to each term in the above conjunction and using (8), we get

$$\begin{aligned} 1 - y_A + 1 - y_{CS} \geq 1 \\ 1 - y_M + 1 - y_{CS} \geq 1 \end{aligned} \tag{17}$$

which can be rearranged to the two inequalities

$$\begin{aligned} y_A + y_{CS} \leq 1 \\ y_M + y_{CS} \leq 1 \end{aligned} \tag{18}$$

Finally, it is important to emphasize that the way one models an MINLP can have a great impact on the performance of the MINLP algorithm. A simple example is the logical constraint $x - Uy \leq 0$, where large values of U reduce the lower bound in the LP-based branch-and-bound search. This phenomenon has been widely recognized in the field of integer programming (e.g., see Williams, 1988; Nemhauser and Wolsey, 1988). The above procedure based on systematically deriving constraints from propositional logic will generally tend to produce well-posed constraints. For example, one logical constraint that may arise in a multiperiod problem states that installing a given unit ($z = 1$) implies that the unit can operate in any of the m time periods ($y_i = 1$, $i = 1, 2, ..., m$):

$$\sum_{i=1}^{m} y_i - mz \leq 0 \tag{19}$$

This constraint simply states that if $z = 0$, no operation of the unit is possible in the m time periods, while if $z = 1$, operation in any of the m periods is possible. While (19) is a "legitimate" constraint, it turns out that its equivalent representation by the set of inequalities

$$y_i - z \leq 0 \qquad i = 1, ..., m \qquad (20)$$

which can be derived from the logic proposition "If unit operates in period 1, in period 2, ..., or in period m, then select unit," is a much more effective way to model the above logic condition. This is because its relaxation with continuous variables y_i yields extreme points, all with 0–1 values, which greatly reduces the computations in the branch-and-bound solution method.

The observations on modeling discussed above have led to the theoretical study in integer linear programming of facets of 0–1 polytopes for which algorithms are starting to emerge. These algorithms can systematically generate these types of constraints and reformulate a "badly" posed problem to improve its LP relaxation. (See, for example, Crowder *et al.*, 1983) for unstructured 0–1 linear problems, Van Roy and Wolsey (1987) and Balas *et al.* (1993) for MILP problems, and Schrijver (1986) for a review of the theoretical concepts.)

In MINLP, however, there is the additional complication that nonlinearities can often be formulated in many different, but equivalent, ways; and, as expected, this can have a great impact on the performance of MINLP algorithms, particularly with respect to the nonconvexities of the nonlinear constraints.

In general, the three major empirical guidelines for a "good" MINLP formulation are as follows:

1. Keep the problem as linear as possible.
2. Develop a formulation whose NLP relaxation is as tight as possible.
3. If possible, reformulate the MINLP as a convex programming problem.

The motivation behind these guidelines requires some basic understanding of the MINLP algorithms which we will cover in the section on algorithms. It should be noted also that a new modeling and solution framework for discrete optimization is emerging that is based on disjunctive programming (Floudas and Grossmann, 1994). Some of these ideas will be covered later in the paper.

A. EXAMPLE OF A MINLP MODEL

To illustrate some of the modeling guidelines, we consider in this section the MINLP model for the synthesis of heat-exchanger networks by Yee and Grossmann (1990), which uses as a basis the superstructure in Fig. 5. The notation that will be used is as follows. Process streams are divided into two sets, set

HP for hot streams represented by index i and set CP for cold streams represented by index j. Index k is used to denote the superstructure stages given by the set ST. Indices HU and CU correspond to the heating and cooling utilities, respectively. Also, the following parameters and variables are used in the formulation:

Parameters:
TIN = inlet temperature of stream
F = heat capacity flow rate
CCU = unit cost for cold utility
CF = fixed charge for exchangers
β = exponent for area cost
Ω = upper bound for heat exchange

TOUT = outlet temperature of stream
U = overall heat transfer coefficient
CHU = unit cost of hot utility
C = area cost coefficient
NOK = total number of stages
Γ = upper bound for temperature difference

Variables:
dt_{ijk} = temperature approach for match (i,j) at temperature location k
$dtcu_i$ = temperature approach for the match of hot stream i and cold utility
$dthu_j$ = temperature approach for the match of cold stream j and hot utility
q_{ijk} = heat exchanged between hot process stream i and cold process stream j in stage k
qcu_i = heat exchanged between hot stream i and cold utility
qhu_j = heat exchanged between hot utility and cold stream j
$t_{i,k}$ = temperature of hot stream i at hot end of stage k
$t_{j,k}$ = temperature of cold stream j at hot end of stage k
z_{ijk} = binary variable to denote existence of match (i,j) in stage k
zcu_i = binary variable to denote that cold utility exchanges heat with stream i
zhu_j = binary variable to denote that hot utility exchanges heat with stream j

With the above definitions, the formulation can now be presented.

(a) *Overall heat balance for each stream:* An overall heat balance is needed to ensure sufficient heating or cooling of each process stream. The constraints specify that the overall heat transfer requirement of each stream must equal the sum of the heat it exchanges with the other process streams at each stage plus the exchange with the utility streams:

$$(\text{TIN}_i - \text{TOUT}_i)F_i = \sum_{k \in \text{ST}} \sum_{j \in \text{CP}} q_{ijk} + qcu_i \quad i \in \text{HP}$$
$$(\text{TOUT}_j - \text{TIN}_j)F_j = \sum_{k \in \text{ST}} \sum_{i \in \text{HP}} q_{ijk} + qhu_j \quad j \in \text{CP}$$
(21)

(b) *Heat balance at each stage:* An energy balance is also needed at each stage of the superstructure to determine the temperatures. Note that for the two-stage superstructure shown in Fig. 5, three temperatures, t_k, are required. Temperatures for the streams are highest at temperature location $k = 1$ and lowest at $k = 3$.

By assuming isothermal mixing, no variables are required for the flows, thus avoiding the introduction of bilinear constraints for the heat balances:

$$(t_{i,k} - t_{i,k+1})F_i = \sum_{j \in CP} q_{ijk} \quad k \in ST, \quad i \in HP$$

$$(t_{j,k} - t_{j,k+1})F_j = \sum_{i \in HP} q_{ijk} \quad k \in ST, \quad j \in CP$$

(22)

(c) *Assignment of superstructure inlet temperatures:* Fixed supply temperatures (TIN) of the process streams are assigned as the inlet temperatures to the superstructure. In Fig. 5, for hot streams, the superstructure inlet corresponds to temperature location $k = 1$, while for cold streams, the inlet corresponds to location $k = 3$.

$$TIN_i = t_{i,1}$$

$$TIN_j = t_{j,NOK+1}$$

(23)

(d) *Feasibility of temperatures:* Constraints are also needed to specify a monotonic decrease of temperature at each successive stage. In addition, a bound is set for the outlet temperatures of the superstructure at the respective stream's target temperature. Note that the outlet temperature of each stream at its last stage does not necessarily correspond to the stream's target temperature since utility exchanges can occur at the outlet of the superstructure.

$$t_{i,k} \geq t_{i,k+1} \quad k \in ST, \quad i \in HP$$

$$t_{j,k} \geq t_{j,k+1} \quad k \in ST, \quad j \in CP$$

$$TOUT_i \leq t_{i,NOK+1} \quad i \in HP$$

$$TOUT_j \geq t_{j,1} \quad j \in CP$$

(24)

(e) *Hot and cold utility load:* Hot and cold utility requirements are determined for each process stream in terms of the outlet temperature in the last stage and the target temperature for that stream. The utility heat load requirements are determined as follows:

$$(t_{i,NOK+1} - TOUT_i)F_i = qcu_i \quad i \in HP$$

$$(TOUT_j - t_{j,1})F_j = qhu_j \quad j \in CP$$

(25)

(f) *Logical constraints:* Logical constraints and binary variables are needed to determine the existence of process match (i,j) in stage k and also any match involving utility streams. The 0–1 binary variables are represented by z_{ijk} for process stream matches, zcu_i for matches involving cold utility and zhu_j for matches involving hot utility. An integer value of 1 for any binary variable designates that the match is present in the optimal network. The constraints then are as follows:

$$q_{ijk} - \Omega z_{ijk} \leq 0 \quad i \in HP, \quad j \in CP, \quad k \in ST$$

$$qcu_i - \Omega zcu_i \leq 0 \quad i \in HP \quad (26)$$

$$qhu_j - \Omega zhu_j \leq 0 \quad j \in CP$$

$$z_{ijk}, zcu_i, zhu_j = 0,1$$

(g) *Calculation of approach temperatures:* The area requirement of each match will be incorporated in the objective function. Calculation of these areas requires that approach temperatures be determined. To ensure feasible driving forces for the exchangers selected in the optimization procedure, the binary variables are used to activate or deactivate the following constraints for approach temperatures:

$$dt_{ijk} \leq t_{i,k} - t_{j,k} + \Gamma(1 - z_{ijk}) \quad k \in ST, \quad i \in HP, \quad j \in CP$$

$$dt_{ijk+1} \leq t_{i,k+1} - t_{j,k+1} + \Gamma(1 - z_{ijk}) \quad k \in ST, \quad i \in HP, \quad j \in CP$$

$$dtcu_i \leq t_{i,NOK+1} - TOUT_{CU} + \Gamma(1 - zcu_i) \quad i \in HP \quad (27)$$

$$dthu_j \leq TOUT_{HU} - t_{j,1} + \Gamma(1 - zhu_j) \quad j \in CP$$

Note that these constraints can be expressed as inequalities because the cost of the exchangers decreases with higher values for the temperature approaches dt. Also, the role of the binary variables in the constraints is to ensure that nonnegative driving forces exist for a selected match. When a match (i,j) occurs in stage k, $z_{ijk} = 1$ and the constraint becomes active so that the approach temperature is properly calculated. However, when the match does not occur, $z_{ijk} = 0$, and the contribution of the upper bound Γ on the right-hand side renders the constraint inactive. Note that the upper bounds can be set to zero for the utility exchangers since, for the data given, all the temperature differences are always positive. Also, one can specify a minimum approach temperature so that in the network, the temperature between the hot and cold streams at any point of any exchanger will be at least TMAPP:

$$dt_{ijk} \geq TMAPP \quad (28)$$

(h) *Objective function:* Finally, the objective function can be defined as the annual cost for the network. The annual cost involves the combination of the utility cost, the fixed charges for the exchangers, and the area cost for each exchanger. LMTD, which is the driving force for a countercurrent heat exchanger, is approximated using the Chen approximation (1987):

$$\text{LMTD} \sim [(dt1 * dt2) * (dt1 + dt2)/2]^{1/3}$$

This approximation is used to avoid the numerical difficulties of the LMTD equation when the approach temperatures for both sides of the exchanger are equal. Furthermore, when the driving force on either side of the exchanger equals zero, the driving force will be approximated to zero. The objective function is defined as follows:

$$\min \sum_{i \in HP} CCU\, qcu_i + \sum_{j \in CP} CHU\, qhu_j$$

$$+ \sum_{i \in HP} \sum_{j \in CP} \sum_{k \in ST} CF_{ij} z_{ijk} + \sum_{i \in HP} CF_{i,CU} zcu_i + \sum_{j \in CP} CF_{j,HU} zhu_j$$

$$+ \sum_{i \in HP} \sum_{j \in CP} \sum_{k \in ST} C_{ij} [q_{ijk}/(U_{ij}[(dt_{ijk}\, dt_{ijk+1})(dt_{ijk} + dt_{ijk+1})/2]^{1/3})]^{\beta_{ik}}$$

$$+ \sum_{i \in HP} C_{i,CU} [qcu_i/(U_{i,CU}[(dtcu_i)(TOUT_i - TIN_{CU}) \quad (29)$$

$$\{dtcu_i + (TOUT_i - TIN_{CU})\}/2]^{1/3})]^{\beta_{i,CU}}$$

$$+ \sum_{j \in CP} C_{HU,j} [qhu_j/(U_{HU,j}[(dthu_j)(TIN_{HU} - TOUT_j)$$

$$\{dthu_j + (TIN_{HU} - TOUT_j)\}/2]^{1/3})]^{\beta_{j,HU}}$$

where

$$\frac{1}{U_{ij}} = \frac{1}{h_i} + \frac{1}{h_j}$$

The MINLP model for the synthesis problem consists of minimizing the objective function in (29) subject to the feasible space defined by eqs. (21)–(28). The continuous variables (t, q, qhu, qcu, dt, $dtcu$, $dthu$) are non-negative and the discrete variables z, zcu, zhu are 0–1. The advantage of this model is that the constraints (21)–(28) are all linear. The nonlinearities have all been placed in the objective function (29). However, it should be noted that since these terms are nonconvex, the MINLP may lead to local optimal solutions.

It should be noted that the simplifying assumption of isothermal mixing at the stage outlets for the stream splits is rigorous for the case when the network to be synthesized does not involve stream splits. For structures where splits are present, the assumption may lead to an overestimation of the area cost since it will restrict trade-offs of area between the exchangers involved with the split streams. In this case, one possibility is to refine the temperatures by introducing flow variables in the selected network structure and perform the corresponding optimization through an NLP model similar to the one given above.

Finally, an interesting feature of the MINLP model is that it is possible to add constraints to avoid generating structures with stream splits. This is accomplished simply by requiring that not more than one match be selected for every stream at each stage; that is,

$$\sum_{i \in HP} z_{ijk} \leq 1 \quad j \in CP, k \in ST$$
$$\sum_{j \in CP} z_{ijk} \leq 1 \quad i \in HP, k \in ST$$
(30)

VI. MINLP Algorithms

A. BASIC ALGORITHMS

While there is a vast body of literature on LP, NLP, and integer LP with special structures, MINLP has received much less attention. In part, the explanation for this is that MINLP has been traditionally regarded as a very difficult problem because it is an NP-hard problem (Nemhauser and Wolsey, 1988) that is prone to combinatorial explosion. In our view, however, it is a mistake to regard these problems as "unsolvable." The applications for MINLP are extremely rich; moreover, with current methods and technology, one can already solve problems of significant size and complexity. Furthermore, with advances in new algorithms and computer architectures, it is reasonable to assume that over the next decade we will see increases in the order of magnitude of sizes of problems that can be currently solved. Several examples of successful solution to optimality of integer programs of very large size can be found, for instance, in Crowder *et al.* (1983), Van Roy and Wolsey (1987), and Balas *et al.* (1993).

Our objective in this section is to provide a general overview of the basic MINLP algorithms, emphasizing their fundamental ideas and properties.

The most basic form of an MINLP problem when represented in algebraic form is as follows (equations are temporarily excluded):

$$\min Z = f(x,y)$$
$$\text{s.t. } g_j(x,y) \leq 0 \quad j \in J \quad \text{(P1)}$$
$$x \in X, \quad y \in Y$$

where $f(\cdot)$, $g(\cdot)$ are convex, differentiable functions, and x and y are the continuous and discrete variables, respectively. The set X is commonly assumed to be a compact set, e.g. $X = \{x \mid x \in \mathbf{R}^n, Dx \leq d, x^L \leq x \leq x^U\}$; the discrete set Y

is a polyhedral set of integer points, $Y = \{y \mid y \in \mathbf{Z}^m, Ay \leq a\}$, which in most applications is restricted to 0–1 values, $y \in \{0,1\}^m$. In most applications of interest the objective and constraint functions $f(\cdot)$, $g(\cdot)$ are linear in y (e.g., fixed cost charges and logic constraints).

Methods that have addressed the solution of problem (P_1) include the branch-and-bound method (BB) (Gupta and Ravindran, 1985; Nabar and Schrage, 1991; Borchers and Mitchell, 1992), the generalized Benders decomposition (GBD) method (Geoffrion, 1972), the outer-approximation (OA) method (Duran and Grossmann, 1986c; Yuan et al., 1988; Fletcher and Leyffer, 1994), the LP/NLP-based branch-and-bound method (Quesada and Grossmann, 1992), and the extended cutting-plane (ECP) method (Westerlund and Pettersson, 1992).

There are three basic NLP subproblems that can be considered for problem (P_1).

(a) NLP relaxation:

$$\min Z_{LB}^k = f(x,y)$$

$$\text{s.t.} \quad g_j(x,y) \leq 0 \quad j \in J$$

$$x \in X, \quad y \in Y_R \quad \text{(NLP1)}$$

$$y_i \leq \alpha_i^k \quad i \in I_{FL}^k$$

$$y_i \leq \beta_i^k \quad i \in I_{FU}^k$$

where Y_R is the continuous relaxation of the set Y, and I_{FL}^k, I_{FU}^k are index subsets of the integer variables y_i, $i \in I$, which are restricted to lower and upper bounds, α_i^k, β_i^k, at the kth step of a branch-and-bound enumeration procedure. Note that $\alpha_i^k = \lfloor y_i^l \rfloor$, $\beta_i^k = \lceil y_i^m \rceil$, $l < k$, $m < k$, where are y_i^l, y_i^m noninteger values at a previous step.

Also note that if $I_{FU}^k = I_{FL}^k = \emptyset$, $k = 0$, then (NLP1) corresponds to the continuous NLP relaxation of (P1); otherwise, it corresponds to the kth step in a branch-and-bound search. Also, the optimal objective function Z_{LB}^0 provides an absolute lower bound to (P1); for $m \geq k$, the bound is only valid for $I_{FL}^k \subseteq I_{FL}^m$, $I_{FU}^k \subseteq I_{FL}^m$.

(b) NLP subproblem for fixed y^k:

$$\min Z_U^k = f(x,y^k)$$

$$\text{s.t.} \quad g_j(x,y^k) \leq 0 \quad j \in J \quad \text{(NLP2)}$$

$$x \in X$$

which clearly yields an upper bound Z_U^k to (P1), provided (NLP2) has a feasible solution (which may not always be the case).

(c) Feasibility subproblem for fixed y^k.

$$\min_{x \in X} \left[\sum_{j \in J} \max[0, \{g_j(x, y^k)\}^p] \right]^{1/p} \quad \text{(NLPF)}$$

which for the 1-norm ($p = 1$) leads to

$$\min \sum_{j \in J} s_j$$

$$\text{s.t. } g_j(x, y^k) \leq s_j \quad j \in J \quad \text{(NLPF-1)}$$

$$x \in X, \quad s_j \in R^1, \quad j \in J$$

where s_j are slack variables.

For the infinity-norm ($p = \infty$) problem (NLPF) yields,

$$\min u$$

$$\text{s.t. } g_j(x, y^k) \leq u \quad j \in J \quad \text{(NLPF-}\infty\text{)}$$

$$x \in X, \quad u \in R^1$$

The new predicted values y^K (or (y^K, x^K)) are obtained from a cutting-plane MILP problem that is based on the K points, (x^k, y^k), $k = 1, ..., K$ generated at the K previous steps:

$$\min Z_L^K = \alpha \quad \text{(M-MIP)}$$

$$\left. \begin{array}{l} \text{s.t. } \alpha \geq f(x^k, y^k) + \nabla f(x^k, y^k)^T \begin{bmatrix} x - x^k \\ y - y^k \end{bmatrix} \\ g_j(x^k, y^k) + \nabla g_j(x^k, y^k)^T \begin{bmatrix} x - x^k \\ y - y^k \end{bmatrix} \leq 0 \quad j \in J^k \end{array} \right\} k = 1, ..., K$$

$$x \in X, \quad y \in Y, \quad \alpha \in R^1$$

where Z_L^K yields a valid lower bound to problem (P1). This bound is nondecreasing with the number of linearization points K.

The different methods can be classified according to the use of the subproblems (NLP1), (NLP2), and (NLPF) and the specific specialization of the MILP (M-MIP) (see Fig. 10). Note that in cases (b) and (d), (NLPF) is used if infeasible subproblems are found.

1. Branch and Bound

The BB method (Gupta and Ravindran, 1985; Nabar and Schrage, 1991; Borchers and Mitchell, 1992) starts by solving first the continuous NLP relaxation. If all discrete variables take discrete values, the search is stopped. Otherwise, it performs a tree search in the space of the integer variables y_i, $i \in I$,

(a) Branch and bound

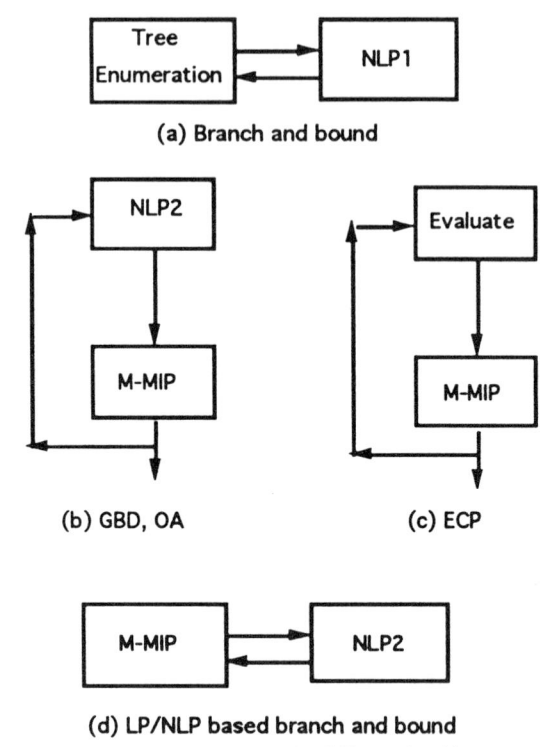

(b) GBD, OA

(c) ECP

(d) LP/NLP based branch and bound

FIG. 10. Major steps in the different algorithms.

and solves a sequence of relaxed NLP subproblems of the form (NLP1) which yield lower bounds; fathoming of nodes occurs when the lower bound exceeds the current upper bound, or when all integer variables y_i take on discrete values. This method is attractive only if the NLP subproblems are relatively inexpensive to solve, or only few of them need to be solved. This could happen either when the dimensionality of the discrete variables is low or when the continuous NLP relaxation of (P1) is tight.

2. Outer Approximation

The OA method (Duran and Grossmann, 1986c; Yuan et al., 1988; Fletcher and Leyffer, 1994) arises when NLP subproblems (NLP2) and MILP master problems (M-MIP) with $J^k = J$ are solved successively in a cycle of iterations to generate the points (x^k, y^k). The (NLP2) subproblems yield an upper bound that corresponds to the best current solution, $\mathrm{UB}^K = \min(Z_U^k)$. The master problems (M-MIP) yield a non-decreasing sequence of lower bounds (Z_L^K since linearizations are accumulated as seen in (M-MIP). The cycle of iterations is stopped when the lower and upper bounds are within a specified tolerance. Also, if infeasible NLP subproblems are found, the feasibility problem (NLPF) is solved to provide the point x^k (commonly NLPF$-\infty$). The OA method generally requires relatively few cycles or major iterations. It trivially converges in one iteration if $f(x,y)$ and $g(x,y)$ are linear. It is also important to note that the MILP master problem need not be solved to optimality. In fact, given the upper bound UB^K and a tolerance ϵ, it is sufficient to generate the new (y^K, x^K) by solving

$$\min Z^K = 0 \cdot \alpha \qquad \text{(M-MIPF)}$$

$$\text{s.t. } \alpha \leq \text{UB} - \epsilon$$

$$\left.\begin{array}{l} \alpha \geq f(x^k,y^k) + \nabla f(x^k,y^k)^T \begin{bmatrix} x-x^k \\ y-y^k \end{bmatrix} \\ g(x^k,y^k) + \nabla g(x^k,y^k)^T \begin{bmatrix} x-x^k \\ y-y^k \end{bmatrix} \leq 0 \end{array}\right\} k = 1, \ldots, K$$

$$x \in X, \quad y \in Y, \quad \alpha \in R^1$$

3. Generalized Benders Decomposition

The GBD method (Geoffrion, 1972) is similar to the OA method. The difference arises in the definition of the MILP master problem (M-MIP). In the GBD method, only active inequalities are considered $J^k = \{j/g_j(x^k, y^k) = 0\}$ and the set $x \in X$ is disregarded. As shown in Quesada and Grossmann (1992), (M-MIP) reduces to a problem projected in the y-space:

$$\min Z_L^K = \alpha$$

$$\text{s.t. } \alpha \geq f(x^k,y^k) + \nabla_y f(x^k,y^k)^T (y-y^k)$$

$$+ (\mu^k)^T [g(x^k,y^k) + \nabla_y g(x^k,y^k)^T (y-y^k)] \quad k \in \text{KFS} \qquad \text{(M-GBD)}$$

$$(\lambda^k)^T [g(x^k,y^k) + \nabla_y g(x^k,y^k)^T (y-y^k)] \leq 0 \quad k \in \text{KIS}$$

$$y \in Y, \quad \alpha \in R^1$$

where KFS is the set of feasible subproblems (NLP2) and KIS is the set of infeasible subproblems whose solution is given by (NLPF$-\infty$). Also, | KFS ∪ KIS | = K. As has been shown by Duran and Grossmann (1986c), the lower bounds of the OA method are greater than or equal to those of the GBD method. For this reason, GBD commonly requires a larger number of cycles or major iterations. As the number of 0–1 variables increases, this difference becomes more pronounced; therefore user-supplied constraints must be added to the master problem to strengthen the bounds (Sahinidis and Grossmann, 1991a). As has been shown by Sahinidis and Grossmann (1991b), GBD converges in one iteration once the optimal (x^*, y^*) has been found when problem (P1) has zero integrality gap. Also, Turkay and Grossmann (1994) have proved that performing one Benders iteration on the MILP master of OA is equivalent to a GBD iteration.

4. Extended Cutting Plane

The ECP method (Westerlund and Pettersson, 1992), which is an extension of Kelly's cutting plane algorithm for convex NLP (Kelley, 1960), does not rely on the use of NLP subproblems and algorithms. It relies only on the iterative solution of the problem (M-MIP) by successively adding the most violated constraint at the predicted point (x^k, y^k): $J^k = \{\hat{j}|\hat{j} \in \arg\{\max_{j \in J} g_j(x^k, y^k)\}\}$. Convergence is achieved when the maximum constraint violation lies within the specified tolerance. The optimal objective value of (M-MIP) yields a nondecreasing sequence of lower bounds. Note that since the discrete and continuous variables are converged simultaneously, a large number of iterations may be required. Also, the objective must be defined as a linear function.

5. LP/NLP-Based Branch and Bound

This method (Quesada and Grossmann, 1992) avoids the complete solution of (M-MIP) at each major iteration. The basic idea consists of performing an LP-based BB method for (M-MIP), solving NLP subproblems (NLP1) at those nodes in which feasible integer solutions are found. By updating the representation of the master problem in the open nodes of the tree with the addition of the corresponding linearizations, one eliminates the need to restart the tree search. This method can be applied both to OA and GBD methods, and commonly reduces quite significantly the number of nodes enumerated. The tradeoff, however, is that the number of NLP subproblems may increase. Computational experience has indicated that the number of NLP subproblems often remains unchanged. Also, Leyffer (1993) has reported substantial savings with this method.

B. Extensions of MINLP Methods

In this section, we present an overview of some of the major extensions of the methods presented in the previous section.

1. Quadratic Master Problems

For most problems of interest, problem (P1) is linear in y: $f(x,y) = \phi(x) + c^T y$, $g(x,y) = h(x) + By$. When this is not the case Fletcher and Leyffer (1994) suggest including a quadratic approximation to (M-MIPF) of the form

$$\min Z^K = \alpha + \frac{1}{2}\begin{pmatrix}x-x^K\\y-y^K\end{pmatrix}^T \nabla^2 \mathcal{L}(x^K,y^K)\begin{pmatrix}x-x^K\\y-y^K\end{pmatrix}$$

$$\text{s.t. } \alpha \leq UB^K - \epsilon \qquad \text{(M-MIQP)}$$

$$\left.\begin{array}{l}\alpha \geq f(x^k,y^k) + \nabla f(x^k,y^k)^T\begin{bmatrix}x-x^k\\y-y^k\end{bmatrix}\\[2mm] g(x^k,y^k) + \nabla g(x^k,y^k)^T\begin{bmatrix}x-x^k\\y-y^k\end{bmatrix} \leq 0\end{array}\right\} \quad k = 1, \ldots, K$$

$$x \in X, \quad y \in Y, \quad \alpha \in R^1$$

where $\nabla^2 \mathcal{L}(x^K, y^K)$ is the Hessian of the Lagrangian of the last NLP subproblem. Note that Z^K does not predict valid lower bounds in this case. As noted by Ding and Sargent (1992), who developed a master problem similar to M-MIQP, the quadratic approximations can help to reduce the number of major iterations since an improved representation of the continuous space is obtained. This, however, comes at the price of having to solve an MIQP instead of an MILP. Note also that using (M-MIQP) for convex $f(x,y)$ and $g(x,y)$ leads to rigorous solutions since the outer approximations remain valid.

2. Reducing the Dimensionality of the Master Problem in OA

The master problem (M-MIP) can be rather large in the OA method. One option is to keep only the last linearization point, but this may lead to nonconvergence even in convex problems (e.g., see Bremicker *et al.*, 1990). A rigorous reduction of dimensionality without greatly sacrificing the strength of the lower bound can be achieved in the case of the "largely" linear MINLP problem

$$\min Z = a^T w + r(v) + c^T y$$

$$\text{s.t. } Dw + t(v) + Cy \leq 0 \qquad \text{(PL)}$$

$$Fw + Gv + Ey \leq b$$

$$w \in W, \quad v \in V, \quad y \in Y$$

where (w, v) are continuous variables and $r(v)$ and $t(v)$ are nonlinear convex functions. As shown by Quesada and Grossmann (1992), linear approximations to the nonlinear objective and constraints can be aggregated with the following MILP master problem:

$$\min Z_L^K = a^T w + \beta + c^T \quad \text{(M-MIPL)}$$

$$\text{s.t. } \beta \geq r(v^k) + (\lambda^k)^T [Dw + t(v^k) + Cy] - (\mu^k)^T G(v-v^k) \quad k = 1, \ldots, K$$

$$Fw + Gv + Ey \leq b$$

$$w \in W, \quad v \in V, \quad y \in Y, \quad \beta \in R^1$$

Numerical results have shown that the quality of the bounds is not greatly degraded with the above MILP, as might happen if GBD is applied to (PL).

3. Incorporating Cuts

One way to expedite the convergence in the OA and GBD algorithms when the discrete variables in problem (P1) are 0–1 is to introduce the following integer cut, which has as an objective to make infeasible the choice of the previous 0–1 values generated at the K previous iterations (Duran and Grossmann, 1986c):

$$\sum_{i \in B^k} y_i - \sum_{i \in N^k} y_i \leq |B^k| - 1 \quad k = 1, \ldots, K \quad \text{(INTCUT)}$$

where $B^k = \{i \mid y_i^k = 1\}$, $N^k = \{i \mid y_i^k = 0\}$, $k = 1, \ldots, K$. This cut becomes very weak as the dimensionality of the 0–1 variables increases. However, it has the useful feature of ensuring that new 0–1 values are generated at each major iteration. In this way, the algorithm will not return to a previous integer point when convergence is achieved. With the above integer cut, the termination takes place as soon as $Z_L^K \geq UB^K$. Also, in the case of the GBD method it is sometimes possible to generate multiple cuts from the solution of an NLP subproblem in order to strengthen the lower bound (Magnanti and Wong, 1981).

4. Handling of Equalities

For the case when linear equalities of the form $h(x, y) = 0$ are added to (P1), there is no major difficulty since these are invariant to the linearization points. If the equations are nonlinear, however, there are two difficulties. First, it is not possible to enforce the linearized equalities at K points. Second, the nonlinear equations may generally introduce nonconvexities. Kocis and Grossmann (1987) proposed an equality relaxation strategy in which the nonlinear equalities are replaced by the inequalities

$$T^k \nabla h(x^k, y^k)^T \begin{bmatrix} x - x^k \\ y - y^k \end{bmatrix} \leq 0 \quad (31)$$

where $T^k = \{t_{ii}^k\}$, and $t_{ii}^k = \text{sign}(\lambda_i^k)$ in which λ_i^k is the multiplier associated with the equation $h_i(x, y) = 0$. Note that if these equations relax as the inequalities $h(x, y) \leq 0$ for all y, and $h(x, y)$ is convex, this is a rigorous procedure. Otherwise, nonvalid supports may be generated. Also, note that in the master problem of GBD, (M-GBD), no special provision is required to handle equations since these are simply included in the Lagrangian cuts of (M-GBD). However, difficulties similar to those in OA arise if the equations do not relax as convex inequalities.

5. Handling of Nonconvexities

When $f(x,y)$ and $g(x,y)$ are nonconvex, two difficulties arise. First, the NLP subproblems (NLP1), (NLP2), and (NLPF) may not have a unique local optimum solution. Second, the master problem (M-MIP) and its variants (e.g., M-MIPF, M-GBD, M-MIQP) do not guarantee a valid lower bound Z_L^K or a valid bounding representation with which the global optimum may be cut off. Two approaches can be used to address this problem: either assume a special structure in the MINLP problem and rely on methods for global optimization (Floudas and Grossmann, 1994); otherwise, apply a heuristic strategy to try to reduce the effect of nonconvexities as much as possible. We will describe only the second approach here, with the objective of reducing the effect of convexities at the level of the MILP master problem.

Viswanathan and Grossmann (1990) proposed introducing slacks in the MILP master problem to reduce the likelihood of cutting off feasible solutions. This master problem (augmented penalty/equality relaxation) (APER) has the following form:

$$\min Z^K = \alpha + \sum_{k=1}^{K} [w_p^k p^k + w_q^k q^k] \quad \text{(M-APER)}$$

$$\left. \begin{array}{l} \text{s.t. } \alpha \geq f(x^k,y^k) + \nabla f(x^k,y^k)^\mathrm{T} \begin{bmatrix} x-x^k \\ y-y^k \end{bmatrix} \\ \\ T^k \nabla h(x^k,y^k)^\mathrm{T} \begin{bmatrix} x-x^k \\ y-y^k \end{bmatrix} \leq p^k \\ \\ g(x^k,y^k) + \nabla g(x^k,y^k)^\mathrm{T} \begin{bmatrix} x-x^k \\ y-y^k \end{bmatrix} \leq q^k \end{array} \right\} \quad k = 1, \ldots, K$$

$$\sum_{i \in B^k} y_i - \sum_{i \in N^k} y_i \leq |B^k| - 1 \quad k = 1, \ldots, K$$

$$x \in X, \quad y \in Y, \quad \alpha \in R^1, \quad p^k, q^k \geq 0$$

where w_p^k, w_q^k are weights that are chosen sufficiently large (e.g., 1000 times magnitude of Lagrange multiplier). Note that if the functions are convex, the MILP master problem (M-APER) predicts rigorous lower bounds to (P1) since all the slacks are set to zero. It should also be noted that the program DICOPT++ (Viswanathan and Grossmann, 1990), which is currently the only MINLP solver that is commercially available (as part of GAMS; Brooke et al., 1988), is based on the above master problem. This code also uses the relaxed (NLP1) to generate the first linearization for the above master problem, so the user need not specify an initial integer value. Also, since bounding properties of (M-APER) cannot be guaranteed, the search for nonconvex problems is terminated when there is no further improvement in the feasible NLP subproblems. Clearly, this is a heuristic, but one that works reasonably well in many problems.

It should also be noted that another modification to reduce the undesirable effects of nonconvexities in the master problem is to apply global convexity tests, followed by a suitable validation of linearizations. One possibility is to apply the test to all linearizations with respect to the current solution vector (y^K, x^K) (Kravanja and Grossmann, 1994). The convexity conditions that have to be verified for the linearizations are as follows:

$$\left. \begin{array}{r} f(x^k,y^k) + \nabla f(x^k,y^k)^T \begin{bmatrix} x_K-x^k \\ y_K-y^k \end{bmatrix} - \alpha \leq \epsilon \\ \\ T^k \nabla h(x^k,y^k)^T \begin{bmatrix} x_K-x^k \\ y_K-y^k \end{bmatrix} \leq \epsilon \\ \\ g(x^k,y^k) + \nabla g(x^k,y^k)^T \begin{bmatrix} x_K-x^k \\ y_K-y^k \end{bmatrix} \leq \epsilon \end{array} \right\} \quad k = 1, ..., K-1 \quad \text{(GCT)}$$

where ϵ is a vector of small tolerances (e.g., 10^{-10}). Note that the test is omitted for the last linearizations since these are always valid for the last solution point (y^K, x^K). Based on this test, a validation of the linearizations is performed so that the linearizations for which the above verification is not satisfied are simply dropped out from the master problem. This test relies on the assumption that the solutions of the NLP subproblems are approaching the global optimum, and that the successive validations are progressively defining valid feasibility constraints around the global optimum. Also note that if the right-hand-side coefficients of linearizations are modified to validate the linearization, the test corresponds to the one in the two-phase strategy by Kocis and Grossmann (1988). Since in the latter case the violated linearizations are just shifted to widen the feasible region around the past solutions and not dropped out as in the former case, the nonconvexities tend to affect the MILP more than in the former case.

C. Logic-Based Methods

One current trend is to represent linear and nonlinear discrete optimization problems by models consisting of algebraic constraints, logic disjunctions, and logic relations (Balas, 1985; Beaumont, 1991; Raman and Grossmann, 1993, 1994). In particular, the mixed-integer program (P1) can also be formulated as a generalized disjunctive program, as has been shown by Raman and Grossmann (1994):

$$\min Z = \sum_i \sum_k c_{ik} + f(x)$$

$$\text{s.t. } g(x) \leq 0 \quad \text{(DP1)}$$

$$\bigvee_{i \in D_k} \begin{bmatrix} Y_{ik} \\ h_{ik}(x) \leq 0 \\ c_{ik} = \gamma_{ik} \end{bmatrix} \quad k \in SD$$

$$\Omega(Y) = \text{true}$$

$$x \in R^n, \quad c \in R^m, \quad Y \in \{\text{true, false}\}^m$$

in which Y_{ik} are the Boolean variables that establish whether a given term in a disjunction is true [$h_{ik}(x) \leq 0$] or false [$h_{ik}(x) > 0$], while $\Omega(Y)$ are logical relations assumed to be in the form of propositional logic involving only the Boolean variables. Y_{ik} are auxiliary variables that control the part of the feasible space in which the continuous variables x lie, and the variables c_{ik} represent fixed charges which are activated to a value γ_{ik} if the corresponding term of the disjunction is true. Finally, the logical conditions, $\Omega(Y)$, express relationships between the disjunctive sets. In the context of synthesis problems the disjunctions in (DP1) typically arise for each unit i in the following form:

$$\begin{bmatrix} Y_i \\ h_i(x) \leq 0 \\ c_i = \gamma_i \end{bmatrix} \vee \begin{bmatrix} \neg Y_i \\ B^i x = 0 \\ c_i = 0 \end{bmatrix} \quad (32)$$

in which the inequalities h_i apply and a fixed cost γ_i is incurred if the unit is selected (Y_i); otherwise ($\neg Y_i$), there is no fixed cost and a subset of the x variables is set to zero with the matrix B^i. An important advantage of the above modeling framework is that there is no need to introduce artificial parameters for the "big-M" constraints that are normally used to model disjunctions. In the latter part of the next section, we will discuss methods for solving problems that are posed in logic form.

When the nonlinear discrete optimization problem is formulated as the generalized disjunctive program in (DP1), one can develop a corresponding logic-based branch-and-bound method. The basic difference is that the branching is performed directly on the disjunctions. While in the linear case branch-and-bound algorithms for solving (DP1) can be attractive (e.g., see Beaumont, 1991; Raman and Grossmann, 1994), for the nonlinear case they present some difficulties: first, because it is generally not trivial to develop a valid surrogate for the disjunctions that will effectively bound the relaxed solution (problem (DP1) without disjunctions); and second, because it is often not possible to reduce the dimensionality of the relaxed NLP subproblems. This can cause difficulties in structural flowsheet optimization problems as "dry" units with zero flow must then be handled, which often leads to singularities. Therefore, we consider only the corresponding OA and GBD algorithms for problems expressed in the form of (DP1).

As described in Turkay and Grossmann (1994), for fixed values of the Boolean variables, $Y_{\hat{i}k}$ = true and Y_{ik} = false, the corresponding NLP subproblem is as follows:

$$\min Z = \sum_i \sum_k c_{ik} + f(x)$$

$$\text{s.t. } g(x) \leq 0 \qquad \text{(NLPD)}$$

$$\left\{\begin{array}{l} c_{ik} = \gamma_{\hat{i}} \\ h_{\hat{i}k}(x) \leq 0 \end{array}\right\} \quad \text{for } Y_{\hat{i}k} = \text{true} \qquad k \in SD$$

$$c_{ik} = 0 \quad \text{for } Y_{ik} = \text{false} \qquad i \neq \hat{i}$$

$$x \in R^n, \qquad c \in R^m$$

Note that only constraints corresponding to Boolean variables that are true are imposed. Also, fixed charges γ_{ik} are only applied to these terms. Assuming that K subproblems (NLPD) are solved in which sets of linearizations $l = 1,...,K$ are generated for subsets of disjunction terms $L(ik) = \{l \mid Y^l_{ik} = \text{true}\}$, one can define the following disjunctive OA master problem:

$$\min Z = \sum_i \sum_k c_{ik} + \alpha$$

$$\text{s.t. } \left.\begin{array}{l} \alpha \geq f(x^l) + \nabla f(x^l)^T(x - x^l) \\ g(x^l) + \nabla g(x^l)^T(x - x^l) \leq 0 \end{array}\right\} \quad l = 1, ..., K \qquad \text{(MDP1)}$$

$$\bigvee_{i \in D_k} \left[\begin{array}{c} Y_{ik} \\ h_{ik}(x^l) + \nabla h_{ik}(x^l)^T (x - x^l) \leq 0 \qquad l \in L(ik) \\ c_{ik} = \gamma_{ik} \end{array}\right] \quad k \in SD$$

$$\Omega(Y) = \text{true}$$

$$\alpha \in R, \quad x \in R^n, \quad c \in R^m, \quad Y \in \{\text{true, false}\}^m$$

It should be noted that before applying the above master problem, it is necessary to solve various subproblems (NLPD) so as to produce at least one linear approximation of each of the terms in the disjunctions. As shown by Turkay and Grossmann (1994), selecting the smallest number of subproblems amounts to the solution of a set-covering problem. In the context of flowsheet synthesis problems, another way of generating the linearizations in (MDP1) is by starting with an initial flowsheet and suboptimizing the remaining subsystems as in the modeling/decomposition strategy (Kocis and Grossmann, 1989b; Kravanja and Grossmann, 1990).

The above problem (MDP1) can be solved by the methods described by Beaumont (1991) and by Raman and Grossmann (1994). It is also interesting to note that, for the case of flowsheet synthesis problems, Turkay and Grossmann (1994) have shown that if the convex hull representation of the disjunctions in (32) is used to convert (MPD1) into an MILP problem, then assuming $B^i = I$ and converting the logic relations $\Omega(Y)$ into the inequalities $Ay \leq a$, it becomes equivalent to the master problem of the modeling/decomposition strategy:

$$\min Z_L = \alpha$$

s.t.

$$\left\{ \begin{array}{l} \alpha \geq \sum_i c_i + f(x^l) + \nabla f(x^l)^T(x - x^l) \\ g(x^l) + \nabla g(x^l)^T (x - x^l) \leq 0 \end{array} \right\} \quad l = 1, ..., L \quad \text{(MIPDF)}$$

$$\nabla h_i(x^l)^T x \leq [-h_i(x^l) + \nabla h_i(x^l)^T x^l] y_i \quad l \in K_L^i, \quad i \in D_k, \quad k \in SD$$

$$Ay \leq a$$

$$x \in R^n, \quad c \geq 0, \quad y \in \{0, 1\}^m$$

in which the linearizations for the constraints $h_i(x) \leq 0$ are "deactivated" when $y_i = 0$. Also, Turkay and Grossmann (1994) have shown that, while a logic-based GBD method cannot be derived as in the case of the OA algorithm, one can nevertheless determine for the MILP version of the master problem (MIPDF) one Benders iteration, which then yields a sequence similar to the GBD method for the algebraic case. Finally, it should also be clear that slacks can be introduced to (MDP1) and to (MIPDF) to reduce the effect of nonconvexities as in the augmented-penalty MILP master problem (M-APER).

D. COMPUTATIONAL EXAMPLE

In order to provide some insight into the computational performance of the MINLP algorithms described in the last section, we consider the following example problem:

$$\min Z = y_1 + 1.5y_2 + 0.5y_3 + x_1^2 + x_2^2$$

s.t.

$$(x_1 - 2)^2 - x_2 \leq 0$$

$$x_1 - 2y_1 \geq 0$$

$$x_1 - x_2 - 4(1 - y_2) \leq 0$$

$$x_1 - (1 - y_1) \geq 0$$

$$x_2 - y_2 \geq 0$$

$$x_1 + x_2 \geq 3y_3$$

$$y_1 + y_2 + y_3 \geq 1$$

$$0 \leq x_1 \leq 4, \quad 0 \leq x_2 \leq 4$$

$$y_1, y_2, y_3 = 0, 1$$

(33)

Note that the nonlinearities involved in problem (33) are convex. Figure 11 shows the convergence of the OA and the GBD methods to the optimal solution using as a starting point $y_1 = y_2 = y_3 = 1$. The optimal solution is $Z = 3.5$, with $y_1 = 0$, $y_2 = 1$, $y_3 = 0$, $x_1 = 1$, $x_2 = 1$. Note that the OA algorithm requires three major iterations, while GBD requires four, and that the lower bounds of OA are much stronger.

Table II presents a set of 15 test problems that were used in the comparison of GBD with OA as implemented in DICOPT (Kocis and Grossmann, 1989a)

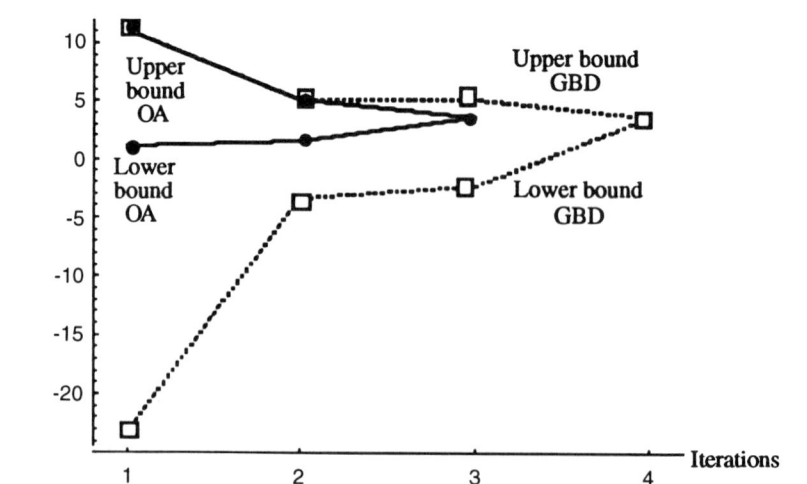

FIG. 11. Progress of iterations of OA and GBD for MINLP in problem (33).

TABLE II
Test Problems for Comparison of GBD, OA, and AP/OA/RP

Problem	Type	Variables Continuous	0–1	Total	Benders	Optimum value DICOPT	DICOPT++
LAZAMY	Nonconvex	5	2	7	−333.89	−333.89	−333.89
HW3	Convex	2	3	5	3.50	3.50	3.50
HW74	Convex	7	3	10	−1.92	−1.92	−1.92
NONCON	Nonconvex	2	3	5	7.67	7.67	7.93
FLEX	Convex	11	4	15	−7.08	−7.08	−7.08
FRENCH	Convex	7	4	11	4.58	4.58	4.58
EX3	Convex	25	8	33	68.01	68.01	68.01
BATCHHW	Convex	10	9	19	106,755.80	106,756.00	106,755.80
TEST1A	Nonconvex	44	12	56	287,169.70	287,168.70	83,637.70
BATCH5	Convex	22	24	46	285,506.50	285,507.00	285,506.50
EX4	Convex	5	25	30	−8.06	−8.06	−8.06
REL1	Nonconvex	20	28	48	294.80	2,449.40	216.50
UTIL	Nonconvex	112	28	140	999.70	1,004.51	999.55
BATCH8	Convex	32	40	72	402,496.70	406,145.00	402,496.70
BATCH12	Convex	40	60	100	4,230,000	2,687,026	2,687,026

and with the augmented penalty/equality relaxation version of OA (AP/OA/RP) as implemented in DICOPT++ (Viswanathan and Grossmann, 1990), which is currently implemented in the modeling system GAMS. The size of these problems is relatively small. As seen in Table III, when there is a smaller number of 0–1 variables (less than 20), the performance of GBD is comparable to that of the OA methods. In fact, DICOPT++ requires more time because of the initialization step of solving the relaxed NLP subproblem. For a larger number of 0–1 variables (last five problems), the requirements of GBD greatly increase while the OA algorithms require a similar number of major iterations (between 3 and 5). Also, from Table II it can be seen that DICOPT (original OA) is most likely to fail to find the global optimum in nonconvex problems (e.g., REL1, UTIL), while DICOPT++ exhibits the greatest robustness in finding near-optimal solutions. It should also be noted that the CPU time in DICOPT is commonly the smallest because it often terminates prematurely owing to the higher lower bounds mistakenly predicted in nonconvex problems.

The solution of significantly larger MINLP problems has been reported in the literature. For instance, Viswanathan and Grossmann (1990) reported the solution with DICOPT++ of the superstructure optimization of the HDA process (see Section VIII, Fig. 17) which involved 13 0–1 variables, 709 continuous variables, and 719 constraints. DICOPT++ obtained the solution in three major iterations requiring 482 seconds on an IBM-3090. Also, Viswanathan and Gross-

TABLE III
TEST PROBLEMS FOR COMPARISON OF GBD, OA, AND OP/OA/RP

Problem	Iterations			CPU Time		
	Benders	DICOPT	DICOPT++	Benders	DICOPT	DICOPT++
LAZAMY						
HW3	3	2	1	4.54	1.40	2.26
HW74	4	3	3	14.73	2.24	9.76
NONCON	4	2	4	3.48	1.58	4.32
FLEX	2	2	1	1.52	1.00	0.42
FRENCH	4	3	3	3.73	2.94	4.31
EX3	8	5	4	8.34	6.19	7.03
BATCHHW	8	3	5	10.94	4.00	45.56
TEST1A	5	3	3	5.62	3.68	34.60
BATCH5	2	2	4	14.97	11.88	93.96
EX4	67	4	3	766.88	33.62	26.94
REL1	70	3	5	394.81	41.51	100.87
UTIL	19	2	3	71.31	3.92	20.78
BATCH8	63	3	3	532.17	9.84	41.99
BATCH12	>66*	5	4	2527.20	141.61	108.58

*Search stopped after 66 iterations.

mann (1993) reported the solution with DICOPT++ of the optimal feed tray and number of plates in a distillation column for the separation of methanol and water. The model involved 115 0–1 variables, 1683 continuous variables, and 1919 constraints. DICOPT++ converged in seven major iterations and required about 70 minutes on an HP-9000/730 workstation. It is also interesting to note that significantly larger MILP problems have been solved in the area of scheduling (e.g., see Sahinidis and Grossmann, 1991a; Pinto and Grossmann, 1994).

VII. Solution Strategies for MINLP Synthesis Problems

In principle, the solution of an MINLP synthesis model can be obtained with the algorithms described in the previous section (e.g., OA or GBD algorithms). There are, however, three major difficulties that have to be addressed when attempting to solve these problems:

- Zero flows in process flowsheets
- Large size of MINLP problems
- Nonconvexities

A. HANDLING ZERO FLOWS

The modeling/decomposition (M/D) strategy proposed by Kocis and Grossmann (1989b) is largely motivated by the need to simplify the solution of the NLP and MILP problems. It reduces the undesirable effect of nonconvexities and eliminates the optimization of "dry units" with zero flows which are temporarily turned off in the superstructure. The solution of the NLP is simplified by optimizing only the particular flowsheet at hand, instead of optimizing it within the superstructure, as implied by problem (NLP2). The MILP solution is simplified by incorporating an approximation to the particular flowsheet only at each iteration. Finally, the effect of nonconvexities is reduced by special modeling techniques.

In the M/D strategy the basic idea is to first recognize that a flowsheet superstructure can be viewed as a network consisting of two types of nodes: interconnection nodes (splitters and mixers) and process unit nodes (reactors, separators). In brief, the modeling is then performed as follows. Since interconnection nodes play a crucial role in defining the flowsheet structures and since they exhibit well defined equations, special modeling techniques can be applied to these nodes. In particular, splitters and mixers that imply the choice of a single alternative can be modeled through linear constraints, thus avoiding the nonconvexities associated with the use of split fractions (see (41)–(42) in

Section VII.C). When multiple choices are possible, one can, in fact, develop valid linear outer approximations that properly bound the nonconvex solution space in the MILP master problem. As for the process unit nodes, the mass balances are expressed in terms of component flows rather than in terms of fractional compositions. Finally, the right-hand side in the linearizations of the process units are modified to ensure that nonzero flows are attained when the 0–1 variable is set to zero.

(a) Superstructure

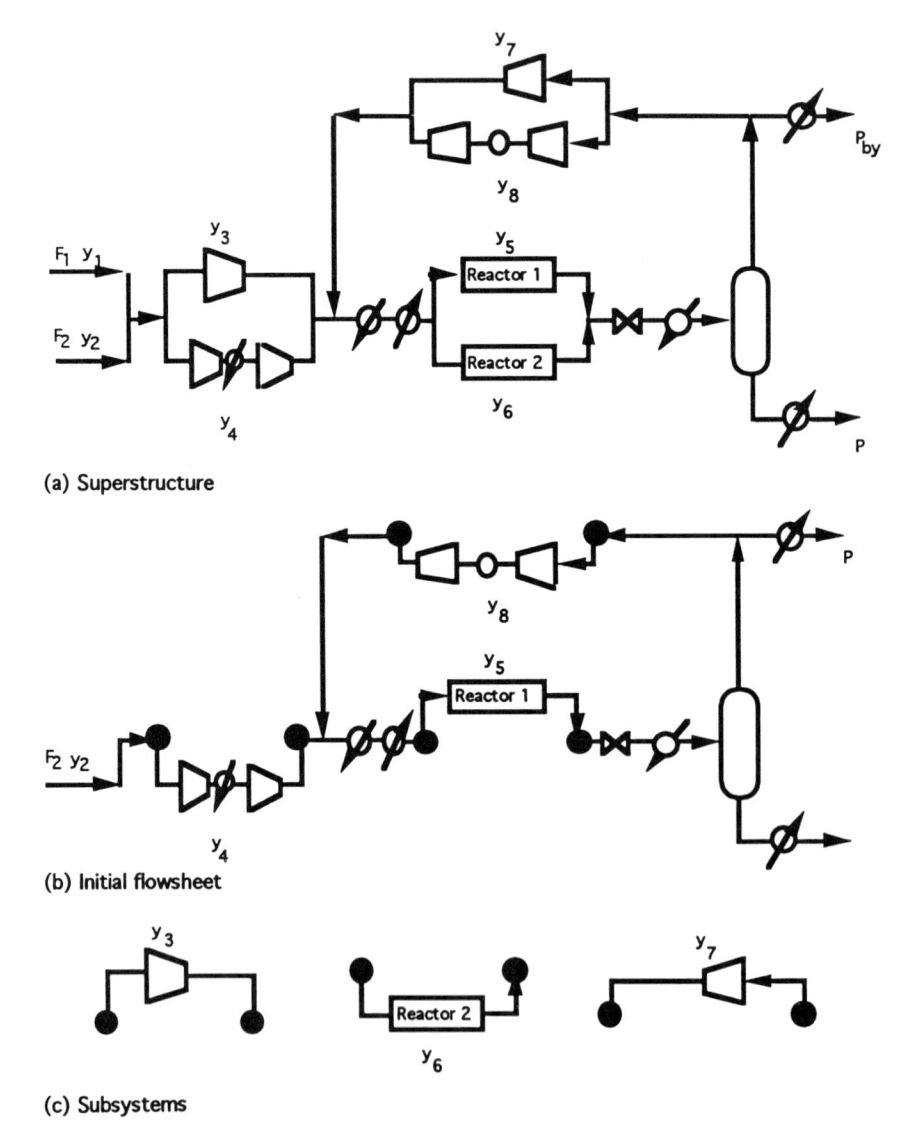

(b) Initial flowsheet

(c) Subsystems

FIG. 12. Initial flowsheet and subsystems in modelling/decomposition strategy.

As for the decomposition part of this strategy, the idea is as follows. Suppose we start by optimizing a particular flowsheet structure. It is clear that we are able to obtain linear approximations for the master problem of the existing process units. The question is, then, how to generate an approximation of the "deleted" units in the superstructure. This can actually be accomplished by suboptimizing groups of units that are tied with existing interconnection nodes. Since prices (i.e., multipliers) and nonzero flows are available at these nodes, these can be used to suboptimize the nonexisting units "as if they were to exist" in the superstructure. This provides not only nonzero flow conditions, but also points that are often good for approximating these units. An example of how a superstructure can be decomposed into subsystems to be suboptimized is given in Fig. 12 for the structural optimization of a process flowsheet. In this way, by optimizing the first flowsheet structure and suboptimizing the groups of nonexisting units, it suffices in subsequent iterations, to optimize the specific flowsheet that is generated in order to update the MILP. This has two desirable effects: it solves the NLP for each specific flowsheet, and it reduces the size of the MILP since only linearizations of existing units are incorporated at each iteration. This strategy has been automated in the flowsheet synthesizer PROSYN by Kravanja and Grossmann (1990), where the heat integration is handled through an extension of the constraints proposed by Duran and Grossmann (1986a), where area is accounted for with a variable HRAT. It is also important to note that if the MINLP were to be formulated as the generalized disjunctive programming problem (DP1), the logic-based algorithms would essentially reduce to the modeling/decomposition strategy, as has been discussed by Turkay and Grossmann (1994).

B. Large Size of MINLP Problems

There are several ways to avoid the solution of large MINLP synthesis problems. One is to rely on a targeting approach in which the solution of the MINLP is avoided by considering a sequence of design targets that give rise to aggregated models which are simpler to solve. As an example, consider the superstructure for heat exchanger networks in Fig. 4 (Section IV). Instead of solving directly the MINLP associated with this superstructure, consider the following stategy by Floudas *et al.* (1986), which has been implemented in the code MAG-NETS by Ciric (1986) and which mimics the pinch design approach (Linnhoff, 1993):

(1) Determine the minimum utility solving the LP transshipment problem (see Fig. 7, Section IV).
(2) Given the energy target in step 1, determine the fewest number of units solving the MILP version of the transshipment model.

(3) Given the energy target of step 1 and the units predicted in step 2, determine the areas and flows of the superstructure in Fig. 4.

Conceptually, the disadvantage of this approach is that it is not entirely rigorous as the energy consumption and the number of units and areas are not optimized simultaneously. On the other hand, the advantage is that it involves the solution of simpler and smaller problems, which are computationally manageable.

Another approach that addresses the reduction of size of the MINLP is the state space approach by Bagajewicz and Manousiouthakis (1992). The basic idea of this strategy is to partition the synthesis problem into two major subsystems, the distribution network and the state space operator. The objective in the former is to make the decisions related to the distribution of flows in the superstructure, while the objective in the latter is to perform the optimization for the decisions selected in the distribution network. At the level of the state space operator one can consider the process either in its detailed level or simply as a pinch-based targeting model. While this strategy has the advantage of reducing the size of the MINLP, it is unclear how to develop automated procedures based on this approach.

A third option to avoid solving a large MINLP problem is to perform preliminary screening to eliminate some of the alternatives that are embedded in the superstructure, thus reducing the size of the MINLP problem. Daichendt and Grossmann (1994a,b) have proposed a design strategy for rigorously integrating preliminary screening and MINLP optimization. The strategy considers the synthesis problem as one in which the superstructure with its corresponding MINLP model is given. Rather than solving the entire MINLP model, the objective is to perform preliminary screening to rigorously eliminate a subset of suboptimal alternatives from the original superstructure to yield a *reduced superstructure*. The corresponding reduced MINLP model is to have the same global optimal solution as the original MINLP model.

The original superstructure is assumed to consist of a set of units $j \in U$, in which binary variables y_j are associated with the existence of these units and continuous variables d_j represent the corresponding sizes of these units. Additional continuous variables are represented by the vector x, corresponding to the state variables of the model, and by the vector z, corresponding to the continuous decision variables that incur linear costs in the objective function (e.g., raw material flows and energy consumption).

The MINLP problem (P1), corresponding to the original superstructure, is assumed to have the following form:

$$\min C = \sum_{j \in U} [a_j y_j + f_j(d_j)] + \sum_{k \in K} c_k z_k \qquad \text{(PP1)}$$

$$\text{s.t. } h_n(y,d,x,z) \leq 0, \qquad n \in N$$

$$d_j \leq d_j^{UB} y_j, \qquad j \in U$$

$$y \in Y, \qquad d \in D, \qquad z \in Z, \qquad x \in X$$

where C is the cost function consisting of a nonlinear fixed charge function for the structural and design variables, y_j and d_j, and linear operating costs and revenues for the continuous variables z_k. The equations for mass and energy balances and inequalities for specifications are represented by the constraints $h_n(y,d,x,z) \leq 0$, $n \in N$. The inequality $d_j \leq d_j^{UB} y_j$, $j \in U$, is a logical condition that will force the design variable d_j to be zero when y_j is zero, d_j^{UB} being a valid upper bound to d_j. Finally, Y is a set of pure 0–1 constraints, $Y = \{y \mid Ay \leq a, y \in \{0,1\}^m\}$; and D, X, and Z are bounded sets in the space of positive real variables.

The strategy for preliminary screening relies on the use of an aggregated model that generally has a dimensionality less than or equal to that of the original model, and that may aggregate the units of the original model into subtasks. The aggregated model bounds the original model: The costs determined by the aggregated model underestimate the costs determined by the original model; the feasible region defined by the aggregated model overestimates the feasible region defined by the original model; and the aggregated model is formulated as a problem that can be solved to global optimality, i.e., as an MILP or convex MINLP problem (PP2), which is defined as follows:

The continuous variables of the aggregated model, $\hat{d}$, $\hat{x}$, and $\hat{z}$, are subvectors of the continuous variables d, x, and z of the original problem:

$$d = \begin{bmatrix} \hat{d} \\ \bar{d} \end{bmatrix}, \quad x = \begin{bmatrix} \hat{x} \\ \bar{x} \end{bmatrix}, \quad z = \begin{bmatrix} \hat{z} \\ \bar{z} \end{bmatrix}$$

where the variables with the bar are the ones excluded from the model. The binary variables of the aggregated model, $\hat{y}$, obey the following logic equivalence relationship (when treated as Boolean variables):

$$\hat{y}_i \Leftrightarrow \bigvee_{j \in U(i)} y_j, \quad i \in \hat{U} \tag{34}$$

where $\Leftrightarrow$ and $\bigvee$ represent the logical equivalence and OR operator, and $U(i) \subseteq \hat{U}$ are subsets of units in the superstructure that are aggregated into subtasks $i \in \hat{U}$. As an example, consider the case of the HENS problem. Hot and cold stream *matches* are represented by the binary variables $\hat{y}_{h,c}$, $h \in H$, $c \in C$. These correspond to aggregated variables of the binary variables $y_{h,c,k}$ that represent the *units* for matches between streams h and c in stage k of the superstructure. It should be noted that when no unit aggregation takes place, $U = \hat{U}$, $|U(i)| = 1$, and $\hat{y}_i$ has a one-to-one correspondence with y_j.

The constraints of the aggregated model are linear combinations of subsets of the constraints of the original problem, $N(l) \subseteq N$:

$$r_l(\hat{y},\hat{d},\hat{x},\hat{z}) = \sum_{n \in N(l)} \lambda_n^l h_n(y,d,x,z), \quad l \in L \tag{35}$$

where L is the set of constraints of the aggregated model and λ_n^l are scalars for the linear combinations. To ensure bounding properties, the nonlinear fixed-charge cost functions of the original model will be *underestimated* by linear fixed-charge cost functions:

$$\sum_{i \subseteq \hat{U}} [b_i \hat{y}_i + g_i(\hat{d}_i)] \leq \sum_{j \subseteq U} [a_j y_j + f_j(d_j)] \qquad (36)$$

Because the aggregated model must rigorously bound the original model, it must be solvable to global optimality. In this paper the aggregated model is formulated as an MILP problem (PP2) having the following form:

$$\min \hat{C} = \sum_{i \in \hat{U}} [b_i \hat{y} + g_i(\hat{d}_i)] + \sum_{k \in \hat{K}} c_k \hat{z}_k \qquad (PP2)$$

$$\text{s.t. } r_l(\hat{y}, \hat{d}, \hat{x}, \hat{z}) \leq 0, \qquad l \in L$$

$$\hat{d}_i \leq \hat{d}_i^{UB} \hat{y}_i, \qquad i \in \hat{U}$$

where $\hat{K} \subseteq K$ since $\hat{z}$ is a subvector of z. The sets $\hat{Y}, \hat{D}, \hat{X}, \hat{Z}$ are such that the feasible space $\hat{F}$ of (PP2) contains the feasible space of (PP1) projected in the space $(\hat{y}, \hat{d}, \hat{x}, \hat{z})$, i.e., $F^p \subseteq \hat{F}$, where the projection of (PP1) is given by

$$F^p = \{\hat{y}, \hat{d}, \hat{x}, \hat{z} \mid (y, d, x, z) \in F\} \qquad (37)$$

Also, since the cost is underestimated, as given in (36), it follows that

$$\hat{C}^* \leq C^* \qquad (38)$$

where C^* and $\hat{C}^*$ are the optimal solutions to (PP1) and (PP2), respectively. Thus, since (PP2) can be solved to global optimality because of its formulation as an MILP (or convex MINLP) problem, its solution provides a lower bound to (PP1).

It should be noted that deriving an aggregated model of the form of problem (PP2) is, in general, a nontrivial task, as it requires a basic understanding of the process as well as a characterization of the nature of the functions that are involved. Thus, problem (PP2) should be viewed as a conceptual framework rather than a recipe.

The solution to (PP2) yields a valid lower bound to the optimal solution of the original MINLP problem (PP1). The efficiency of the preliminary screening is clearly related to the tightness of the over- and underestimations; i.e., the closer the representation of the aggregated model is to the original model, the greater is the potential for problem size reduction. The purpose of the aggregated model is to generate successive solutions to (PP2) through integer cuts, whose costs are constrained to be less than or equal to the cost of a base-case design. The base-case design is a particular selection of units whose cost is determined using the original model. Since the base-case design is a feasible solution to (PP1), it provides a valid upper bound to the global optimal solution of (PP1) as well as a valid upper bound for the optimal solution for each iteration of (PP2), because (PP2) is always an underestimation of (PP1). Thus, when no additional feasible solution can be found to (PP2), preliminary screening is terminated.

Those units or subtasks that are not selected in any of the solutions are excluded from the superstructure. Those units that are selected in all solutions are permanently fixed in the superstructure; or, if the original units are aggregated into subtasks, constraints are imposed on the units that must be selected. Only those combinations of remaining units or subtasks that are feasible solutions to (PP2) are considered as alternatives for (PP1). The exclusion of certain units and the fixing of other units or combinations of units produces the reduced superstructure.

The base-case design can be determined either from the first solution to (PP2) or by selecting units based on heuristics. These are then fixed in (PP1), which is solved as an NLP problem. The closer the solution of the base-case design is to the global optimal solution of (PP1), the fewer solutions that result from (PP2). Thus, the efficiency, but not the rigorousness, of the preliminary screening procedure is dependent on the determination of a good base-case design. It should be noted also that work by the authors is currently under way to integrate preliminary screening and MINLP optimization within a hierarchical decomposition framework.

C. Handling Nonconvexities

One of the difficulties in the application of NLP and MINLP optimization techniques in process synthesis has been the fact that these problems are often nonconvex, and hence give rise to multiple local solutions. Although such techniques as simulated annealing might help to locate global or near-optimal solutions, they have the drawback that, aside from not being rigorous, they may require a very large number of trials, which in the long term will not help for handling process models that are relatively expensive to evaluate. Two major strategies for handling nonconvexities in synthesis problems are the reformulation of (MINLP) to an MILP model and the rigorous global optimization of nonlinear problems that exhibit special structure. We briefly describe these strategies here.

In order to derive an MILP approximation to problem (MINLP) (Papoulias and Grossmann, 1983c), the continuous variables x are partitioned as follows:

$$x = \begin{bmatrix} z^d \\ x^c \end{bmatrix} \qquad (39)$$

in which z^d is the vector of operating conditions that gives rise to the nonlinearities (e.g. pressures, temperatures, split fractions, conversions, etc.), and x^c is a vector of material, heat and power flow variables. In this way, given a fixed value of z^d, the nonlinear equations, which are assumed to be only a function of x, reduce to a subset of linear equations; that is,

$$h(x) = 0 \Rightarrow Ex^c = e \qquad (40)$$

in which the matrix of coefficients E and the right-hand side e are functions of z^d, $E(z^d)$, $e(z^d)$.

Since, in general, more than one fixed value for the variables z^d are considered, this requires the introduction of the additional 0–1 variables y^d to represent the potential selection of the discrete operating conditions. In this way, the general form of the MILP approximation will be as follows:

$$\min C = a_1^T y + a_2^T y^d + b^T x^c$$
$$\text{s.t. } E_1 y^d + E_2 x^c = e \qquad \text{(MAPP)}$$
$$D_1 y + D_2 y^d + D_3 x^c \leq d$$
$$y, y^d = 0, 1 \quad x^c \geq 0$$

It should be noted that the derivation of the above problem generally requires the disaggregation of the vector of continuous variables x^c in terms of the discretized conditions. To illustrate this point more clearly, consider the simple splitter shown in Fig. 13. The corresponding mass balance equations for each component i are as follows:

$$f_i^1 = \eta f_i^{in} \qquad (41)$$
$$f_i^2 = f_i^{in} - f_i^1 \qquad (42)$$

where η is the split fraction for outlet stream 1. Note that Eq. (41) is nonlinear (in fact bilinear), and despite its simplicity it is a major source of nonconvexities and numerical difficulties.

Now assume that we consider N discrete values of η, η_k, $k = 1, 2, ..., N$. Then if we disagggregate the flow for the inlet stream as $f_i^{in,k}$, $k = 1, 2, ..., N$, and introduce the 0–1 variables $y^{d,k}$, $k = 1, 2, ..., N$, equations (41) and (42) can be replaced by the linear constraints

$$f_i^1 = \sum_{k=1}^{N} \eta_k f_i^{in,k} \qquad (43)$$
$$f_i^{in,k} - U y^{d,k} \leq 0 \quad k = 1, 2, ..., N \qquad (44)$$
$$\sum_{k=1}^{N} y^{d,k} = 1 \qquad (45)$$
$$f_i^2 = f_i^{in} - f_i^1 \qquad (46)$$

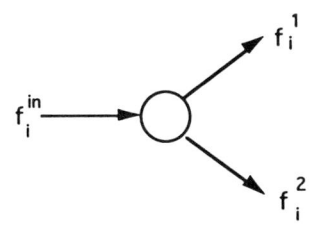

FIG. 13. Stream splitter.

While the nonlinearities are eliminated, it is clear the number of discrete and continuous variables is increased as well as the number of constraints. Also, in the general case the definition of the matrix of coefficients and the right-hand sides of problem (MAPP) requires an *a priori* evaluation or simulation of nonlinear models.

Extensive reviews on global optimization can be found in Horst (1990) and Horst and Tuy (1990). In this section we present a summary of a global optimization method that has been developed by Quesada and Grossmann for solving nonconvex NLP problems which have the special structure that they involve linear fractional and bilinear terms. It should be noted that global optimization has clearly become one of the new trends in optimization and synthesis, and active workers involved in this area include Floudas and Visweswaran (1990), Swaney (1990), Manousiouthakis and Sourlas (1992), and Sahinidis (1993).

The special class of NLP problems considered by Quesada and Grossmann (1995b) has been motivated by heat exchange networks and separations problems. These can be represented in general as follows:

$$\min g_0$$
$$\text{s.t. } g_l \leq 0 \quad l = 1, \ldots, L \quad \text{(NLP)}$$

where

$$g_l = \sum_{i \in I} \sum_{j \in J} c_{ijl} \frac{x_i}{y_j} - \sum_{i \in I'} \sum_{j \in J'} c_{ijl} x_i y_j + h_l(x, y, z), \quad l = 0, 1, \ldots, L$$

$$x^L \leq x \leq x^u$$

$$y^L \leq y \leq y^u$$

$$z \in Z$$

As shown above, the objective function and the constraints generally involve linear fractional and bilinear terms corresponding to the two summation terms, while the last term $h_l(x, y, z)$ is assumed to correspond to a convex function. This type of problem arises, for instance, in the optimization of heat-exchanger networks (linear area cost, arithmetic mean driving force, isothermal mixing; see Quesada and Grossmann, 1993) and in the case of synthesis of sharp separation networks with mixed products (see Quesada and Grossmann, 1995a). The difficulty involved in solving these NLP optimization problems is that a straightforward application of common local search methods is generally not rigorous. Conventional NLP algorithms can not only produce local solutions that are suboptimal, but also may fail to find a feasible solution owing to the nonconvexities of the constraints.

In the method proposed by Quesada and Grossmann (1995b), the main idea is to replace the bilinearities and linear fractional terms by valid under- and over-estimators which will yield a convex NLP (or LP) whose solution provides a lower bound to the global optimum. Consider, for instance, fractional terms with positive coefficients. By introducing the variables r_{ij}, we can express the fractional term as the constraint

$$r_{ij} y_j \geq x_i \quad i \in I, j \in J \qquad (47)$$

which is nonconvex. Valid linear over-estimators, which were suggested by McCormick (1976) for this constraint, are given by,

$$x_i \leq y_j^u r_{ij} + r_{ij}^L y_j - y_j^u r_{ij}^L \quad i \in I, j \in J \qquad (48)$$

$$x_i \leq y_j^L r_{ij} + r_{ij}^u y_j - y_j^L r_{ij}^u \quad i \in I, j \in J \qquad (49)$$

where x_i^L, x_i^u, y_j^L, y_j^u, r_{ij}^L, r_{ij}^u are valid lower and upper bounds of the variables. In addition, Quesada and Grossmann (1995b) showed that the nonlinear convex constraint

$$r_{ij} \geq \frac{x_i}{y_j^u} + x_i^L \left(\frac{1}{y_j} - \frac{1}{y_j^u} \right) \quad i \in I, j \in J \qquad (50)$$

can be used as a valid underestimator. The interesting feature of (50) is that it is a stronger constraint than (48) and (49), provided r_{ij}^L, r_{ij}^u are given by the bounds of x_i and y_j. In fact, when these bounds are obtained by the optimization of individual variables in (NLP), it is also possible to generate projected bounding constraints which can serve to tighten the representation of the NLP.

The proposed method then consists in reformulating problem (NLP) in terms of valid linear and nonlinear bounding constraints such as in (48)–(50), giving rise to a convex NLP (or LP) problem which predicts valid lower bounds to the global optimum. If there is a difference between the current upper and lower bounds, the idea is to partition the feasible region by performing a spatial branch-and-bound search. The major steps in the global optimization algorithm by Quesada and Grossmann (1995b) for NLP problems involving linear fractional and bilinear terms are as follows:

Step 0. Initialization step

(a) Set the upper bound to $f^* = \infty$; the tolerance ϵ is selected.

(b) Bounds over the variables involved in the nonconvex terms are obtained. For this purpose, specific subproblems can be solved or a relaxation of the original problem is used. Update the upper bound f^*.

(c) Define space W_0 as a valid relaxation of the feasible region in the space of the nonconvex variables. The branch-and-bound search will be conducted over W_0. The list F is initially defined as the region W_0.

(d) Construct a convex underestimator problem (CU_L) by replacing the nonconvex terms in the original problem with additional variables and introducing valid convex approximations of these nonconvex terms. Valid constraints that were not present in the original problem because they were redundant can be included to tighten the convex relaxation.

Step 1. Convex underestimator problem

(a) Solve problem CU_L over the relaxed feasible region W_0. The solution corresponds to a valid lower bound (f^L) of the global optimum. The actual objective function is evaluated if this is a feasible solution; otherwise, the original problem is solved using the convex solution as the initial point. Update the upper bound.

(b) If $(f^* - f^L) \leq \epsilon f^*$, stop; the global solution corresponds to f^*.

Step 2. Partition

From the list F, consider a subregion W_j (generally the region with the smallest f^L is selected) and divide it into two new subregions W_{j+1} and W_{j+2}; these are added to the list F and subregion W_j is deleted from F.

Step 3. Bounding

(a) Solve problem CU_L for the two new subregions.

(b) If the solutions are feasible, evaluate the actual objective function. Otherwise, the original nonconvex problem can be solved according to a given criterion.

Step 4. Convergence

Delete from list F any subregion with $(f^* - f^L) \leq \epsilon f^*$. If list F is empty, stop; the global optimum is f^*. Otherwise, go to step 2.

The global optimization algorithm described above uses a spatial branch-and-bound procedure (steps 2 to 4). Like many branch-and-bound methods, the algorithm consists of a set of branching rules, together with upper bounding and lower bounding procedures.

The branching rules include the node selection rule, the branching variable selection, and the level at which the variable is branched. A simple branching strategy is as follows. The node with the smallest lower bound is the node selected to branch on, and two new nodes are generated using constraints of the type

$$x_i \geq x_i^* \quad \text{and} \quad x_i \leq x_i^* \tag{51}$$

Different strategies can be used to do the branching. These include generating more than two nodes from a parent node, using different types of branching constraints or different node selection rules. For the latter, we can use some type of degradation function similar to the one used in branch-and-bound procedures for MILP problems.

Additional criteria used in branch-and-bound algorithms for MILP problems can be extrapolated to the global optimization case. These include the fixing of variables, tightening of bounds, range reduction, etc. (see Sahinidis, 1993). One main difference between the branch-and-bound procedure used for binary variables and the spatial branch-and-bound search used here is the fact that it might be necessary to branch more than once on the same variable. When the selection rule involves more than one variable within a small range, it is often useful to branch on a variable that has not been used previously, even though it may not be the first candidate.

Information of the convex underestimator problem can be employed to select the branching variables. At this point, only the difference between the convex solution and the actual value of the functions is used. It is also possible to consider dual or second-order information or to generate small selection subproblems (Swaney, 1990). With respect to the upper bound, there are two cases. In the first case, when the feasible region of the original problem is convex, the evaluation of the original objective function at the solution of the convex underestimator problem often provides a good upper bound. In the second case, when the feasible region is nonconvex, it is sometimes necessary to obtain an upper bound through a different procedure, since the solution of the convex underestimator problems might be infeasible for the original problem. In some particular cases, it may be better to use a specialized heuristic to obtain a good upper bound. In general, however, it is necessary to solve the original nonconvex problem to generate an upper bound. As pointed out in Quesada and Grossmann (1995a,b) the solution of the convex underestimator problem provides a good initial point to the nonconvex problem.

VIII. Applications

As was described in the review of previous work, over the last ten years MINLP optimization models have been reported for the synthesis of process flowsheets, heat-exchanger networks, separation sequences, reactor networks, utility plants, and design of batch processes. Rather than describing in detail each of these works, we will briefly highlight several examples from our research group at Carnegie Mellon to illustrate the capabilities and the current limitations of the MINLP approach.

First, we present a small example from the MINLP model by Yee *et al.* (1990b) for the synthesis of a heat-exchanger network for the stream data given in Table IV. If the MINLP model (21)–(29) is solved with two stages (see Fig. 5) and with a code such as DICOPT++ (Viswanathan and Grossmann, 1990), we obtain the design given in Fig. 14. Note that that design requires neither heating nor cooling. On the other hand, the network involves stream splitting which is not always attractive from a practical point of view as this requires the additional investment of a control valve and a potentially more complex operation. One can easily generate a network structure with no stream splitting by adding the inequalities that allow at most one heat exchange for each stream in each stage. The resulting solution is shown in Fig. 15. Note that the new structure requires heating and cooling, although in small amounts. Nevertheless, the network now consists of four instead of three units, and the cost penalty for not having stream splits is of the order of 15% ($94,268/year vs. $82,491/year). This example shows that by imposing discrete constraints, one can synthesize several network structures with which one can assess different designs with other criteria such as complexity, operability, etc.

As a second example we present the synthesis of a 10-stream problem (5 hot and 5 cold), using the program SYNHEAT (Bolio *et al.*, 1993) which implements the MINLP method by Yee *et al.* (1990b) with DICOPT++. The size of the MINLP model is 135 0–1 variables, 376 continuous variables, and 536 constraints. The solution, which is shown in Fig. 16, required about 5 minutes on a VAX-6325. It should be noted that the solution to this problem was simplified by the various exchangers that are assigned only to utilities.

As a third example, consider the HDA process studied extensively by Douglas (1988). The superstructure for this process is shown in Fig. 17, which is based on a preliminary qualitative analysis of alternatives described in Douglas (1988). Given the basic options considered for the selection of reactors and the use of membrane separators, as well as a restricted set of alternatives for the

TABLE IV
DATA FOR ONE-HOT/TWO-COLD STREAM PROBLEM

Stream	TIN (K)	TOUT (K)	F_{cp} (kW/K)	h (kW/h m K)	Cost ($/kW-year)
H1	440	350	22	2.0	—
C1	349	430	20	2.0	—
C2	320	368	7.5	0.67	—
S1	500	500	—	1.0	120
W1	300	320	—	1.0	20

Note. Minimum approach of temperatures (EMAT) = 1K. Exchanger cost = 8,600 + 670 (Area)$^{0.83}$.

C1
349
(20)

↓

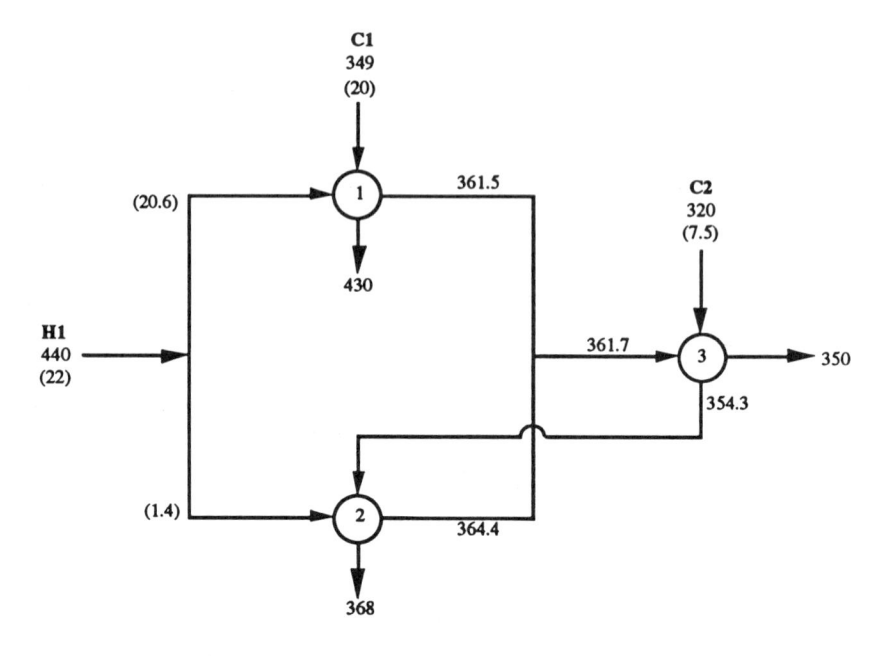

```
         361.5
(20.6) ──┬──(1)──────┐                C2
         │            │                320
         │           430               (7.5)
         ↓                              ↓
H1                      361.7          
440 ────────────────────(3)──────► 350
(22)                    │
                     354.3
        ┌───────────────┘
(1.4)   │
 ──────(2)
        │   364.4
        ↓
       368
```

Total Heat Exchangers Area = 182.9 m2

Utilities:
  Heaters heat load =   0 Kw
  Coolers heat load =   0 Kw

Costs:
  Investment = $ 82,491.6 per year

Total   = $ 82,491.6 per year

FIG. 14. Optimal network with no constraints on split streams.

separation and recycle, it is a relatively simple task to develop the superstructure representation, which has embedded close to 200 different flowsheet configurations. In this case, the simplified nonlinear models were used to model the problem as an MINLP, which involves 13 0-1 variables, 672 continuous variables, and 678 constraints (140 nonlinear equations, 567 linear equations, 71 linear inequalities). Using the starting configuration shown in Fig. 18, the optimal solution, which is shown in Fig. 19, was obtained with both the M/D strategy and with the AP/OA/ER algorithm. Note that the profit is increased from $4,814,000/year to $5,887,000/year. The M/D strategy required 2 minutes of CPU time (IBM-3083), while AP/OA/ER required 8 minutes; both took two major iterations. This example shows that computational savings can be achieved with the M/D strategy and that the economic improvements obtained in structural optimization can be very large.

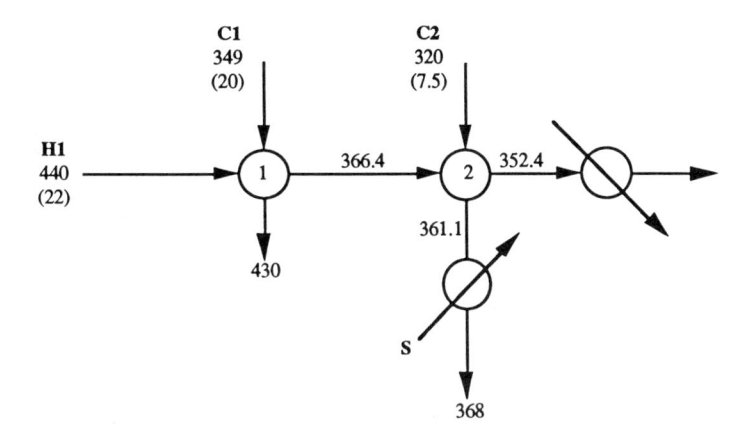

```
              C1              C2
              349             320
              (20)            (7.5)
                │               │
                ▼               ▼
   H1        ┌───┐   366.4   ┌───┐   352.4
   440 ─────▶│ 1 │──────────▶│ 2 │──────────⊘────▶
   (22)      └───┘           └───┘
                │           361.1 ↗
                ▼               ⊘
               430              │
                                ▼
                                S
                                │
                                ▼
                               368
```

Total Heat Exchangers Area = 165.4 m2

Utilities:
    Heaters heat load = 52.1 Kw
    Coolers heat load = 52.1 Kw

Costs:

    Utilities   = $ 7, 293.4 per year
    Investment = $ 86, 975.8 per year

FIG. 15. Network structure with no stream splits.

As a fourth example, consider the simultaneous optimization and heat integration of the flowsheet superstructure in Fig. 12 which has 16 embedded flowsheet alternatives. The M/D strategy implemented in PROSYN was applied to this problem. The resulting MINLP formulation contains 293 constraints, 279 continuous variables, and 8 binary variables. This example problem was solved with PROSYN for the three following cases: (a) MINLP optimization with no heat integration, (b) simultaneous MINLP optimization and heat integration using the model by Duran and Grossmann, (c) simultaneous NLP optimization and heat integration with HEN costs for the optimal structure obtained in case (b). For cases (a) and (b), the OA/ER algorithm was terminated based on the progress of the NLP solutions, since higher bounds on the profit were obtained from the MILP master with the proposed deactivation scheme for the linearizations of the splitter in the recycle. The OA/ER algorithm requires two NLP subproblems to confirm that the initial flowsheet in case (a) is the optimum. In case (b) it requires three NLP subproblems to find the structure in Fig. 19. This clearly indicates that the quality of the information supplied to the MILP master problem by the M/D strategy is good.

228   IGNACIO E. GROSSMANN

Total Heat Exchangers Area = 106.7 m2

Utilities:
  Heaters heat load = 420 Kw
  Coolers heat load = 1,810 Kw

Costs:
  Utilities   = $ 70,600 per year
  Investment = $ 32,027 per year

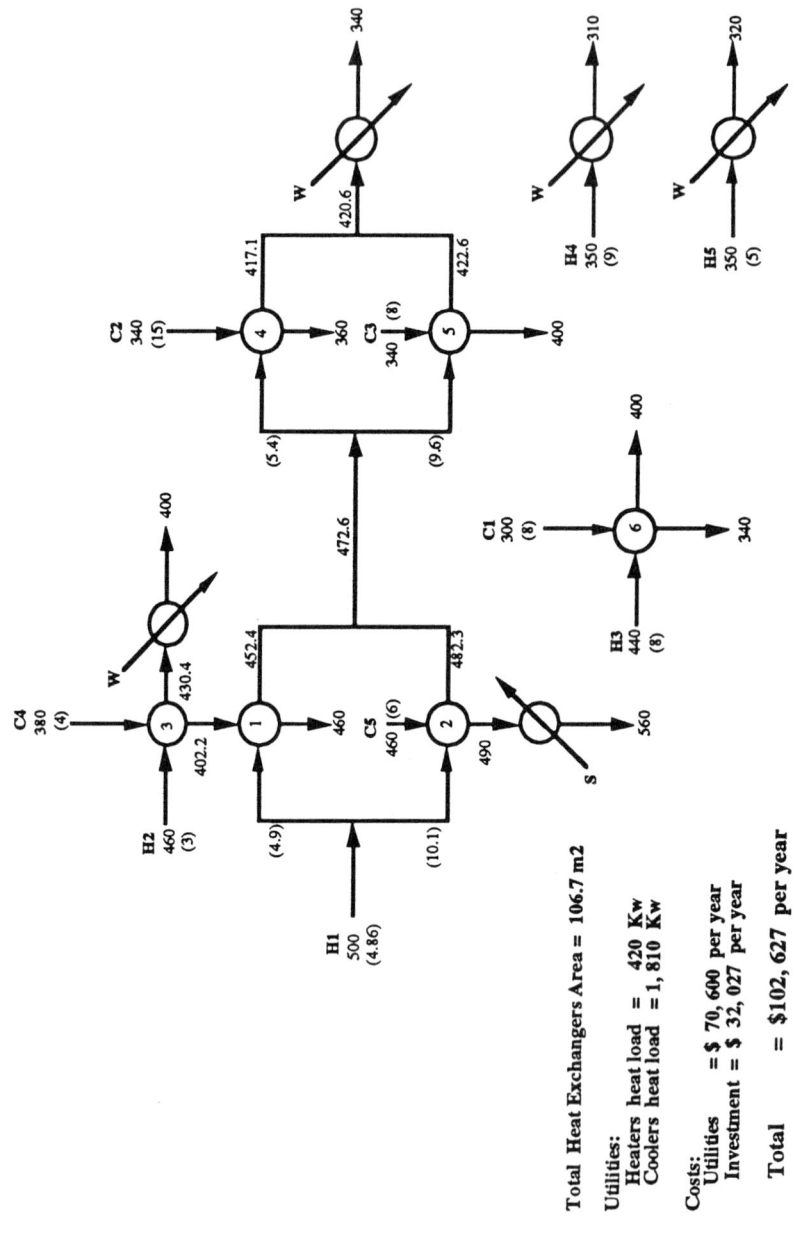

  Total   = **$102,627 per year**

FIG. 16. Synthesis of a 10-stream problem with SYNHEAT.

MIXED-INTEGER OPTIMIZATION TECHNIQUES   229

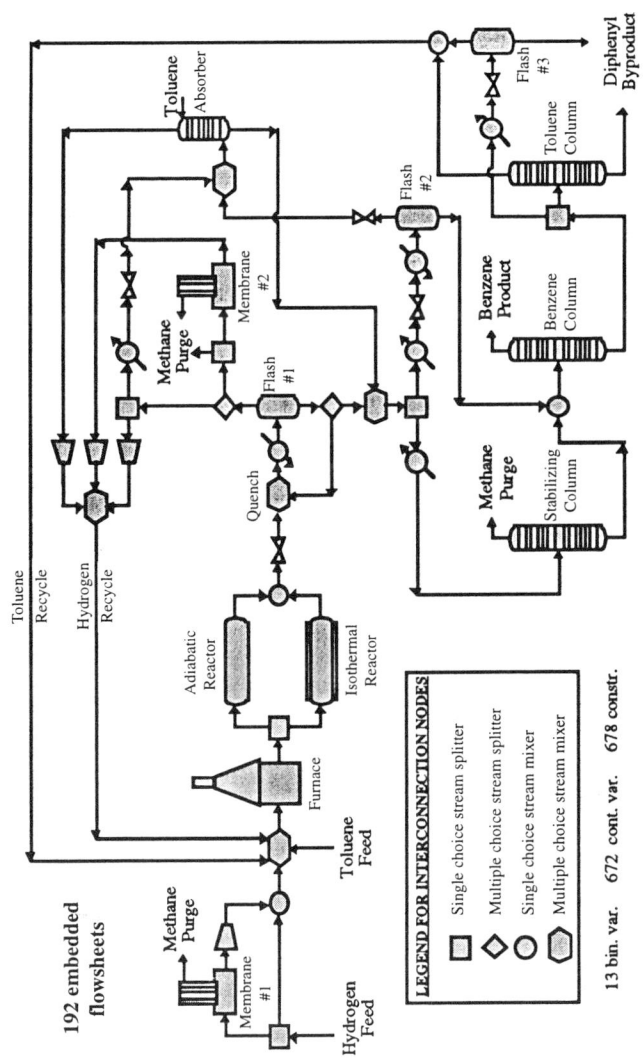

FIG. 17. Superstructure of HDA process.

13 bin. var.   672 cont. var.   678 constr.

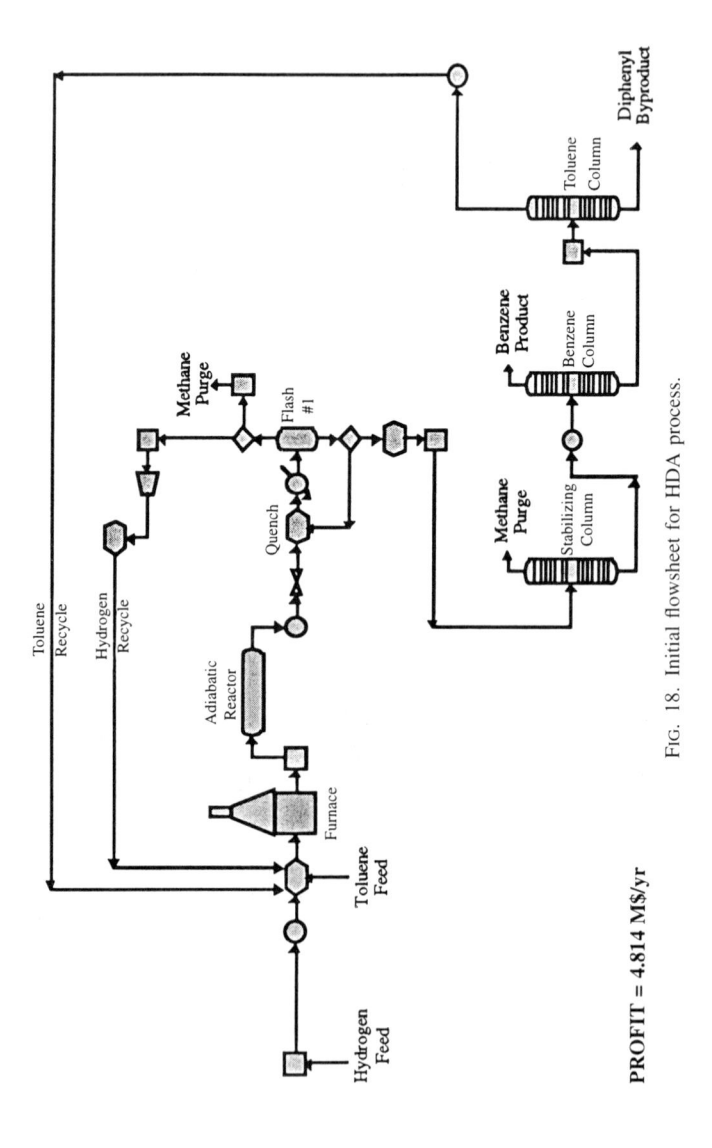

FIG. 18. Initial flowsheet for HDA process.

PROFIT = 4.814 M$/yr

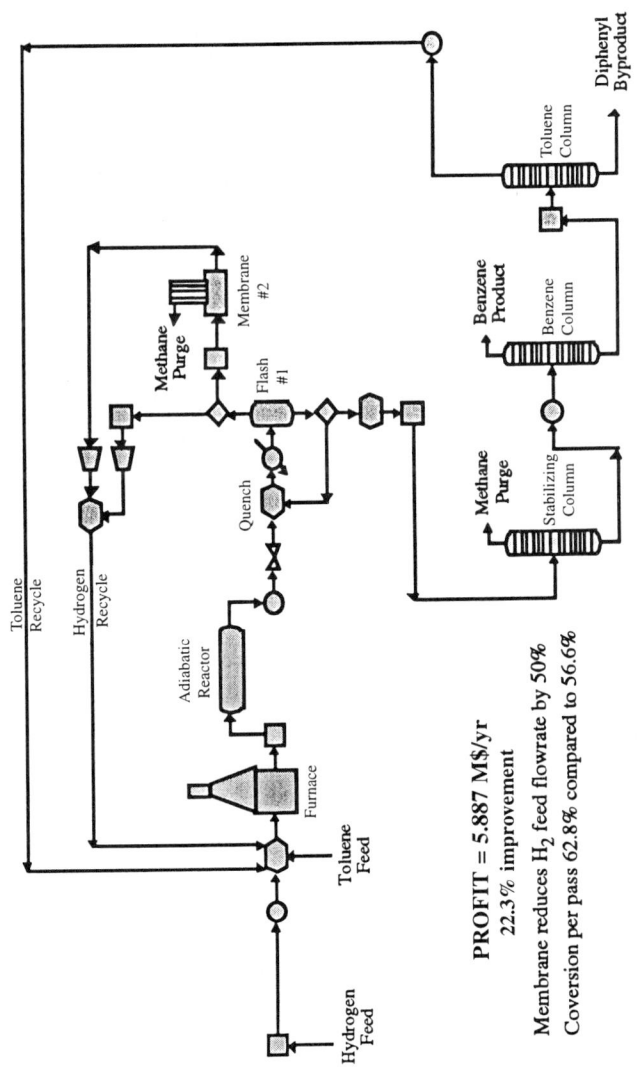

FIG. 19. Optimal flowsheet from HDA superstructure.

PROFIT = 5.887 M$/yr
22.3% improvement
Membrane reduces H$_2$ feed flowrate by 50%
Coversion per pass 62.8% compared to 56.6%

First consider case (a) when only the MINLP optimization of the superstructure is performed without heat integration. The optimal flowsheet is $y^k = \{1,0,0,1,1,0,0,1\}$ with annual profit of 794,000 $/year. It utilizes the cheaper feedstock F1, two-stage feed compression, cheap reactor R1 with low conversion, and a two-stage compressor for the recycle. If costs of the HEN (which are quite significant) are subsequently calculated and added to the profit, the result is a loss of 1,192,000 $/year. When heat integration is simultaneously performed in the MINLP optimization of the superstructure (Duran and Grossmann, 1986b), the results are much better at first glance. The optimal flowsheet in Fig. 20 yields an annual profit of 3,403,000 $/year (2,609,000 $/year more than for the nonintegrated flowsheet). The difference in the new flowsheet lies in the selection of single-stage compressors for the feed and the recycle. Also, almost all operating conditions change significantly since the trade-offs between heat integration (consumption of steam and cooling water), electricity, and consumption of feedstock are now appropriately established. As seen in Table V, energy is recovered within the process, so no expensive heating utility is required. The overall conversion of B is increased from 58.3 to 63.7%, and the reactor operates at 2.5 MPa instead of 7.05 MPa as in case (a). Because of the relatively small vertical driving forces and the gas–gas matches, however, the HEN costs are very high, so that annual profit when these costs are added to the expenses reduces $-292,000$ $/year. In order to consider the HEN costs, the approach suggested by Kravanja and Grossmann (1990) was applied, yielding a profit of 1,679,000 $/year. As can be seen from Table V, the operating conditions again undergo considerable changes. The most significant differences are a further increase in the overall conversion to 66.04%, elimination of the preheat of the reactor feed (gas–gas matches with small temperature driving forces), and

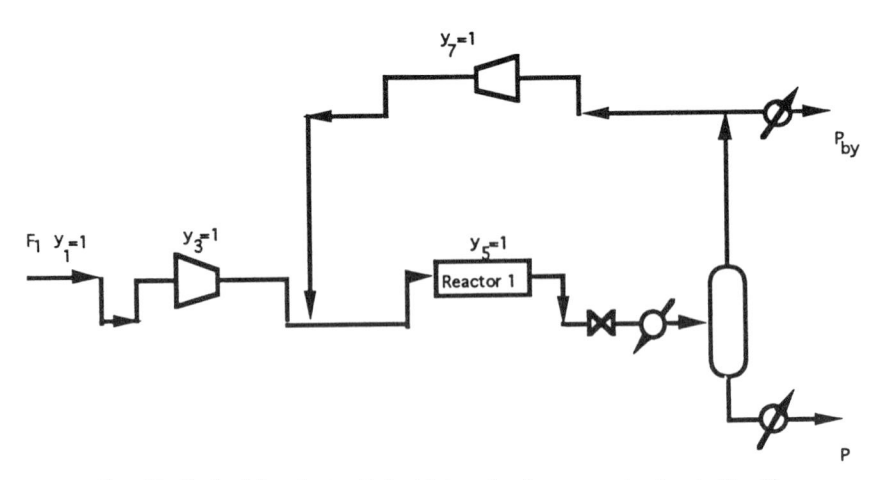

FIG. 20. Optimal flowsheet with heat integration from superstructure in Fig. 12.

TABLE V
TECHNICAL AND ECONOMIC RESULTS FOR EXAMPLE OF FIG. 20

	MINLP only	Heat integration Duran–Grossman	Heat integration HEN costs
Flows, kg-mol/s			
F1	6.176	5.648	5.451
F2	0	0	0
Purge rate, %	14.5	14.6	19.7
Reactor			
P_{in}, MPa	7.048	2.500	4.377
P_{out}, MPa	6.343	2.250	3.939
T_{out}, K	378	430	419
T_{in}, K	332	379	356
Conversion of B Per pass, %	25.5	25.4	29.4
Volume, m^3	55.7	49.1	67.7
Flash separation			
P, MPa	6.343	2.250	4.377
T_{out}, K	378	310	310
Utilities			
Electricity, MW	3.718	1.798	2.78
Heating, steam, 10^9 MJ/year	0.114	0	0
Cooling, water, 10^9 MJ/year	1.566	0.834	1.05
Other			
Overall conversion of B, %	58.92	63.7	66.04
Load of HEN, MW	54.9	71.5	48.0
Earnings, 10^3 $/year			
Product	8000	8000	8000
Byproduct	1513	1341	1309
Expenses, 10^3 $/year			
Feedstock capital investment	4632	4236	4088
HEN	1986	3695	1173
Other	1131	659	925
Electricity compress	948	459	709
Heating utility	912	0	0
Cooling utility	1096	584	735
Annual profit, 10^3 $/year			
Without HEN costs	794	3403	2852
With HEN Costs	**−1192**	**−292**	**1679**

selection of the reactor pressure at 4.377 MPa, which lies between the pressures of cases (a) and (b). Note that the HEN costs are significantly reduced (see Fig. 21) while other capital and utility costs increased (electricity and cooling) to yield a profit increase of 2,871,000 $/year when compared to case (a) where no heat integration was considered, and an increase of 1,971,000 $/year compared to case (b). This clearly shows the importance of anticipating the heat integration in the synthesis stage with both utility and investment costs.

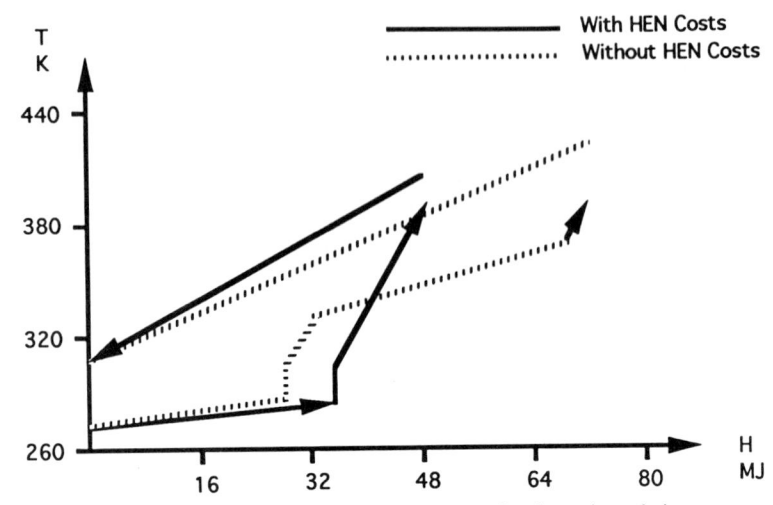

FIG. 21. Composite heating and cooling curves for alternative solutions.

As the next example, we consider the synthesis of a network of sharp separators for the specifications given in Fig. 22. The objective function considered is simply a linear function of the total flows. As shown in Quesada and Grossmann (1995a), the optimization of the corresponding superstructure corresponds to a nonconvex NLP with bilinear constraints (flows times compositions and flows times splits)—one for which a special LP underestimator function can be derived. This method, which has been automated in the program GLOBESEP developed by Bolio et al. (1994), produced the global optimum configuration given in Fig. 23. The global optimum solution with an objective value of 159.48 was determined by analyzing a total of seven nodes in the branch and bound search, solving seven LP underestimators and six NLPs. The global optimum was confirmed within a tolerance of less than 0.1%. It is interesting that the

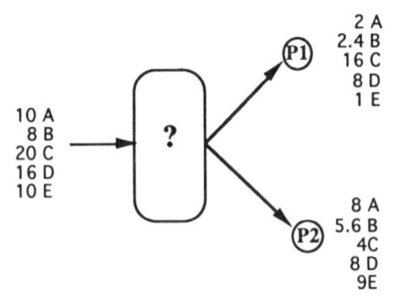

FIG. 22. Synthesis of 5-component separation network.

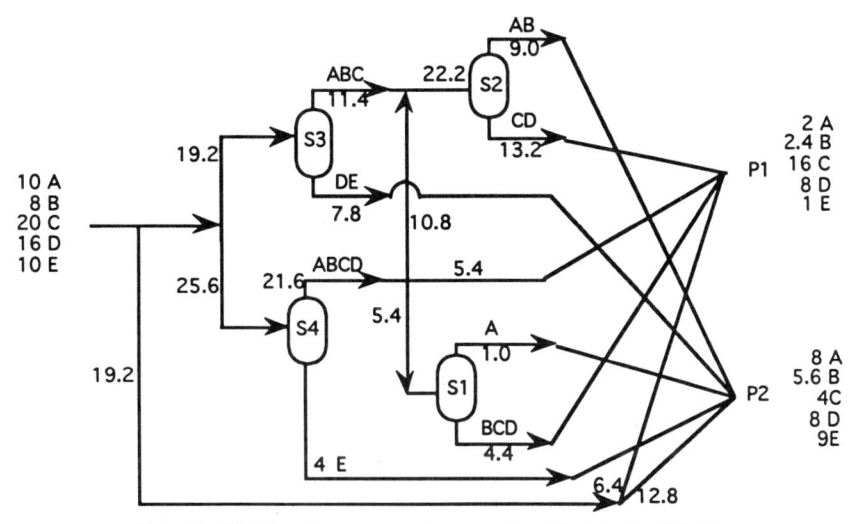

FIG. 23. Global optimum separation network with objective 159.48.

initial lower bound that is predicted is rather tight, and that there is at least a second local solution with objective 162.36. This problem, which involved 225 continuous variables, was solved with GLOBESEP in less than 10 seconds of CPU time on an IBM RS6000 using the GAMS/MINOS optimization software.

As the final example, we consider the preliminary screening procedure by Daichendt and Grossmann (1994a,b) that was applied to the MINLP model and superstructure for heat-integrated distillation sequences of Floudas and Paules (1988). The superstructure is given in Fig. 24. The problem involves the separation of a simple ternary mixture into 98%-pure components using sharp splits and adiabatic columns. The problem was aggregated into an MILP as described in the previous section. The base-case design is determined by the initial solution to the MILP, yielding a lower bound to the global optimal solution of the MINLP. Furthermore, the actual costs corresponding to the selection of units and operating conditions can readily be calculated, yielding an upper bound to the global optimal solution. The DICOPT++ implementation for solving directly the MINLP involves 18 binary variables, 77 continuous variables, and 143 constraints. This can be compared to the aggregated MILP model, which involves 18 binary variables, 117 continuous variables, and 225 constraints. The base-case solution consists of columns 1 and 3, with the distillate of column 3 (376 K) heat-integrated with the bottoms of column 1 (361 K). Low-pressure steam (421 K) is used to heat the reboiler of column 3 (396 K), while the less expensive exhaust steam (373 K) is used to supply the additional heat required

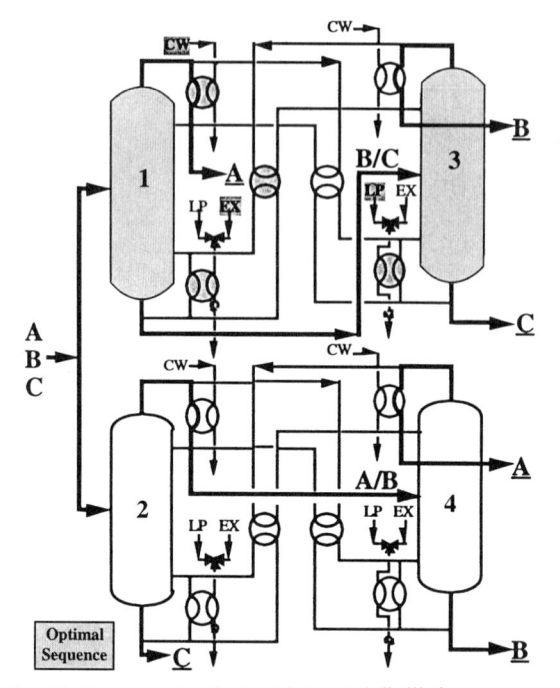

FIG. 24. Superstructure for heat-integrated distillation sequences.

for the reboiler of column 1. Cooling water (305–325 K) is used for the condenser of column 1 (342 K). The optimal sequence is shaded in Fig. 24. The lower bound to the global optimal solution is $464,500/year. The upper bound is determined to be $466,800/yr. Therefore, the global optimal solution is bounded to within 0.5%.

Solving this sequence as an NLP yields an optimal solution of $466,600/year. No feasible alternative solution to this problem could be found during preliminary screening, so the base-case design corresponds to the optimal configuration. In contrast, the full MINLP when solved with DICOPT++ converged to a suboptimal solution of $763,600/year (64% more expensive) using default NLP parameter settings, and $502,100/year (8% more expensive) using the modified NLP parameter settings. Although the DICOPT++ implementation does not converge to the same solution, we note that determining a base-case design and performing preliminary screening yielded a reduction of 41% compared to the computation time required with DICOPT++ using the default settings (8.1 s), and 51% using the modified settings (9.8 s). This example indicates the extent to which preliminary screening can be used to obtain tight bounds to the global optimal solution of the original MINLP problem, thus yielding a measure of confidence to the goodness of the solution.

IX. Concluding Remarks

We have presented an overview of algorithmic methods for process synthesis which rely on MINLP techniques. From this overall review, it is clear that considerable progress has been made over the last decade. Heuristics and thermodynamic targets, which have been dominant in the past, offer some valuable insights and motivate higher level representations for process synthesis. However, they do not provide a systematic framework for the modeling and decision making in process synthesis. In contrast, the mathematical programming approach offers a formalization and a basic modeling framework for simultaneous optimization in which trade-offs and interactions are systematically accounted for. Futhermore, the mathematical programming approach is more suitable for producing automated tools owing to the general mathematical representations that are used. At this time, we have reached a stage in which the modeling and solution of MINLP problems have become fact—especially with the increased computational power that is now available. Furthermore, the mathematical programming approach has increasingly been adopted by researchers in process systems over the last five years. This is not to say, of course, that MINLP optimization techniques are without their difficulties. Several of the outstanding problems that require attention for future work are the following:

(1) *Integrating approaches:* Clearly, a major question that remains is how to combine the heuristic search, mathematical programming and targeting approaches in such a way that the integration is conceptually consistent and rigorous, while yet exploiting the strengths of each approach. At the root of this problem lies the question of how to perform preliminary screening to eliminate alternatives, but within a rigorous multilevel design framework. In addition, the point below on aggregated models needs to be addressed to support this integration.

(2) *Developing superstructure representations at various levels of abstraction:* We should not view general design targets or aggregated models and fairly detailed process superstructures as different representations that are unrelated. Rather, they should be viewed as representations at different levels of abstraction in which the feasible space of the target model contains the feasible space of the superstructure. It should be interesting and useful to systematically aggregate detailed superstructure models, and vice-versa, to decompose target models that can be mapped to detailed flowsheets.

(3) *Generating novel superstructures for process flowsheets:* While for most homogeneous systems it is possible to postulate a general superstructure for analyzing the alternatives of interest, it is largely unclear how to systematically generate superstructures for process flowsheets involving different types of processing tasks or nonlinearities which give rise to infeasibilities that are

nontrivial to predict (e.g., azeotropic distillation). Furthermore, there is a need for considering representations that may lead to process intensification (e.g., integration of separation and reaction) and that may accommodate more readily the choice of different chemistries.

(4) *Finding new applications for process synthesis:* Most of the research work in synthesis has been aimed at conceptual flowsheets of continuous processes. There is a clear need to integrate the selection of chemical pathways in process synthesis (Douglas, 1990; Knight and McRae, 1993) as well as to consider the design of molecules (Maranas and Floudas, 1993). For example, the piping layout design (Guirardello and Swaney, 1993) has received limited attention, despite its importance. Also, work in the synthesis of batch processes is starting to emerge (Klossner and Rippin, 1984; Reklaitis, 1990; Papageorgaki and Reklaitis, 1990; Voudouris and Grossmann, 1993; Charalambides *et al.,* 1993).

(5) *Reducing combinatorial search:* In order to make mathematical programming tools useful for industrial problems, it is important that large combinatorial problems be handled effectively (this, of course, also includes large continuous models). The challenge here lies in accomplishing this goal without sacrificing optimality. New developments in MINLP algorithms are required for methods that can exploit more explicitly the logic and qualitative knowledge of a synthesis problen, making effective use of new ideas in polyhedral theory and new developments in advanced computer architectures.

(6) *Developing global optimization methods:* In most cases, the MINLP or NLP models for process synthesis involve nonconvexities in the continuous variables, which may give rise to several local or suboptimal optima. The difference between these solutions may be largely due to the expanded space of alternatives in these problems. Automating synthesis procedures that will sometimes produce poor solutions is obviously not satisfactory. Therefore, we need to develop global optimization methods that are both relevant to process synthesis problems and, preferably, rigorous in nature.

(7) *Handling of rigorous models:* In order to make synthesis techniques more applicable and relevant to industry, it is important that these techniques be either capable of explicitly handling complex process models or at least in some sense compatible with these models (e.g., in terms of bounding properties).

(8) *Synthesizing of process systems with multiple objectives:* Given the importance and potential impact of early design decisions in synthesis, it is highly desirable to develop computational frameworks that allow the evaluation and determination of trade-offs for a number of different attributes. Besides traditional economic measures, these include operability, safety, and environmental aspects.

Although these questions remain largely unanswered at this point, there is no doubt that over the next decade we will see some exciting developments along these lines.

Acknowledgments

Most of the optimization developments for process synthesis described in this paper have been due to the following Ph.D. students at Carnegie Mellon: Soterios Papoulias, Marco Duran, Chris Floudas, Mark Shelton, Gary Kocis, Terry Yee, Ramesh Raman, Mark Daichendt, Ignacio Quesada and Metin Turkay. These efforts have been complemented by collaborations with Zdravko Kravanja, J. Viswanathan and Truls Gundersen. The author is grateful to Eastman Chemical Company and to the Engineering Design Research Center for financial support of this work.

References

Achenie, L. K. E., and Biegler, L. T. "Algorithmic Synthesis of Chemical Reactor Networks Using Mathematical Programming," *Ind. Eng. Chem. Fundam.* **25**, 621 (1986).

Achenie, L. K. E., and Biegler, L. T. "Developing Targets for the Performance Index of a Chemical Reactor Network: Isothermal Systems," *Ind. Eng. Chem. Res.* **27**, 1811 (1988).

Andrecovich, M. J., and Westerberg, A. W. "A Simple Synthesis Method Based on Utility Bounding for Heat-Integrated Distillation Sequences," *AIChE J.* **31**, 363 (1985a).

Andrecovich, M. J., and Westerberg, A. W. "An MILP Formulation for Heat-Integrated Distillation Sequence Synthesis," *AIChE J.* **31**, 1461 (1985b).

Bagajewicz, M. J., and Manousiouthakis, V. "Mass/Exchange Network Representation of Distillation Networks," *AIChE J.* **38**(11), 1769 (1992).

Balakrishna, S., and Biegler, L. T. "Constructive Targeting Approaches for the Synthesis of Chemical Reactor Networks," *Ind. Eng. Chem. Res.* **31**, 300 (1992a).

Balakrishna, S., and Biegler, L. T. "Targeting Strategies for the Synthesis and Energy Integration of Nonisothermal Reactor Networks," *Ind. Eng. Chem. Res.* **31**, 2152 (1992b).

Balas, E. "Disjunctive Programming and a Hierarchy of Relaxations for Discrete Optimization Problems," *SIAM J. Alg. Disc. Methods* **6**, 466–486 (1985).

Balas, E., Ceria, S., and Cornuejols, G. "A Lift-and-Project Cutting Plane Algorithm for Mixed 0–1 Programs," *Math. Program.* **58**, 295–324 (1993).

Beaumont, N. "An Algorithm for Disjunctive Programs," *Eur. J. Oper. Res.* **48**, 362–371 (1991).

Bolio, B., Daichendt, M., Iyer, R., Yee, T., and Grossmann, I. E. "SYNHEAT: An Interactive Program for Heat Exchanger Network Synthesis." Carnegie-Mellon University, Pittsburgh, PA, 1993.

Bolio, B., Quesada, I., and Grossmann, I. E. "GLOBESEP: An Interactive Program for the Global Optimization of Separation Networks." Carnegie-Mellon University, Pittsburgh, PA, 1994.

Borchers, B., and Mitchell, J. E. "An Improved Branch and Bound Algorithm for Mixed Integer Nonlinear Programs," TIMS/ORSA Meeting (1992).

Bremicker, J. F., Papalambros, P. Y., and Loh, H. T. "Solution of Mixed-Discrete Structural Optimization with a New Sequential Linearization Model," *Comput. Struct.* **37**, 451–461 (1990).

Brooke, A., Kendrick, D., and Meeraus, A. "GAMS—A User's Guide." Scientific Press, Palo Alto, CA, 1988.
Carlberg, N. A., and Westerberg, A. W. "Temperature-Heat Diagrams for Complex Columns. 2. Method for Side Strippers and Enrichers," *Ind. Eng. Chem. Res.* **28,** 1379 (1989a).
Carlberg, N. A., and Westerberg, A. W. "Temperature-Heat Diagrams for Complex Columns. 3. Underwood's Method for the Petlyuk Configuration," *Ind. Eng. Chem. Res.* **28,** 1386 (1989b).
Cavalier, T. M., and Soyster, A. L. "Logical Deduction via Linear Programming," IMSE Working Paper 87-147. Department of Industrial and Management Systems Engineering, Pennsylvania State University, University Park, 1987.
Cavalier, T. M., Pardalos, P. M., and Soyster, A. L. "Modelling and Integer Programming Techniques applied to Propositional Calculus," *Comput. Oper. Res.* **17**(6), 561–570 (1990).
Cerda, J., and Westerberg, A. W. "Synthesizing Heat Exchanger Networks Having Restricted Stream/Stream Matches Using Transportation Problem Formulations," *Chem. Eng. Sci.* **38,** 1723 (1983).
Charalambides, M. S., Shah, N., and Pantelides, C. C. "Optimal Batch Process Synthesis," Paper No. 153c, AIChE Meeting, St. Louis, MO (1993).
Chitra, S. P., and Govind, R. "Synthesis of Optimal Serial Reactor Structures for Homogeneous Reactions. Part I: Isothermal Reactors," *AIChE J.* **31,** 177 (1985a).
Chitra, S. P., and Govind, R. "Synthesis of Optimal Serial Reactor Structures for Homogeneous Reactions. Part II: Nonisothermal Reactors," *AIChE J.* **31,** 185 (1985b).
Ciric, A. R., and Floudas, C. A. "Heat Exchanger Network Synthesis without Decomposition," *Comput. Chem. Eng.* **15,** 385–396 (1991).
Clocksin, W. F., and Mellish, C. S., "Programming in Prolog," Springer–Verlag, Heidelberg, 1981.
Colberg, R. D., and Morari, M. "Area and Capital Cost Targets for Heat Exchanger Network Synthesis with Constrained Matches and Unequal Heat Transfer Coefficients," *Comput. Chem. Eng.* **14,** 1 (1990).
Colmenares, T. R., and Seider, W. D. "Heat and Power Integration of Chemical Processes," *AIChE J.* **33,** 898 (1987).
Crowder, H., Johnson, E. L., and Padberg, M. "Solving Large-Scale Zero-One Linear Programming Problems," *Oper. Res.* **31,** 803–834 (1983).
Daichendt, M. M., and Grossmann, I. E. "Preliminary Screening for the MINLP Synthesis of Process Systems. I. Aggregation and Decomposition Techniques," *Comput. Chem. Eng.* **18,** 663 (1994a).
Daichendt, M. M., and Grossmann, I. E. "Preliminary Screening for the MINLP Synthesis of Process Systems. II. Heat Exchanger Networks," *Comput. Chem. Eng.* **18,** 679 (1994b).
Dhole, V. R., and Linnhoff, B. "Distillation Column Targets," *Proc. Eur. Symp. Comput. Aided Process Design (Escape-1)* (1992).
Ding, M., and Sargent, R. W. H. "A Combined SQP and Branch and Bound Algorithm for MINLP Optimization," Internal Report. Centre for Process Systems Engineering, Imperial Colleges London, 1992.
Doherty, M. F., and Cardarola, G. A. "Design and Synthesis of Homogeneous Azeotropic Distillations. 3. The Sequencing of Columnds for Azeotropic and Extractive Distillation," *Ind. Eng. Chem. Fundam.* **24,** 474–485 (1985).
Douglas, J. M. "A Hierarchical Decision Procedure for Process Synthesis," *AIChE J.* **31,** 353 (1985).
Douglas, J. M. "Conceptual Design of Chemical Processes." McGraw-Hill, New York, 1988.
Douglas, J. M. "Synthesis of Multistep Reaction Processes," *in* "Foundations of Computer-Aided Design" (J. J. Siirola, I. E. Grossmann, and G. Stephanopoulos, eds.). Cache-Elsevier, Amsterdam, 1990.

Duran, M. A., and Grossmann, I. E. "Simultaneous Optimization and Heat Integration of Chemical Processes," *AIChE J.* **32,** 123 (1986a).
Duran, M. A., and Grossmann, I. E. "A Mixed-Integer Nonlinear Programming Algorithm for Process Systems Synthesis," *AIChE J.* **32,** 592 (1986b).
Duran, M. A., and Grossmann, I. E. "An Outer-Approximation Algorithm for a Class of Mixed-integer Nonlinear Programs," *Math. Program.* **36,** 307 (1986c).
El-Halwagi, M., and Manousiouthakis, V. "Synthesis of Mass Exchange Networks," *AIChE J.* **35**(8), 1233 (1989a).
El-Halwagi, M., and Manousiouthakis, V. "Design and Analysis of Mass Exchange Networks with Multicomponent Targets," Paper 137f, AIChE Meeting San Francisco (1989b).
El-Halwagi, M., and Manousiouthakis, V. "Automatic Synthesis of Mass-Exchanger Networks with Single Component Targets," *Chem. Eng. Sci.* **45**(9), 2813 (1990).
Eliceche, A. M., and Sargent, R. W. H. "Synthesis and Design of Distillation Sequences," *Inst. Chem. Eng. Symp. Ser.* **61,** 1–22 (1986).
Fletcher, R., and Leyffer, S. "Solving Mixed Integer Nonlinear Programs by Outer Approximation," *Math. Program.* **66,** 327 (1994).
Floquet, P., Pibouleau, L., and Domenach, S. "Mathematical Programming Tools for Chemical Engineering Process Design Synthesis," *Chem. Eng. Process.* **23,** 1 (1988).
Floudas, C. A. "Separation Synthesis of Multicomponent Feed Streams into Multicomponent Product Stream," *AIChE J.* **33,** 540–550 (1987).
Floudas, C. A., and Aggarwal, A. "A Decomposition Strategy for Global Optimum Search in the Pooling Problem," *ORSA J. Comput.* **2,** 225–235 (1990).
Floudas, C. A., and Grossmann, I. E. "Algorithmic Approaches to Process Synthesis: Logic and Global Optimization," in "Foundations of Computer-Aided Process Design" (M. F. Doherty and L. T. Biegler eds.). Snowmass, CO, 1994.
Floudas, C. A., and Paules, G. E., IV "A Mixed-Integer Nonlinear Programming Formulation for the Synthesis of Heat-Integrated Distillation Sequences," *Comput. Chem. Eng.* **12,** 531 (1988).
Floudas, C. A., and Visweswaran, V. "A Global Optimization Algorithm (GOP) for Certain Classes of Nonconvex NLPs-I Theory," *Comput. Chem. Eng.* **4,** 1397–1417 (1990).
Floudas, C. A., Ciric, A. R., and Grossmann, I. E. "Automatic Synthesis of Optimum Heat Exchanger Network Configurations," *AIChE J.* **32,** 276 (1986).
Fonyó, Z. "Thermodynamic Analysis of Rectification. I. Reversible Model of Rectification," *Int. Chem. Eng.* **14,** 18 (1974a).
Fonyó, Z. "Thermodynamic Analysis of Rectification. II. Finite Cascade Models," *Int. Chem. Eng.* **14,** 203 (1974b).
Fonyó, Z., and Mizsey, P. "A Global Approach to the Synthesis and Preliminary Design of Integrated Total Flowsheets," Annual AIChE Meeting, Chicago (1990).
Friedler, F., Tarjan, K., Huang, Y. W., and Fan, L. T. "An Accelerated Branch and Bound Method for Process Synthesis," presented at the 4th World Congress of Chemical Engineering, Karlsruhe (1991).
Friedler, F., Tarjan, K., Huang, Y. W., and Fan, L. T. "Graph Theoretic Approach to Process Synthesis: Axioms and Theorems," *Chem. Eng. Sci.* **47,** 1973–1988 (1992).
Friedler, F., Tarjan, K., Huang, Y. W., and Fan, L. T. "Graph Theoretic Approach to Process Synthesis: Polynomial Algorithm for Maximal Structure Generation," *Comput. Chem. Eng.* **17,** 929–942 (1993).
Geoffrion, A. M. "Generalized Benders Decomposition," *J. Optim. Theory Appl.* **10**(4), 237–260 (1972).
Glasser, D., Hildebrandt, D., and Crowe, C. "A Geometric Approach to Steady Flow Reactors: The Attainable Region and Optimization in Concentration Space," *Ind. Eng. Chem. Res.* **26,** 1803 (1987).

Glavic, P., Kravanja, Z., and Homsak, M. "Heat Integration of Reactors. I. Criteria for Placement of Reactors into the Flowsheet," *Chem. Eng. Sci.* **43**, 593 (1988).

Gomez-Muñoz, A., and Seader, J. D. "Synthesis of Distillation Trains by Thermodynamic Analysis," *Comput. Chem. Eng.* **9**, 311 (1985).

Grossmann, I. E. "Mixed-Integer Programming Approach for the Synthesis of Integrated Process Flowsheets," *Comput. Chem. Eng.* **9**, 463 (1985).

Grossmann, I. E. "MINLP Optimization Strategies and Algorithms for Process Synthesis," *in* "Foundations of Computer-Aided Design" (J. J. Siirola, I. E. Grossmann, and G. Stephanopoulos, eds.). Cache-Elsevier, Amsterdam, 1990a.

Grossmann, I. E. "Mixed-Integer Nonlinear Programming Techniques for the Synthesis of Engineering Systems," *Res. Eng. Des.* **1**, 205 (1990b).

Grossmann, I. E., and Daichendt, M. M. "New Trends in Optimization-based Approaches for Process Synthesis," *in* Proceedings of Process Systems Engineering, Korea (1994).

Guirardello, R., and Swaney, R. E. "Process Piping Layout Design and Optimization," Paper No. 151g, AIChE Meeting, St. Louis, MO (1993).

Gundersen, T., and Grossmann, I. E. "Improved Optimization Strategies for Automated Heat Exchanger Network Synthesis through Physical Insights," *Comput. Chem. Eng.* **14**, 925 (1990).

Gundersen, T., and Naess, L. "The Synthesis of Cost Optimal Heat Exchanger Networks. An Industrial Review of the State of the Art," *Comput. Chem. Eng.* **12**, 503 (1988).

Gupta, O. K., and Ravindran, V. "Branch and Bound Experiments in Convex Nonlinear Integer Programming," *Manage. Sci.* **31**(12), 1533–1546 (1985).

Hendry, J. E., and Hughes, R. R. "Generating Separation Process Flowsheets," *Chem. Eng. Prog.* **68**, 69 (1972).

Hendry, J. E., Rudd, D. F., and Seader, J. D. "Synthesis in the Design of Chemical Processes," *AIChE J.* **19**, 1 (1973).

Hildebrandt, D., Glasser D., and Crowe, C. "Geometry of the Attainable Region Generated by Reaction and Mixing: With and without Constraints," *Ind. Eng. Chem. Res.* **29**, 49 (1990).

Hlavacek, V. "Synthesis in the Design of Chemical Processes," *Comput. Chem. Eng.* **2**, 67–75 (1978).

Horst, R. "Deterministic Method in Constrained Global Optimization: Some Recent Advances and Fields of Application," *Nav. Res. Logistics* **37**, 433–471 (1990).

Horst, R., and Tuy, T. "Global Optimization: Deterministic Approaches." Springer-Verlag, Berlin and New York, 1990.

Ichikawa, A., and Fan, L. T. "Optimal Synthesis of Process Systems," Necessary Condition for Optimal System and Its Use in Synthesis of Systems," *Chem. Eng. Sci.* **28**, 357 (1973).

Kakhu, A. I., and Flower, J.R. "Synthesising Heat-Integrated Distillation Sequences Using Mixed Integer Programming," *Chem. Eng. Res. Des.* **66**, 241 (1988).

Kelley, J. E., Jr., 'The Cutting-Plane Method for Solving Convex Programs," *J. SIAM* **8**, 703–712 (1960).

Kirkwood, R. L., Locke, M. H., and Douglas, J. M. "A Prototype Expert System for Synthesizing Chemical Process Flowsheets," *Comput. Chem. Eng.* **12**, 329 (1988).

Klossner, J., and Rippin, D. W. T. "Combinatorial Problems in the Design of Multiproduct Batch Plants-Extensions to Multiplant and Partly Parallel Operations," AIChE Meeting, San Francisco (1984).

Knight, J. R., and Doherty, M. F. "Optimal Design and Synthesis of Homogeneous Azeotropic Distillation Sequences," *Ind. Eng. Chem. Res.* **28**, 564–572 (1989).

Knight, J. R., and McRae, G. J. "An Approach to Process Integration based on the Choice of the System Chemistry," Paper No. 153d, AIChE Meeting, St. Louis, MO (1993).

Kocis, G. R., and Grossmann, I. E. "Relaxation Strategy for the Structural Optimization of Process Flow Sheets," *Ind. Eng. Chem. Res.* **26**, 1869 (1987).

Kocis, G. R., and Grossmann, I. E. "Global Optimization of Nonconvex Mixed-Integer Nonlinear Programming (MINLP) Problems in Process Synthesis," *Ind. Eng. Chem. Res.* **27,** 1407 (1988).

Kocis, G. R., and Grossmann, I. E. "Computational Experience with DICOPT Solving MINLP Problems in Process Systems Engineering," *Comput. Chem. Eng.* **13,** 307 (1989a).

Kocis, G. R., and Grossmann, I. E. "A Modelling and Decomposition Strategy for the MINLP Optimization of Process Flowsheets," *Comput. Chem. Eng.* **13,** 797 (1989b).

Koehler, J., Aguirre, P., and Blass, E. "Evolutionary Thermodynamic Synthesis of Zeotropic Distillation Sequences," *Gas Separ. Purif.* **6,** 4153 (1992).

Kokossis, A. C., and Floudas, C. A. "Synthesis of Isothermal Reactor-Separator-Recycle Systems," *Chem. Eng. Sci.* **46,** 1361–1383 (1991).

Kravanja, Z., and Grossmann, I. E. "PROSYN, An MINLP Process Synthesizer," *Comput. Chem. Eng.* **14,** 1363 (1990).

Kravanja, Z., and Grossmann, I. E. "Recent Developments in PROSYN—A Topology MINLP Synthesizer," *Comput. Chem. Eng.* (1994).

Lang, Y. D., Biegler, L. T., and Grossmann, I. E. "Simultaneous Optimization and Heat Integration with Process Simulators," *Comput. Chem. Eng.* **12,** 311 (1988).

Laroche, L., Bekiaris, N., Andersen, H. W., and Morari, M. "Homogeneous Azeotropic Distillation: Separability and Flowsheet Synthesis," *Ind. Eng. Chem. Res.* **31,** 2190–2209 (1992).

Lee, K. F., Masso, A. H., and Rudd, D. F. "Branch and Bound Synthesis of Integrated Process Designs," *Ind. Eng. Chem. Fundam.* **9,** 48 (1970).

Leyffer, S. "Deterministic Methods for Mixed-Integer Nonlinear Programming," Ph.D. Thesis, Department of Mathematics and Computer Science, University of Dundee, Dundee (1993).

Linnhoff, B. "Pinch Analysis—A State of the Art Overview," *Trans. Inst. Chem. Eng.* **71**(a), 503–522 (1993).

Linnhoff, B., and Hindmarsh, E. "The Pinch Design Method of Heat Exchanger Networks," *Chem. Eng. Sci.* **38,** 745 (1983).

Linnhoff, B. et al., "User Guide on Process Integration for the Efficient Use of Energy." Inst. Chem. Eng., Rugby, 1982.

Magnanti, T. L., and Wong, R. T. "Acclerated Benders Decomposition: Algorithm Enhancement and Model Selection Criteria," *Oper. Res.* **29,** 464–484 (1981).

Mahalec, V., and Motard, R. L. "Procedures for the Initial Design of Chemical Processing Systems," *Comput. Chem. Eng.* **1,** 57 (1977a).

Mahalec, V., and Motard, R. L. "Evolutionary Search for an Optimal Limiting Process Flowsheet," *Comput. Chem. Eng.* **1,** 149 (1977b).

Manousiouthakis, V., and Sourlas, D. "A Global Optimization Approach to Rationally Constrained Rational Programming," *Chem. Eng. Commun.* **115,** 127–147 (1992).

Maranas, C. D., and Floudas, C. A. "Global Minimum Potential Energy Conformation of Small Molecules," Paper No. 151d, AIChE Meeting, St. Louis, MO (1993).

Masso, A. H., and Rudd, R. D. "The Synthesis of System Designs. II: Heuristic Structuring," *AIChE J.* **15,** 10 (1969).

McCormick, G. P. "Computability of Global Solutions to Factorable Nonconvex Programs: Part I. Convex Underestimating Problems," *Math. Program.* **10,** 146–175 (1976).

Nabar, S. and Schrage, L. "Modelling and Solving Nonlinear Integer Programming Problems," presented at Annual AIChE Meeting, Chicago (1991).

Nath, R., and Motard, R. L. "Evolutionary Synthesis of Separation Processes," *AIChE J.* **27,** 578 (1981).

Nemhauser, G. L., and Wolsey, L. A. "Integer and Combinatorial Optimization." Wiley (Interscience), New York, 1988.

Nishida, N., Stephanopoulos, G., and Westerberg, A. W. "A Review of Process Synthesis," *AIChE J.* **27,** 321 (1981).

Omtveit, T., and Lien, K. "Reactor System Design Revisited," Paper No. 153b, AIChE Meeting, St. Louis, MO (1993).
Papageorgaki, S., and Reklaitis, G. V. "Optimal Design of Multipurpose Batch Plants. I. Problem Formulation," *Ind. Eng. Chem. Res.* **29**, 2054 (1990).
Papalexandri, K. P., Pistikopoulos, E. N., and Floudas, C. A. "Mass Exchange Networks for Waste Minimization: A Simultaneous Approach," Paper No. 153e, AIChE Meeting, St. Louis, MO (1993).
Papoulias, S. A., and Grossmann, I. E. "A Structural Optimization Approach in Process Synthesis. Part I. Utility Systems," *Comput. Chem. Eng.* **7**, 695 (1983a).
Papoulias, S. A., and Grossmann, I. E. "A Structural Optimization Approach in Process Synthesis. Part II. Heat Recovery Networks," *Comput. Chem. Eng.* **7**, 707 (1983b).
Papoulias, S. A., and Grossmann, I. E. "A Structural Optimization Approach in Process Synthesis. Part III. Total Processing Systems," *Comput. Chem. Eng.* **7**, 723 (1983c).
Petlyuk, F. B., Platonov, V. M., and Slavinski, D. M. "Thermodynamically Optimal Method for Separating Multicomponent Mixtures," *Int. Chem. Eng.* **5**, 555 (1965).
Pinto, J. M., and Grossmann, J. E., "Optimal Cyclic Scheduling of Multistage Continuous Multiproduct Plants," *Comput Chem. Eng.* **18**, 797–816 (1994).
Powers, G. J., "Heuristic Synthesis in Process Development," *Chem. Eng. Prog.* **68**, 88 (1972).
Quesada, I., and Grossmann, I. E. "An LP/NLP Based Branch and Bound Algorithm for Convex MINLP Optimization Problems," *Comput. Chem. Eng.* **16**, 937–947 (1992).
Quesada, I., and Grossmann, I. E. "Global Optimization Algorithm for Heat Exchanger Networks," *Ind. Eng. Chem. Res.* **32**, 487 (1993).
Quesada, I., and Grossmann, I. E. "Global Optimization Algorithm of Process Networks with Multicomponent Flows," *Comput. Chem. Eng.* **19**, 1219–1242 (1995a).
Quesada, I., and Grossman, I. E. *J. Global Optim.* **6**, 39–76 (1995b).
Raman, R., and Grossmann, I. E. "Relation Between MILP Modelling and Logical Inference for Chemical Process Synthesis," *Comput. Chem. Eng.* **15**, 73 (1991).
Raman, R., and Grossmann, I. E. "Symbolic Integration of Logic in Mixed Integer Linear Programming Techniques for Process Synthesis," *Comput. Chem. Eng.* **17**, 909 (1993).
Raman, R., and Grossmann, I. E. "Modelling and Computational Techniques for Logic Based Integer Programming," *Comput. Chem. Eng.* **18**, 563–578 (1994).
Rathore, R. N. S., van Wormer, K. A., and Powers, G. J. "Synthesis Strategies for Multicomponent Separation Systems with Energy Integration," *AIChE J.* **20**, 491 (1974a).
Rathore, R. N. S., van Wormer, K. A., and Powers, G. J. "Synthesis of Distillation Systems with Energy Integration," *AIChE J.* **20**, 940 (1974b).
Reklaitis, G. V. "Progress and Issues in Computer-Aided Batch Process Design," *in* "Foundations of Computer-Aided Design" (J. J. Siirola, I. E. Grossmann, and G. Stephanopoulos, eds.). Cache-Elsevier, Amsterdam, 1990.
Rigg, T. J. "Synthesis of Separation Systems," A Preliminary Report for the Degree of Doctor of Philosophy in Chemical Engineering, University of Wisconsin-Madison (1991).
Rippin, D. W. T. "Introduction: Approaches to Chemical Process Synthesis," *in* "Foundations of Computer-Aided Design" (J. J. Siirola, I. E. Grossmann, and G. Stephanopoulos, eds.). Cache-Elsevier, Amsterdam 1990.
Rudd, D. F., "The Synthesis of System Designs. I. Elementary Decomposition Strategy," *AIChE J.* **14**, 343 (1968).
Rudd, D. F., and Watson, C . C. "Strategy of Process Engineering." Wiley, New York, 1968.
Sahinidis, N. V. "Accelerating Branch and Bound in Continuous Global Optimization," Paper MA 36.2, TIMS/ORSA Meeting, Phoenix, AZ (1993).
Sahinidis, N. V., and Grossmann, I. E. "MINLP Model for Cyclic Multiproduct Scheduling on Continuous Parallel Lines," *Comput. Chem. Eng.* **15**, 85 (1991a).
Sahinidis, N. V., and Grossmann, I. E. "Convergence Properties of Generalized Benders Decomposition," *Comput. Chem. Eng.* **15**, 481 (1991b).

Sargent, R. W. H., and Gaminibandara, K. "Introduction: Approaches to Chemical Process Synthesis," *in* "Optimization in Action" (L. C. W. Dixon, ed.). Academic Press, London, 1976.
Schrijver, A. "Theory of Linear and Integer Programming." Wiley, New York, 1986.
Seader, J. D., and Westerberg, A. W. "A Combined Heuristic and Evolutionary Strategy for Synthesis of Simple Separation Sequences," *AIChE J.* **23,** 951 (1977).
Shelton, M. R., and Grossmann, I. E. "Optimal Synthesis of Integrated Refrigeration Systems. I: Mixed-integer Programming Model," *Comput. Chem. Eng.* **10,** 445 (1986a).
Shelton, M. R., and Grossmann, I. E. "Optimal Synthesis of Integrated Refrigeration Systems. II: Implicit Enumeration Algorithm," *Comput. Chem. Eng.* **10,** 461 (1986b).
Siirola, J. J., and Rudd, D. F. "Computer-Aided Synthesis of Chemical Process Designs," *Ind. Eng. Chem. Fundam.* **10,** 353 (1971).
Siirola, J. J., Powers, G. J., and Rudd, D. F. "Synthesis of System Designs, III: Toward a Process Concept Generator," *AIChE J.* **17,** 677 (1971).
Smith, R., and Linnhoff, B. "The Design of Separators in the Context of Overall Processes," *Chem. Eng. Res. Des.* **66,** 195 (1988).
Stephanopoulos, G., and Westerberg, A. W. "Studies in Process Synthesis, II. Evolutionary Synthesis of Optimal Process Flowsheets," *Chem. Eng. Sci.* **31,** 195 (1976).
Stichlmair, J., Fair, J., and Bravo, J. L. "Separation of Azeotropic Mixtures via Enhanced Distillation," *Chem. Eng. Prog.* **85,** 63–69 (1989).
Swaney, R. E. "Global Solution of Algebraic Nonlinear Programs," Paper No.22f, AIChE Meeting, (1990).
Terranova, B. E., and Westerberg, A. W. "Temperature-Heat Diagrams for Complex Columns. 1. Intercooled/Interheated Distillation Columns," *Ind. Eng. Chem. Res.* **28,** 1374 (1989).
Turkay, M., and Grossmann, I.E. "A Logic Based Outer-Approximation Algorithm for MINLP Optimization of Process Flowsheets," AIChE Annual Meeting, San Francisco (1994).
Umeda, T., Hirai, A., and Ichikawa, A. "Synthesis of Optimal Processing System by an Integrated Approach," *Chem. Eng. Sci.* **27,** 795 (1972).
Umeda, T., Harada, T., and Shiroko, K. "A Thermodynamic Approach to the Synthesis of Heat Integration Systems in Chemical Processes," *Comput. Chem. Eng.* **3,** 273 (1979).
Van Dongen, D. B., and Doherty, M. F. "Design and Synthesis of Homogeneous Azeotropic Distillation. 1. Problem Formulation for Single Column," *Ind. Eng. Chem. Fundam.* **24,** 454 (1985).
Van Roy, T. J., and Wolsey, L. A. "Solving Mixed Integer Programs by Automatic Reformulation," *Oper. Res.* **35,** 45–57 (1987).
Viswanathan, J., and Grossmann, I. E. "A Combined Penalty Function and Outer-Approximation Method for MINLP Optimization," *Comput. Chem. Eng.* **14,** 769 (1990).
Viswanathan, J., and Grossmann, I. E. "Optimal Feed Locations and Number of Trays for Distillation Columns with Multiple Feeds," *Ind Eng. Chem. Res.* **32,** 2942–2949 (1993).
Visweswaran, C., and Floudas, C. A. "A Global Optimization Algorithm (GOP) for Certain Classes of Nonconvex NLPs. II. Application of Theory and Test Problems," *Comput. Chem. Eng.* **14**(2), 1419–1434 (1990).
Voudouris, V. T., and Grossmann, I. E. "Optimal Synthesis of Multiproduct Batch Plants with Cyclic Scheduling and Inventory Considerations," *Ind. Eng. Chem. Res.* **32,** 1962 (1993).
Wahnschafft, O. M., Jurain, T. P., and Westerberg, A. W. "SPLIT: A Separation Process Designer," *Comput. Chem. Eng.* **15,** 565–581 (1991).
Wahnschafft, O. M., LeRudulier, J. P., Blania, P., and Westerberg, A. W. "SPLIT: II. Automated Synthesis of Hybrid Liquid Separation Systems," *Comput. Chem. Eng.* **16,** S305–S312 (1992).
Wahnschafft, O. M., Wareck, J. S., and Ahmad, S. "Distillation System Synthesis Based on Thermodynamic Targeting," Paper No. 152f, AIChE Meeting, St. Louis, MO (1993).
Wehe, R. R., and Westerberg, A. W. "An Algorithmic Procedure for the Synthesis of Distillation Sequences with Bypass," *Comput. Chem. Eng.* **11,** 619–627 (1987).

Westerberg, A. W. "The Synthesis of Distillation Based Separation," *Comput. Chem. Eng.* **9,** 421 (1985).

Westerlund, T., and Pettersson, F. "A Cutting Plane Method for Solving Convex MINLP Problems," Report 92-124-A. Process Design Laboratory, Abo Akademi, 1992.

Williams, H. P. "Model Building in Mathematical Programming," Wiley, Chichester, 1988.

Yee, T. F., Grossmann, I. E., and Kravanja, Z. "Simultaneous Optimization Models for Heat Integration. I. Energy and Area Targeting," *Comput. Chem. Eng.* **14,** 1151 (1990a).

Yee, T. F., and Grossmann, I. E. "Simultaneous Optimization Models for Heat Integration. II: Heat Exchanger Network Synthesis," *Comput. Chem. Eng.* **14,** 1165 (1990b).

Yee, T. F., Grossmann, I. E., and Kravanja, Z. "Simultaneous Optimization Models for Heat Integration. III. Optimization of Process Flowsheets and Heat Exchanger Networks," *Comput. Chem. Eng.* **14,** 1185 (1990c).

Yuan, X., Zhang, S., Pibouleau, L., and Domenech, S. "Une Méthode d'optimisation Nonlinéare en Variables pour la Conception de Procèdés," *Oper. Res.* **22,** 331 (1988).

Chemical Reactor Network Targeting and Integration: An Optimization Approach

Subash Balakrishna* and Lorenz T. Biegler

Chemical Engineering Department
Carnegie Mellon University
Pittsburgh, Pennsylvania

I. Introduction	248
II. Geometric Concepts for Attainable Regions	250
III. Reactor Network Synthesis: Isothermal Systems	254
A. The Segregated-Flow Approximation	254
B. Sufficiency Conditions for Segregated-Flow Networks	256
C. Optimization Formulations for Reactor Synthesis	258
D. Example Problems	262
IV. Reactor Network Synthesis: Nonisothermal Systems	265
A. Nonisothermal Model Formulation	265
B. Reactor Extensions	269
V. Energy Integration of Reactor Networks	274
A. Model Formulation for Energy Integration	274
B. Extensions from the Targeting Model	279
C. Energy-Integration Example	280
VI. Simultaneous Reaction, Separation, and Energy System Synthesis	283
A. Combined Reaction–Separation Model	284
B. Unified Formulation for Optimal Energy Utilization	290
C. Example Problem	292
VII. Summary and Conclusions	295
Acknowledgments	298
References	298

While synthesis strategies are well developed for energy integration and separation systems, relatively little work has been done in synthesizing reactor networks. This is due to the complex and nonlinear behavior of the reacting system, coupled with the combinatorial aspects inherent in all synthesis problems. This paper provides a brief summary of work to date in this area, focusing on targeting approaches for reactor network synthesis.

*Currently with Mobil Corporation, Beaumont, Texas.

As with energy integration, reactor network targeting seeks to describe the performance of the network without its explicit construction. Once this description has been obtained, a network is then determined that is guaranteed to match this target. To achieve these objectives, we rely on recent concepts of attainable regions and extend these to simple optimization formulations. With these formulations, network targets can be achieved for isothermal and nonisothermal systems with complex kinetics.

Moreover, these optimization formulations can easily be coupled to other process systems such as heat integration and separation sequences. As a result, they provide a framework for integrated process design. This strategy is illustrated with detailed examples. Finally, limitations of the current approach are summarized and topics for future work are outlined.

I. Introduction

Synthesis of chemical reactor networks can be defined by the following problem statement:

Given the reaction stoichiometry and rate laws, a desired objective and system constraints, what is the optimal reactor network structure? What is the flow pattern of this network? Where should mixing or segregatation occur in this network? Where should heating and cooling be applied in this network?

Despite significant research, both in reactor modeling and analysis and in the design of specific reactors, relatively little work has been reported in reactor network synthesis. As noted in the process synthesis review of Nishida *et al.* (1981), other areas of process synthesis, including heat integration and separation synthesis, have advanced much more than reactor networks. Several reasons account for this. First, reacting systems are typically more difficult to model, and they generally have more diverse elements than energy or separation systems. This is typified by an important (and expensive) experimental component. Moreover, given the resource constraints in process development, there is often little opportunity to find an optimal reactor network by developing a detailed reactor model or by investigating many possible alternatives.

Nevertheless, as noted by Douglas (1988), the reactor system is often the heart of the chemical process, as it dictates the downstream processes (e.g., separation and waste treatment) and strongly influences both the recycle and separation structures as well as the energy network. Despite this, the general approach is to design the reactor system "in isolation" and then to design the remaining subsystems. As will be seen in this paper, these "sequential" approaches are clearly suboptimal, and large improvements in the overall process can be made through process integration. It is also hoped that these case studies will encourage the development of more detailed models for reactor targeting at the design stage.

Previous work in reactor network synthesis can be summarized from various sources. First, most textbooks on reactor design (e.g., Levenspiel, 1962; Kramers and Westerterp, 1963; Froment and Bischoff, 1979; Fogler, 1992) present graphical and heuristic approaches that emphasize the effects of mixing for various reaction orders and heating for exothermic and endothermic reactions. These methods are then used to guide the selection of ideal reactors [e.g., plug flow (PFR) and continuous stirred tank (CSTR) reactors]. However, these approaches are usually limited to single reactions or simple series/parallel cases. While these heuristics can also be applied to more complex problems, a quantitative evaluation of trade-offs is often required and considerably more effort is introduced if an optimal network is desired.

Early studies that applied systematic optimization approaches can be classified as optimal control studies and superstructure optimization. Studies based on the former include the use of adjoint variables and sensitivities by Horn and Tsai (1967), the development of adjoint networks for reacting streams by Jackson (1968), and its extension to include CSTRs by Ravimohan (1971). Paynter and Haskins (1970) attempted to formulate the selection of CSTRs and PFRs through optimal controls for axial dispersion (ADR) models, and Waghmere and Lim (1981) related CSTR/PFR selection to optimal profiles in batch and semibatch reactor systems. Achenie and Biegler (1986) modified the approach of Jackson to use ADRs as well in order to aid in CSTR/PFR selection. Later, they developed an optimal control formulation (Achenie and Biegler, 1988) based on transfer from segregated flow to maximum mixing compartments. In addition to these general network studies, there are numerous optimal control studies related to specific reactor systems (e.g., Glasser *et al.*, 1987; Jackson, 1968; Dyson and Horn, 1967; Narasimhan, 1969).

The structural optimization of superstructures can probably be traced to Aris (1961), who applied dynamic programming to a series of reactors. This approach was advanced and summarized by Hartmann and Kaplick (1990) and extended through optimization of serial recycle reactors by Chitra and Govind (1985). More efficient uses of nonlinear programming to solve superstructure problems were also developed by Pibouleau *et al.* (1988) and Achenie and Biegler (1990). Among the superstructure approaches, several studies by Kokossis and Floudas (1989, 1990, 1991) applied sophisticated mixed-integer nonlinear programming (MINLP) strategies to a large reactor network. Modeled as CSTRs or a series of sub-CSTRs that represented PFRs, this MINLP problem was capable of handling arbitrary kinetics for both isothermal and nonisothermal cases. In addition, these authors demonstrated the interaction of this network with other parts of the process flowsheet. Finally, they also incorporated stability constraints within the MINLP problem in order to avoid the selection of unstable network structures.

While these optimization-based approaches have yielded very useful results for reactor networks, they have a number of limitations. First, proper problem definition for reactor networks is difficult, given the uncertainties in the process and the need to consider the interaction of other process subsystems. Second, all of the above-mentioned studies formulated nonconvex optimization problems for the optimal network structure and relied on local optimization tools to solve them. As a result, only locally optimal solutions could be guaranteed. Given the likelihood of extreme nonlinear behavior, such as bifurcations and multiple steady states, even locally optimal solutions can be quite poor. In addition, superstructure approaches are usually plagued by the question of completeness of the network, as well as the possibility that a better network may have been overlooked by a limited superstructure. This problem is exacerbated by reaction systems with many networks that have identical performance characteristics. (For instance, a single PFR can be approximated by a large train of CSTRs.) In most cases, the simpler network is clearly more desirable.

For the above reasons, we instead develop in this paper an approach based on reactor network targeting. Here, we develop simplified optimization formulations that accurately predict network performance prior to their construction. This approach is based on geometric concepts for attainable regions (AR) of the reactor network initially developed by Glasser *et al.* (1987). In Section II, we describe and summarize the geometric properties that relate to the attainable region. In Section III, these are applied to develop optimization formulations for isothermal systems, and Section IV extends these formulations to nonisothermal system. Sections V and VI then describe the integration of reactor targeting optimization problems with energy and separation systems, respectively. Finally, Section VII summarizes the paper and outlines areas for future work.

II. Geometric Concepts for Attainable Regions

For chemical reactor networks, the attainable region concept was first defined by Horn (1964), who noted that

> [V]ariables such as recycle flow rate and composition of the product form a space which in general can be divided into an attainable region and a nonattainable region. The attainable region corresponds to the totality of physically possible reactors. . . . Once the border is known the optimum reactor corresponding to a certain environment can be found by simple geometric considerations.

To illustrate this concept, consider the attainable region for the following series reaction:

$$A \to B \to C$$

which can be defined in the space of concentrations for A and B as shown in Fig. 1. Here, we assume a fixed feed and initial temperature and trajectories that are determined entirely by the state equations for concentration. (This is true in steady state for isothermal or adiabatic systems.) With this attainable region, we clearly see that point F and the line segment GH represent the maximum concentration of B and maximum selectivity of B to C, respectively. These can be achieved by the reactor networks needed to construct the attainable region boundary. Moreover, if a more complex objective represented in terms of C_A and C_B yields an interior optimum point, then again this point can be achieved by any linear combination of the boundary structures.

We construct the attainable region by noting that the concentration space is a vector field with a rate vector (e.g., in Fig. 1, $dC_B/dC_A = R_B/R_A$) defined at each point. Moreover, we are not restricted to concentration space, but can consider any other variable that satisfies a linear conservation law (e.g., mass fractions, residence time, energy, and temperature—for constant heat capacity and density). The attainable region is an especially powerful concept; once it is known, performance of the network can often be determined without the network itself.

More recently, Glasser *et al.* (1987) developed geometric properties of the attainable region along with a constructive approach for determining this region. They defined the necessary conditions for the attainable region as follows:

(1) The attainable region (AR) must be convex. Any point created by a linear combination of two points in the AR must be in the AR, as it can be

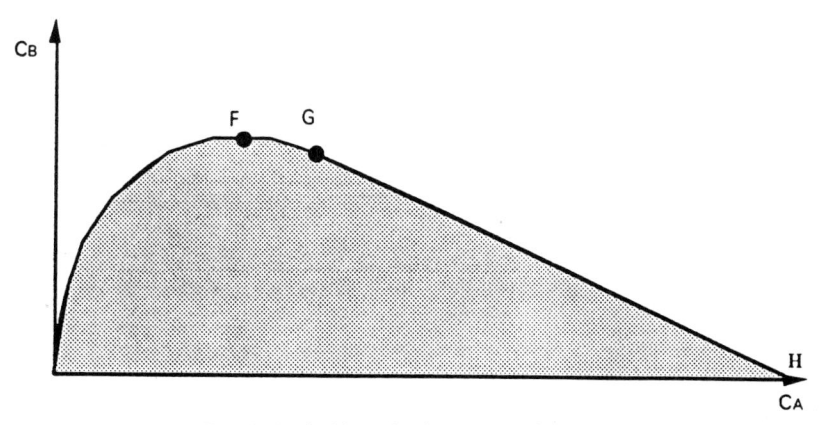

FIG. 1. Attainable region in concentration space.

created by mixing these two points. Moreover, this property ensures that the AR cannot be extended by further mixing.
(2) No reaction vector in the AR boundary can point out of the AR. If this were the case, the AR could be extended further by PFR reactors, which have trajectories that are always tangent to the rate vectors.
(3) No reversed reaction vector in the complement of the AR can point back into the AR. Note that a CSTR can be represented in the AR by a line with ends at the feed and outlet concentrations, with the rate vector at the CSTR outlet collinear with this line. Thus, this condition ensures that the AR cannot be extended further by a CSTR.

These properties hold for all dimensions and are, in fact, stronger than the simple exclusion of CSTRs, PFRs, and mixing. Hildebrandt (1989) proved that an AR closed to further extension by PFRs and CSTRs is also closed to extension by recycle PFRs (RRs) as long as the AR is not constrained in concentration. Hildebrandt *et al.* (1990) also showed how these properties could be applied to systems with nonconstant densities and heat capacities.

These concepts can also be shown to apply to the more general case where the reacting system is described by

$$dC/d\tau = R(C) + q(C_0 - C)$$

where C_0 is an arbitrary concentration and $q(\tau)$ represents the sidestream flowrate. In addition to the limiting cases of PFRs ($q = 0$), CSTRs ($dC/d\tau = 0$), and pure mixing ($q = \infty$), we can also consider differential sidestream reactors (DSRs), the maximum mixed case discussed by Zwietering (1959). Finally, Hildebrandt and Feinberg (1992) have shown that the AR itself consists of PFR surfaces and straight-line segments. If the intersection of these surfaces and straight lines—called a *connector*—is smooth, it defines a DSR with a feed point at C_0.

If the problem can be modeled in two concentration dimensions, the connector is simply a point and this line–surface intersection can be represented by a simple CSTR at C_0. Thus, DSRs are not necessary for the construction of a two-dimensional AR. In fact, two-dimensional systems allow for a straightforward algorithm for the construction of the AR, based on tracing alternating PFR and CSTR trajectories until the necessary conditions are satisfied. Here, we begin with a PFR trajectory from the feed point and trace it to its equilibrium point. Any nonconvexities in the region are then filled in with straight-line segments to form a convex hull. If any rate vectors point out of the straight line segments, we know that the AR can be extended by a CSTR; we thus find a CSTR with feed on the PFR trajectory that extends the AR the most. Any nonconvexities in this trajectory are again filled in with line segments, and we continue until there are no further rate vectors pointing out from the straight-line segments. From the outlet of the (boundary) CSTR, we then trace the PFR trajectory to its equilibrium point, continue to fill in nonconvexities, and apply the above steps until the necessary AR conditions are satisfied.

Glasser *et al.* (1987) and Hildebrandt *et al.* (1990) demonstrated this two-dimensional approach on a number of small reactor network problems, with better results than previously reported. Moreover, Omtveit and Lien (1993) were able to consider higher-dimensional problems as well through projections in concentration space that allow a complete two-dimensional representation. These projections were accomplished through the principle of reaction invariants (Fjeld *et al.*, 1974) and the imposition of system specific constraints.

For three dimensions, the AR construction algorithm is similar to the one described above—with the added possibility that we can find a (one-dimensional) connector on the AR that is described by a DSR. Glasser *et al.* (1992) defined conditions under which DSRs appear on the AR along with a direct method for finding the feed addition rate, q. While the conditions for DSRs appear to occur infrequently, examples have been constructed in the space of conversion, temperature, and residence time where the DSR was a prominent part of the AR. Nevertheless, Hildebrandt and co-workers conclude that most ARs will consist only of CSTR and PFR surfaces. In dealing with n-dimensional problems, Hildebrandt and Feinberg noted that the AR boundary is defined by line segments and PFR trajectories, with at most n structures needed to define a point on the AR boundary and $n + 1$ structures needed to define an interior point of the AR. Thus, for three-dimensional problems, at most three parallel structures (PFRs, CSTRs, DSRs) are needed to define any AR boundary point.

Finally, the AR concept can be extended beyond regions that are closed to further mixing and reaction. Godorr *et al.* (1994) have also considered multirate processes where, for example, a mixture of catalysts will lead to a superposition of two separate reaction mechanisms. These authors also considered operations of mixing, condensing, and boiling as well as separation coupled with reaction. Here, they modified the concept of connectors to accommodate multiple rates and hence create more appropriate attainable regions. However, the graphical procedure for constructing ARs has limitations beyond three dimensions, as it requires an inspection of all the boundary points. Thus, the attainable region concepts need to be combined with more powerful, higher-dimensional search procedures. Hildebrandt and Biegler (1994) presented a discussion of these concepts and the role of optimization formulations for higher-dimensional search and for integration of the reactor targeting procedure with other process subsystems. In the next sections we further develop and review these AR optimization formulations, based on the recent work of Balakrishna and Biegler (1992a, 1993).

III. Reactor Network Synthesis: Isothermal Systems

In this section, we develop a simple and efficient formulation for target-based reactor synthesis for homogeneous, constant-density reacting systems. As described in Section I, previous superstructure and target-based approaches have many limitations. In superstructure-based approaches, the solution is only as rich as the superstructure chosen; moreover, these approaches usually suffer from local and nonunique solutions which are characteristic of reactor networks. By combining AR concepts and optimization formulations, we instead create targets for this network through the solution of simple optimization problems. Unlike the geometric approach to finding an AR, optimization approaches, in principle, do not have a dimensionality limitation.

A. The Segregated-Flow Approximation

Given the reaction stoichiometry and rate laws for an isothermal system, a simple representation for targeting of reactor networks is the segregated-flow model (see, e.g., Zwietering, 1959). A schematic of this model is shown in Fig. 2. Here, we assume that only molecules of the same age, t, are perfectly mixed and that molecules of different ages mix only at the reactor exit. The performance of such a model is completely determined by the residence time distribution function, $f(t)$. By finding the optimal $f(t)$ for a specified reactor network objective, one can solve the synthesis problem in the absence of mixing.

The isothermal formulation for maximizing the performance index in segregated flow is given by

$$\underset{f(t)}{\text{Max}} \quad J(X_{\text{exit}}, \tau)$$

$$\frac{dX_{\text{seg}}}{dt} = R(X_{\text{seg}})$$

$$X_{\text{seg}}(0) = X_0 \quad \quad \text{(P1)}$$

$$X_{\text{exit}} = \int_0^{t_{\text{max}}} f(t) \, X_{\text{seg}}(t) \, dt$$

$$\int_0^{t_{\text{max}}} t \, f(t) \, dt = \tau$$

$$\int_0^{t_{\text{max}}} f(t) \, dt = 1$$

CHEMICAL REACTOR NETWORK TARGETING AND INTEGRATION 255

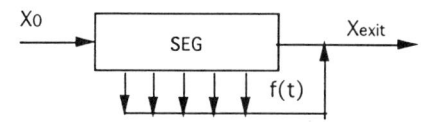

FIG. 2. Segregated-flow model.

Here, X_{seg} is the dimensionless concentration vector (normalized, e.g., by a feed concentration), $R(X)$ is the corresponding rate vector, and X_{exit} is the dimensionless output concentration of the segregated flow system with a residence time τ. We allow the objective function, J, to be specified by the designer as any function of X_{exit} and τ. One can see that the differential equation system can be uncoupled from the rest of the model and solved offline if the dimensionless feed concentration, X_0, is prespecified. Once the vectors X_{seg} are determined, we merely solve for $f(t)$, which satisfies an additional set of linear constraints. If Gaussian quadrature on finite elements is applied to the above model over the domain $[0, t_{max}]$, we get

$$\underset{f_{(i,j)}}{\text{Max}} \quad J(X_{exit}, \tau)$$

$$\sum_i \sum_j w_j f_{(i,j)} \Delta\alpha_{(i)} = 1$$

$$\tau = \sum_i \sum_j w_j f_{(i,j)} t_{(i,j)} \Delta\alpha_{(i)} \tag{P2}$$

$$X_{exit} = \sum_i \sum_j w_j f_{(i,j)} X_{seg\,(i,j)} \Delta\alpha_{(i)}$$

where

- i = Index set of finite elements
- j = Index set of Gaussian quadrature (or collocation) points
- $f_{(i,j)}$ = RTD function at jth quadrature point in ith element (point $[i,j]$)
- $X_{seg(i,j)}$ = Dimensionless concentration at point $[i,j]$
- w_j = Weights of Gaussian quadrature
- $\Delta\alpha_i$ = Length of ith finite element (fixed)

It is clear that the above formulation is linearly constrained in $f_{(i,j)}$ and the solution of (P2) gives us a global optimum in segregated flow for any concave objective function. Moreover, we can reduce this problem to a linear program for both yield and selectivity objective functions by applying suitable transformation techniques (Balakrishna, 1992). Then we can solve the problem by any linear programming algorithm.

The solution to this problem provides a good lower bound for the targeting problem. Also, the segregated flow model is often sufficient for the reactor synthesis problem. The solution to this simple formulation thus could be chosen to represent the first stage in an iterative synthesis approach. Before constructing this procedure, we first develop conditions under which the segregated-flow model is itself sufficient for the reactor synthesis problem.

B. Sufficiency Conditions for Segregated-Flow Networks

The segregated-flow model described by (P2) forms a basis to generate an AR. We now develop conditions for the closure of this space with respect to the operations of mixing and reaction by means of a PFR, a CSTR, or a recycle PFR (RR). Consider the region depicted by the constraints of (P2). Our aim is to develop conditions that can be checked easily for the reaction system in question so that, if these conditions are satisfied, we need to solve only (P2) for the reactor targeting problem. We will analyze these conditions based on PFR trajectories projected into two dimensions. Here, a PFR, which is an n-dimensional trajectory in concentration space and parametric in time, is generated by the solution of the initial value differential equation system in (P1). Figure 3 illustrates a PFR trajectory and its projections in three-dimensional space, where the solid line represents the actual PFR trajectory and the dotted lines represent the projected trajectories.

Consider a reaction system consisting of a set of components $i = 1, 2, ..., n$. Associated with each is a reaction rate r_i. Let $I = \{i\}$ denote the complete set of reacting species. This set can be further classified into

$$I_1 = \{a\} = \{a \in I : r_a \leq 0\}$$

$$I_2 = \{j\} = \{j \in I : r_j \geq 0\}$$

$$I_3 = \{k\} = \{k \in I : r_k >, < 0\}$$

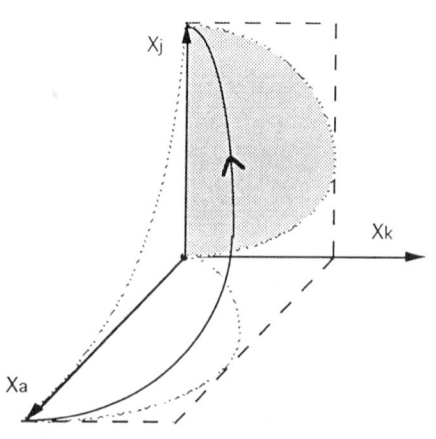

FIG. 3. An example of a PFR trajectory (solid line) and its projections (dotted lines).

In the most common case, our objective function is an explicit function of elements of I_3, which represent products of the reacting system. For completeness of segregated-flow systems, we are able to state the following properties, which are proved in Balakrishna and Biegler (1992a):

Property 1. *If the projected (onto R^2) PFR trajectories are such that $\{X_i : i \in I_1, I_3\}$ encloses a convex region with respect to any element of $\{X_a : a \in I_1\}$ and no two PFR projections meet within the bounds of possible concentrations, then none of these projected (two dimensional) spaces can be extended under the operations of mixing, additional PFRs/CSTRs/RRs, or any combination of these. This convex region implies that the projected curve is a concave function, $\partial^2 X_i/\partial X_a^2 \leq 0$.*

Corollary 1. *For any system that satisfies the conditions of Property 1, any reactor starting from the projected space will have an outlet concentration such that*

$$\frac{\Delta X_i}{\Delta X_a} \geq \frac{r_{i0}}{r_{a0}}$$

where ΔX_i is the change in concentration of component i accomplished by the reactor, ΔX_a is the corresponding change in the concentration of A, and r_{i0} and r_{a0} are reaction rates at the reactor inlet.

On the other hand, if the X_i vs. X_a curves are convex ($\partial^2 X_i/\partial X_a^2 \geq 0$) but still satisfy the condition that no two PFR projections meet, then the inequality above is reversed.

At this point, the utility of this property with respect to (P2) deserves attention. A careful look at P2 reveals that the shaded region in the projected space (for example, the X_j–X_k space) is exactly the projection on the X_j–X_k space of the feasible region of P2. The concave PFR projection defines the concentrations in segregated flow, and the interior is a convex combination of all boundary points created by the *residence time distribution* function. This gives a new interpretation to the residence time distribution as a convex combiner. For any convex objective function to be *maximized,* the solution to the segregated flow model will always lead to a boundary point of the AR.

Since we have closed-form expressions for the rate equations that form the tangents to the trajectories in concentration space (although we do not have closed-form expressions for the trajectories themselves), checking for concavity of the curves is easy. However, it is clear that we require certain convexity properties among the projections; i.e, we need to know *a priori* whether certain projected regions are convex. This task can be simplified by developing certain properties for mutual convexity in the reacting system.

Property 2. *If the projection of any element of $\{X_j\}$ vs. $\{X_a\}$ defines a convex curve and the projection of any element of $\{X_k\}$ vs. $\{X_a\}$ defines a concave* curve, then the projected curves of $\{X_k\}$ vs. $\{X_j\}$ are concave up to the stationary point in this projection, and this stationary point is unique if the X_k–X_a projected curve is strictly concave.

Corollary 2. *If the projections of $\{X_j\}$ vs. $\{X_a\}$ define concave curves and the projections of $\{X_k\}$ vs. $\{X_a\}$ define concave trajectories, then the projected curves of $\{X_k\}$ vs. $\{X_j\}$ are concave beyond the stationary point in this projection, and this stationary point is unique if the X_k–X_a projected curve is strictly concave.*

C. Optimization Formulations for Reactor Synthesis

The properties described above enable us to identify the nature of different projections and can be very useful, as shown in the example problems (Section III.D). It is essential to note, however, that Property 1 is only a sufficient condition, but not always necessary. In other words, the segregated-flow model can be optimal even if these conditions are not satisfied. If the segregated-flow region (for P2) is not sufficient, we generate optimization formulations for extending this region, by superimposing various reactors on the region provided by the linearly constrained formulation. The main idea for this approach is

Given a candidate for the AR, can extensions to this region be generated? If yes, then create the extension and, on its convex hull, check for further extensions that improve the objective function. This procedure is continued until there are no extensions that improve the objective function.

The main insight in this approach is that the residence time distributions (RTDs) lead to convex combinations and the region enclosed by the segregated flow model is always convex. The aim now is to develop an algorithm by which, given a candidate for an AR, we should be able to check whether it can be extended to our advantage. Here, we restrict ourselves to PFR, RR, and CSTR extensions only.

Consider the feasible region from (P2), a convex combination of the concentrations in segregated flow, as the first candidate for the AR. Each combination of the RTDs and the concentrations gives a unique point in the feasible region. The following subproblem can also be embedded within (P2) in order to check whether, starting from any feasible point, a reactor can provide an extension that gives a point outside of the region of P2. Here, we consider a recycle reactor extension, since it includes the PFR and CSTR extensions as special cases. Strictly speaking, recycle reactors do not form the boundary of an attainable region (Hildebrandt, 1989) as any AR extended by an RR can also be extended by a CSTR in the absence of constraints. Nevertheless, we use the RR formulation to simplify the algorithm. Additional formulations for purely CSTR or PFR extensions can also be developed along the same lines.

Recycle Reactor (RR) Extension: If $J_{rr} > J_{P2}$, then the recycle reactor provides an advantageous extension to the AR.

$$\text{Max} \quad J_{rr}(X_{exit})$$

$$X_{P2} = \sum_i \sum_j w_j f_{(i,j)} X_{seg\,(i,j)} \Delta\alpha_{(i)}$$

$$\frac{dX_{rr}}{dt} = R(X_{rr})$$

$$X_{rr}(t=0) = \frac{R_e X_{exit} + X_{P2}}{R_e + 1} \tag{P3}$$

$$X_{exit} = \sum_i \sum_j w_j f_{r(i,j)} X_{rr\,(i,j)} \Delta\alpha_{(i)}$$

$$\sum_i \sum_j w_j f_{(i,j)} \Delta\alpha_{(i)} = 1.0$$

$$\sum_i \sum_j w_j f_{r(i,j)} \Delta\alpha_{(i)} = 1.0$$

$$l \leq X_{exit} \leq u$$

where

J_{rr} = Objective function at the exit of the recycle reactor extension
X_{rr} = Dimensionless concentrations within the RR extension
R_e = Recycle ratio
X_{exit} = Vector of reactor exit concentrations
f_r = Linear combiner of all the concentrations from the plug flow section of the recycle reactor

The first equation describes the concentrations available from segregated flow. The model equations for a recycle reactor starting from any feasible point are described by the second and third equations. The fourth equation gives the concentration at the exit of the recycle reactor. Here, the vectors l and u are lower and upper bounds, respectively, on the exit concentration vector. Thus, if $J_{rr} > J_{P2}$, then the recycle reactor provides an advantageous extension over (P2).

In addition, if any of the projections X_i vs. X_j ($j \in I_2$) are concave and if we wish to eliminate searching in the interior of these projections, we can always include the following inequality, which arises out of geometric arguments (Corollary 1):

$$\frac{(X_{P2}^i - X_{exit}^i)}{(X_{P2}^j - X_{exit}^j)} \leq \frac{R^i(X_{P2})}{R^j(X_{P2})}$$

It is important to note that CSTR/PFR/RR extensions can be applied to any convex candidate region, not just the one defined by (P2). The residence time distribution can be used to generate the convex candidates. A sequence of convex hulls can be generated until the conditions for completeness are satisfied (i.e., there are no further extensions). The synthesis flowchart shown in Fig. 4 illustrates these ideas. In the algorithm, we initially check the possibility of global optimality for (P2). If this solution is suboptimal, a more complex model can be solved to give an updated optimal solution. Thus the new or updated convex

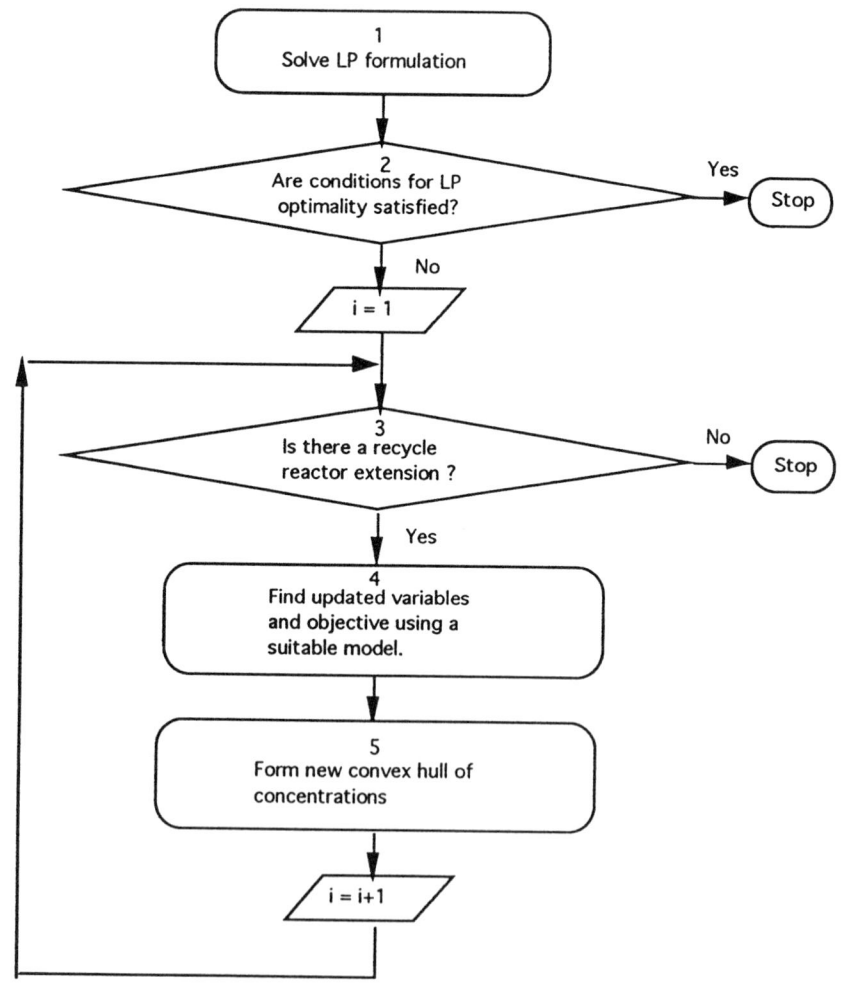

FIG. 4. Flowchart for stagewise synthesis.

hull based on the new concentrations can be generated, and the following subproblem, which represents Box 3 of the flowchart, can be solved.

$$\max_{R_e,\ f_{(i,j)},\ f_{\text{model}(k)}} J(X_{\text{exit}})$$

$$\frac{dX_{\text{rr}}}{dt} = R(X_{\text{rr}})$$

$$X(t=0) = \frac{R_e X_{\text{exit}} + X_{\text{update}}}{R_e + 1} \quad \text{(P4)}$$

$$X_{\text{update}} = \sum_i \sum_j f_{(i,j)} X_{\text{seg }(i,j)} + \sum_k f_{\text{model}(k)} X_{\text{model}(k)}$$

$$X_{\text{exit}} = \sum_i \sum_j f_{r(i,j)} X_{\text{rr}(i,j)}$$

$$\sum_i \sum_j f_{(i,j)} + \sum_k f_{\text{model}(k)} = 1.0$$

In this formulation, $X_{\text{model}(k)}$ is a constant vector and reflects the concentration at the exit in the models previously chosen. A convex combination of this with the segregated-flow region described by (P2) gives the fresh feed point for the recycle reactor we are looking for, X_{update}. Then X_{exit} represents the concentration at the exit of the recycle reactor; and if $J(X_{\text{exit}}) > J(X_{\text{model}(k)})$, the earlier model chosen is insufficient. The control variables essentially are the f's and $f_{\text{model}(k)}$, which are the linear combiners used to provide a convex candidate. A careful look at the formulation reveals that we are checking for completeness of the convex hull of the region found by the model.

Figure 5 gives a geometric interpretation to the solution of (P4). Here, if the solution of (P4) indicates that the objective function can be improved by extending the AR, e.g., based on segregated flow, we consider a more complex model. Thus, the expression for X_{update} automatically includes all the points in the convex hulls attained so far by previous recycle reactor extensions, and this region can be made up of the segregated-flow model as well as favorable recycle reactor extensions from previous solutions of (P4). We continue to check for extensions and terminate when there are no further extensions that improve the objective function. Note that with this approach, the algorithm allows the reactor network to be synthesized readily. It is also clear that this approach shares many characteristics with the geometric approach of Glasser et al. (1987). An important difference is that their approach searches for all possible extensions of candidate ARs and requires checking of an infinite number of points on the convex hull.

Our approach, on the other hand, automatically finds only those extensions that improve the objective function. Since it is an optimization-based approach, it is not limited, in principle, by problem dimensionality or the addition of

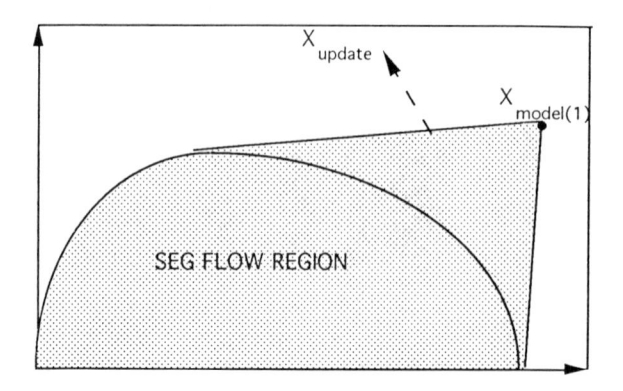

$X_{model(1)}$: Solution to first reactor extension from segregated flow.
X_{update} : Reactor Extension from combined hull of segregated flow and $X_{model(1)}$

FIG. 5. Extension of the convex hull (P4).

constraints, as will be demonstrated by the examples in the next section. We also note that because problems (P2) and (P4) are relatively small and simple optimization problems, AR solutions can be found very quickly.

One potential disadvantage to our optimization formulations, however, is that a particular extension that does not improve the objective function may still enlarge the AR enough so that a point from within this extension can improve the objective function beyond what we started with. This *nonmonotonic* increase in the objective is a limitation of the proposed approach. Also, we note that even though the attainable space of concentrations is always convex, (P4) is not necessarily a convex nonlinear program and therefore will not necessarily find the global optimum that we are seeking. However, termination at local optima is not always serious because the solution of (P2) usually yields an excellent starting point for (P4). (Also, multiple starting points could be tried to improve the likelihood of finding a global optimum for (P4).)

D. EXAMPLE PROBLEMS

We conclude this section with four example problems to illustrate our approach. The first problem satisfies the sufficiency conditions for segregated flow and is easily addressed by our approach. The second and third examples do not satisfy these properties but are readily solved by the algorithm of Fig. 4. Finally, the fourth example illustrates the difference between our optimization formulation and the geometric approach of Glasser *et al.* Several additional problems are also considered in Balakrishna and Biegler (1992a), with results superior to those presented in other articles.

Example 1. The isothermal van de Vusse (1964) reaction involves four species for which the objective is the maximization of the yield of intermediate species B, given a feed of pure A. The reaction network is given by

$$A \xrightarrow{k_1} B \xrightarrow{k_2} C$$
$$\downarrow k_3$$
$$D$$

Here, the reaction from A to D is second-order. The feed concentration is $c_{A0} = 0.58$ mol/liter and the reaction rates are $k_1 = 10$ s^{-1}, $k_2 = 1$, s^{-1} and $k_3 = 1$ liter g-mol^{-1} s^{-1}. The reaction rate vector for components A, B, C, and D, respectively is given in dimensionless form by:

$$R(X) = [-10X_A - 0.29X_A^2, 10X_A - X_B, X_B, 0.29X_A^2]$$

where $X_A = C_A/C_{A0}$, $X_B = C_B/C_{A0}$, and C_A, C_B are the molar concentrations of A and B, respectively. The objective function is the yield of component B. Following the algorithm in Fig. 4, we find that the solution of (P2), which is an LP, gives a globally optimal reactor network. This is because the reaction kinetics satisfy the conditions in Property 1 (i.e., the X_B vs. X_A curves are concave). The verification of concavity is simple, because we just have to verify the monotonicity of the slopes, given in closed form for the reactor synthesis problem. Here, the slope of the PFR trajectories in X_B vs. X_A space is just given by Rate(B) / Rate(A), and a simple program to find the maximum of the second derivative can be solved. The optimal value of the objective function is given by $X_B^{exit} = 0.7636$. This is the globally optimal solution and can be realized by a PFR with a residence time of 0.288 s. A brief comparison with results from the literature is presented below:

Study	Yield	Residence Time
Chitra and Govind (1985)	0.752	0.2551
Achenie and Biegler (1986)	0.7531	0.2965
Achenie and Biegler (1988)	0.757	0.2370
Kokossis and Floudas (1990)	0.752	0.2539
(P2) LP formulation	0.7636	0.2880

The results from the different approaches are nearly the same and differences could be attributed to numerical solutions of the PFR equations. The slightly better objective function predicted by the LP formulation may be due to a better approximation procedure when these differential equations are solved offline.

Example 2: The Trambouze reaction (Trambouze and Piret, 1959) involves four components and has the following reaction scheme:

$$A \xrightarrow{k_1} B \quad A \xrightarrow{k_2} C \quad A \xrightarrow{k_3} D$$

where the reactions are zero-order, first-order and second-order, respectively, with $k_1 = 0.025$ mol liter^{-1} min^{-1}, $k_2 = 0.2$ min^{-1}, $k_3 = 0.4$ liter mol^{-1} min^{-1}. We wish to maximize the selectivity of C to A defined by $X_C / (1 - X_A)$. Here the conditions of (P2) optimality are not satisfied; following the stagewise approach, we arrive at a CSTR extension from the feed point of the segregated-flow model, indicating a single CSTR with a selectivity of 0.5 and residence time of 7.5 s. No further extensions are observed by solving (P3). Achenie and Biegler (1990) observe a selectivity of 0.4999 in a two-CSTR combination. Kokossis and Floudas (1990) report many optimal networks to this problem with the same objective function of 0.5; Glasser et al. (1987) observe that this problem has an infinite number of optimal solutions with a selectivity of 0.5 in a CSTR with variable bypass.

Example 3: The α-pinene problem is a reaction network that consists of five species and has the following reaction network:

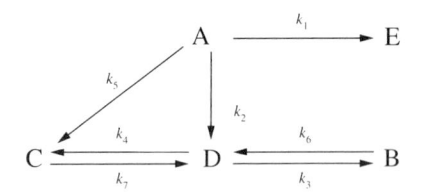

The objective function here is the maximization of the selectivity of C over D, given a feed of pure A.

The reaction vector for the components A, B, C, D, and E, respectively, is given by

$$R(X) = [-k_1 + k_2)X_a - 2k_5X_a^2, -k_6X_b + k_3X_d, k_5X_a^2 + k_4X_d^2$$
$$- k_7X_c, k_2X_a + k_6X_b - k_3X_d - 2k_4X_d^2 + 2k_7X_c, k_1X_a]$$

where $X_i = C_i/C_{A0}$ and $C_{A0} = 1$ mol/liter; $k_1 = 0.33384$ s^{-1}, $k_2 = 0.26687$ s^{-1}, $k_3 = 0.14940$ s^{-1}, $k4 = 0.18957$ liter mol^{-1} s^{-1}, $k_5 = 0.009598$ liter mol^{-1} s^{-1}, $k_6 = 0.29425$ s^{-1}, $k_7 = 0.011932$ s^{-1}.

This problem was solved by applying formulation (P2). Interestingly, the RTD always settles at the upper bound in the time horizon. Thus, we may infer that the selectivity in this problem increases monotonically with age within the segregated flow model. For instance, the selectivity obtained with a t_{max} of 60 s was 1.48. While this approach does not reveal the entire AR, this example corroborates some of the strengths of a target-based perspective that we mentioned earlier. Achenie and Biegler (1988) arrived at an optimal selectivity of 0.2336 by placing a bound on the residence time of 6 s. Kokossis and Floudas (1990), using a superstructure-based approach, arrived at a complex network of a CSTR and a PFR (represented by sub-CSTRs) with recycles and intermediate feeds to get a maximum selectivity of 1.4020. Moreover, (P2) (an LP in this case) is based on a finite approximation to an infinite residence time domain. Thus, from the solution to our targeting model, it is easy to detect monotonicities in the objective function.

Example 4: Here, we revisit the van de Vusse reaction of Example 1 with altered rate constants. The objective function again is the yield of intermediate species B. The rate vector is given by $R(X) = [-X_A - 20X_A^2, X_A - 2X_B, 2X_B, 20X_A^2]$. In this case, the segregated-flow model gives a yield of 0.061. However, the sufficiency conditions for the LP formulation are not satisfied. Using our optimization formulation with recycle reactor extensions (P3), we observe a B yield of 0.069. Our network corresponds to a recycle reactor from the feed point (recycle fraction = 0.772; residence time in the plug section of RR = 0.1005 s) in series with a PFR with a residence time of 0.09 s. Glasser *et al.*, on the other hand, report a yield of 0.071 with a graphical approach. The lower yield obtained in this example can be attributed to a nonmonotonic increase in the objective, as mentioned earlier. It is interesting to note that if we apply the multicompartment approach of Achenie and Biegler (1988) to this problem, a yield of 0.0705 is obtained.

IV. Reactor Network Synthesis: Nonisothermal Systems

In this section, we develop a formulation for the synthesis of nonisothermal reactors, based on the isothermal targeting approach proposed in Section III. The targeting model is based on mixing between different reacting environments and is formulated as a dynamic optimization problem, where the temperature, the feed distribution function, and an exit flow distribution function are the control profiles. The optimization procedure results in the sequential solution of small nonlinear programming problems, where the solution to each NLP generates a component of the reactor network. This provides a constructive technique for the target-based synthesis of nonisothermal reactor networks for any general objective function and process constraints.

A. NONISOTHERMAL MODEL FORMULATION

The targeting model for nonisothermal reactors derives much of its motivation from the targeting methodology described in the previous section. There, the only control profiles are the mixing functions (RTD) and we assumed that mixing does not involve additional costs. In the case of nonisothermal systems, however, temperature is an added profile and we must consider the cost of maintaining this profile. One rather inexpensive technique for exothermic reactions is cold-shot cooling. However, mixing may not always be optimal in the space of concentrations, even if it is desirable in terms of temperature manipulation.

We therefore consider a different reaction flow model as our basic targeting model—one that can address temperature manipulation by feed mixing as well as by external heating or cooling. The model consists of a differential sidestream reactor (DSR), shown in Fig. 6, with a sidestream concentration set to the feed concentration and a general exit flow distribution function. (As mentioned in Section II, the boundary of an AR can be defined by DSRs for higher-dimensional (≥ 3) problems). We term this particular structure a *cross-flow reactor*. By construction, this model not only allows the manipulation of reactor temperature by feed mixing, but often eliminates the need to check for PFR extensions.

Figure 6 shows a schematic of a cross-flow reactor (CFR) with side exits, which we choose as our basic targeting model. Here, X_0 is the dimensionless concentration of the feed entering the reactor network, α is the independent variable denoting time as it progresses along the length of the reactor, and $T(\alpha)$ denotes the temperature as a function of the reactor length. We define $f(\alpha)\, d\alpha$ as the fraction of molecules in the reactor exit that leave between points α and $\alpha + \delta\alpha$ of the reactor (an exit flow distribution function), and $q(\alpha)$ as the probability density function for a molecule entering the system at point α in the reactor. Thus the number of molecules entering between points α and $\alpha + d\alpha$ is given by $q(\alpha)Q_0\, d\alpha$, where Q_0 is the flow rate entering the reactor network. Implicit in the formulation is the assumption of instantaneous mixing between the feed and the mixture in the reactor. In addition, the formulation can easily be extended for variable-density systems, although we consider only constant-density systems in this study.

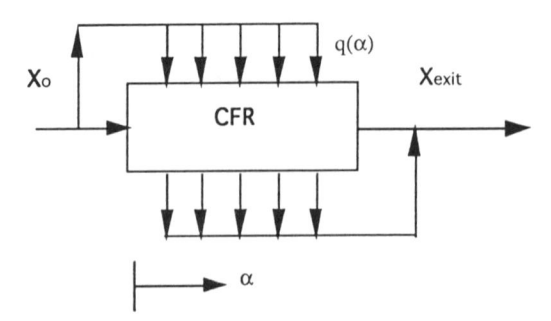

FIG. 6. General cross-flow reactor model.

Clearly, at one extreme—when $q(\alpha)$ is zero throughout the reactor and we have a general $f(\alpha)$—we have the equations for a segregated-flow model. On the other extreme—when $f(\alpha)$ is a Dirac delta exactly at one point and we have a general nonzero $q(\alpha)$—this model reduces to the Zwietering (1959) model of maximum mixedness. Also, we define $Q(\alpha)$ as the flow of molecules at point α. Based on this nomenclature, a differential mass balance on an element $\Delta\alpha$ leads to

$$\frac{dX}{d\alpha} = R(T(\alpha), X) + \frac{q(\alpha)Q_0}{Q(\alpha)} (X_0 - X(\alpha))$$

and with this governing equation in a cross-flow reactor, the mathematical model for maximizing the performance index in cross flow can be derived as follows:

$$\underset{q(\alpha), f(\alpha), T(\alpha)}{\text{Max}} \quad J(X_{\text{exit}}, \tau)$$

$$\frac{dX}{d\alpha} = R(T(\alpha), X) + \frac{q(\alpha)Q_0}{Q(\alpha)} (X_0 - X(\alpha))$$

$$X(0) = X_0$$

$$X_{\text{exit}} = \int_0^\infty f(\alpha) X(\alpha) \, d\alpha \tag{P6}$$

$$\int_0^\infty f(\alpha) \, d\alpha = 1$$

$$\int_0^\infty q(\alpha) \, d\alpha = 1$$

$$\frac{Q(\alpha)}{Q_0} = \int_0^\infty [q(\alpha') - f(\alpha')] \, d\alpha'$$

$$\int_0^\infty \int_0^\alpha [q(\alpha') - f(\alpha')] \, d\alpha' \, d\alpha = \tau$$

Here, the last two equations define the flow rate and the mean residence time, respectively. This formulation is an optimal control problem, where the control profiles are $q(\alpha)$, $f(\alpha)$, and $T(\alpha)$. The solution to this problem will give us a lower bound on the objective function for the nonisothermal reactor network along with the optimal temperature and mixing profiles. Similar to the isothermal formulation (P3), we discretize (P6) based on orthogonal collocation (Cuthrell and Biegler, 1987) on finite elements, as the differential equations can no longer be solved offline. This type of discretization leads to a reactor network more practically achievable than the schematic shown in Fig. 6 with a finite number of mixing points.

In the finite-element discretization, the subscript i denotes the ith finite element, and the subscript j (or k) denotes the jth (or kth) collocation point in any finite element. There are a total of N finite elements and K collocation points ($i = 1, N; j = 1, K$). The state variable X is approximated over each finite element by Lagrange interpolation basis functions ($L_k(\alpha)$) as:

$$X(\alpha) = \sum_{k=0}^{K} X_{ik} L_k(\alpha) \quad \text{for } \alpha_{i0} \leq \alpha \leq \alpha_{i+1,0}$$

and

$$L_k(\alpha) = \prod_{l=0;k}^{K} \left[\frac{\alpha - \alpha_{il}}{\alpha_{ik} - \alpha_{il}}\right]$$

Substitution of this into (P6) leads to the following nonlinear program, the solution of which gives us the optimal control variables at the collocation points.

$$\begin{aligned}
& \underset{\phi_i, f_{ij}, T_i}{\text{Max}} \quad J(X_{\text{exit}}, \tau) \\
& \sum_k X_{ik} L_k'(\alpha_j) - R(X_{ij}, T_{ij}) \Delta\alpha_i = 0 \quad j = 1, K \\
& X(0) = X_0 \\
& X_{i\text{end}} = \sum_k X_{ik} L_k(\alpha_{i+1}) \\
& X_{i,0} = \phi_i X_0 + (1 - \phi_i) X_{(i-1)\text{end}} \\
& X_{\text{exit}} = \sum_i \sum_j X_{ij} f_{ij} \quad \quad \quad \quad \quad \quad \text{(P7)} \\
& \sum_i \sum_j \alpha_{ij} (f_{ij} - q_{ij}) = \tau \\
& \sum_i \sum_j f_{ij} = 1 \\
& \phi_i Q_{i,1} = q_{i,1} Q_0 \\
& Q_{ij} = \sum_{ij} (q_{ij} - f_{ij}) Q_0 \\
& 0 \leq \phi_i \leq 1
\end{aligned}$$

where

$L_k'(\alpha_j)$ = Derivatives of the Lagrange basis polynomials evaluated at α_j
ϕ_i = Ratio of the side inlet flow rate to the bulk flow rate within the reactor after mixing before element i
$\Delta\alpha_i = \alpha_{i+1,0} - \alpha_{i,0}$, the length of each finite element i

f_{ij} = Exit flow distribution at collocation point j in element i (point $[i,j]$)
q_{ij} = Fraction of inlet flow entering at $[i,j]$
T_{ij} = Temperature at $[i,j]$
X_{ij} = Dimensionless concentration at $[i,j]$
$X_{i\text{end}}$ = Concentration at end of ith finite element

Note that ϕ_i in this model is an approximation to

$$\frac{q(\alpha)Q_0}{Q(\alpha^+)} \quad \left(\phi_i = \frac{q_i Q_0}{Q_{i,1}}; \; Q_{ij} = \sum_{ij}(q_{ij} - f_{ij})Q_0\right)$$

and the equations in (P7) result from collocation applied to the differential equation model and Gaussian quadrature applied to the integral expressions. For convenience, the quadrature weights are absorbed into the respective discretized variables in the integration. Figure 7 is a schematic of this discretization.

Balakrishna (1992) has shown that if the finite elements ($\Delta\alpha_i$) are chosen sufficiently small, then (P7) simply reduces to a numerical scheme for solving (P6). Thus (P6) can now be solved as a nonlinear program to obtain the optimal set of f, T, and ϕ over each element. In this model, even though the temperature along the reactor is a control variable, part of the temperature manipulation can be readily accomplished by feed mixing if this is optimal for the reactor. In addition, the cross-flow reactor, by construction, mixes all available points on the reacting segments with the feed point, thus continuously checking for PFR extensions as long as the PFR trajectories are such that their nonconvexities can be enveloped from the feed point. These mixed points correspond geometrically to convex hulls; and since PFRs are already generated from these, there is often no need to check for the PFR extensions from the solution to the cross-flow reactor model.

B. REACTOR EXTENSIONS

The solution to (P7) provides a lower bound to the performance index of the reactor network. In the case of purely isothermal reactors, we derive theoretical

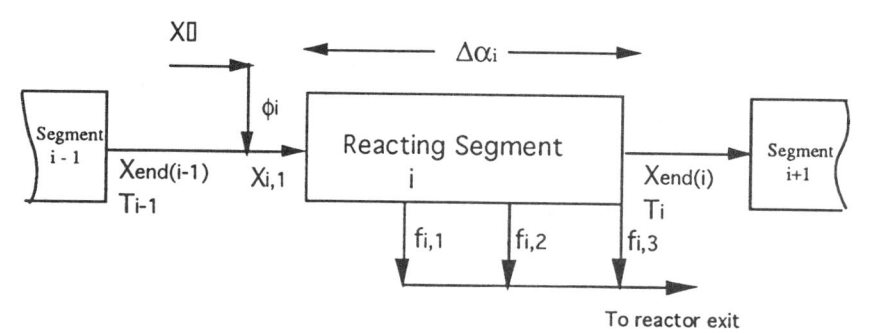

FIG. 7. Reactor representation for discretized cross-flow reactor model.

conditions of sufficiency for the segregated-flow model as in Section III, and we have checks for favorable extensions from the segregated-flow model, which are formulated as nonlinear programs. These results apply directly if the temperature profile can be related to the concentration profile (e.g., adiabatic reactors). However, more general nonisothermal reactors do not lend themselves to easy analysis owing to the assumption of a completely arbitrary temperature profile. Nevertheless, by applying the optimization formulations detailed in Section III, we can now develop techniques for extending the target provided by the CFR model. The constraints of (P7) define the feasible region for any achievable cross-flow reactor. The convex combination of all the concentrations in this region provides the entire region attainable by the cross-flow reactor and mixing, which corresponds to the first candidate for the AR. Based on the convex hull extensions illustrated in Section III, the following subproblem can be solved to check whether a reactor can provide an extension to the candidate AR. Here, we again consider a recycle reactor extension, since it includes the PFR and CSTR extensions as special cases. In the nonisothermal recycle reactor, we assume that the temperature is a control profile along the length of the plug flow section of the recycle reactor. The inlet temperature to the reactor that constitutes the extension will also follow a convex combination rule if intermediate heating or cooling is not permitted.

Recycle Reactor (RR) Extension: If $J_{rr} > J_{P7}$, then the recycle reactor provides an advantageous extension outside of A.

$$\text{Max} \quad J_{rr}(X_{exit}, T_R)$$

$$X_{P7} = \sum_i \sum_j \lambda_{ij} X_{cf}(i,j)$$

$$\frac{dX_{rr}}{dt} = R(X_{rr}, T_{rr})$$

$$X_{rr}(t=0) = \frac{R_e X_{exit} + X_{P7}}{R_e + 1}$$

$$X_{exit} = \sum_i \sum_j f_r(i,j) X_{rr}(i,j) \quad \text{(P8)}$$

$$\sum_i \sum_j \lambda_{ij} = 1.0$$

$$\sum_i \sum_j f_r(i,j) = 1.0$$

$$T_R < T_{max}$$

$$l \leq X_{exit} \leq u$$

Here, J_{rr} is the value of the objective function at the exit of the recycle reactor; J_{P7} is the value of the objective obtained from the solution of (P7). λ_{ij} is the convex combiner of all points available from the CFR model. The variables X_{rr} and R_e represent the concentrations in the recycle reactor extension and the recycle ratio, respectively. X_{exit} is the vector of exit concentrations from the RR reactor. f_r is a linear combiner of all the concentrations from the plug flow section of the recycle reactor.

The next iteration consists of creating the new convex hull of concentrations, which include the concentrations obtained by this extension, and checking for favorable recycle reactor extensions from this new convex hull. At iteration P, this involves the solution of the following nonlinear programming problem:

$$\max_{R_e, \lambda_{ij}, f_{\text{model}(p)}, T_{rr}(t)} J^{(P+1)}(\nu)$$

$$\frac{dX_{rr}}{dt} = R(X_{rr}, T_{rr}(t))$$

$$X_{rr}(t=0) = \frac{R_e X_{exit} + X_{update}}{R_e + 1} \quad \text{(P9)}$$

$$X_{update} = \sum_i \sum_j \lambda_{ij} X_{cf}(i,j) + \sum_{p=1}^{P} f_{\text{model}(p)} X_{\text{model}(p)}$$

$$X_{exit} = \sum_i \sum_j f_r(i,j) X_{rr}(i,j)$$

$$\sum_i \sum_j \lambda_{ij} + \sum_{p=1}^{P} f_{\text{model}(p)} = 1.0$$

Here, $X_{\text{model}(p)}$ is a constant vector and reflects the concentration at the exit at iteration p in the models previously chosen. A convex combination of this with the cross-flow region described by (P7) gives the fresh feed point for the recycle reactor we are looking for, X_{update}. X_{exit} then represents the concentration at the exit of the recycle reactor; and if $J^{(P+1)} > J^{(P)}$, the earlier model chosen is insufficient. The control profiles are $[f, f_{\text{model}(p)}]$ and T_{rr}, which are the linear combiners used to provide a convex candidate and the temperature profile in the recycle reactor, respectively. This procedure is repeated until no improvement in the objective function is observed. Figure 8 illustrates the flowchart for the synthesis procedure.

It is easy to see that with this approach, the reactor network is synthesized readily. Also, because of its stagewise nature, only the simplest reactor model needed is solved. In fact, for many problems, the solution is obtained through just one iteration of the flowchart in Fig. 8, mainly because the variable tem-

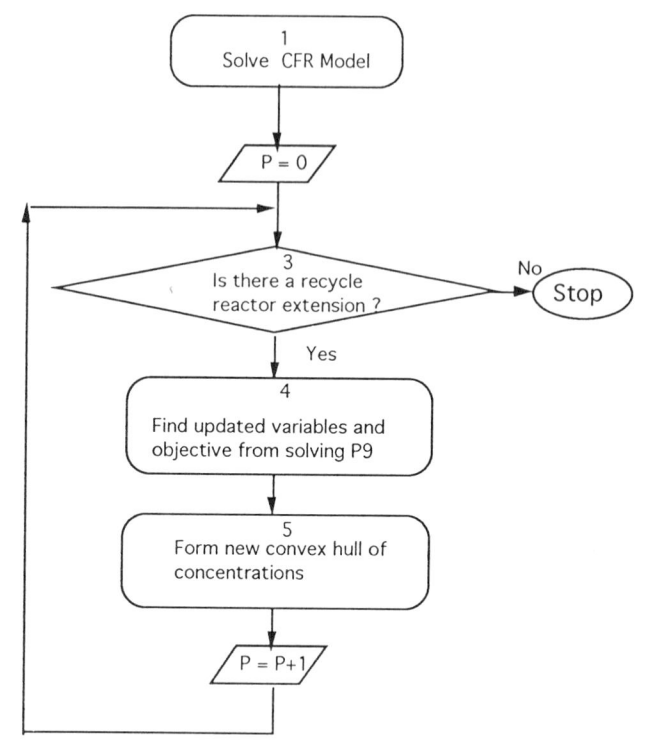

```
                      ┌─────────────────┐
                      │        1        │
                      │  Solve CFR Model│
                      └────────┬────────┘
                               │
                          ╱ P = 0 ╱
                               │
                               ▼
                      ╱╲
                     ╱  ╲        3
            ┌──────╱     ╲──────────────────┐  No
            │     ╱ Is there a recycle ╲────┼──────▶ ( Stop )
            │     ╲ reactor extension ?╱    │
            │      ╲                  ╱
            │       ╲       ╱
            │         Yes
            │          │
            │          ▼
            │  ┌─────────────────┐
            │  │        4        │
            │  │ Find updated variables and
            │  │ objective from solving P9
            │  └────────┬────────┘
            │           ▼
            │  ┌─────────────────┐
            │  │        5        │
            │  │ Form new convex hull of
            │  │ concentrations
            │  └────────┬────────┘
            │           │
            │     ╱ P = P+1 ╱
            │           │
            └───────────┘
```

FIG. 8. Flowchart for reactor synthesis.

perature profile allows such a large area for the AR. It can be shown that this algorithm will converge to the optimal solution as long as *nonmonotonic* extensions to the AR are not encountered (as in Example 3). The examples solved in Balakrishna and Biegler (1992b), as well as the one shown next, suggest that this restriction does not seem to be limiting for the applicability of the proposed approach.

Example 5: Here, we maximize the conversion in the catalytic oxidation of sulfur dioxide in fixed-bed reactors, which has been investigated by Lee and Aris (1963). The reaction and the kinetics are described as follows:

$$SO_2 + {}^1\!/_2 O_2 = SO_3$$

$$R(g,t) = 3.6 \times 10^6 \left[\exp\left\{ 12.07 - \frac{50}{1 + 0.311t} \right\} \frac{\{2.5 - g\}^{0.5}\{3.46 - 0.5g\}}{\{32.01 - 0.5g\}^{1.5}} \right.$$
$$\left. - \exp\left\{ 22.75 - \frac{86.45}{1 + 0.311t} \right\} \frac{g\{3.46 - 0.5g\}^{0.5}}{\{32.01 - 0.5g\}\{2.5 - g\}^{0.5}} \right]$$

where g is defined as the number of moles of SO_3 formed per unit mass of mixture and t is defined as $(T - T_0)/J$, where T is the temperature, T_0 is 310 K (fresh feed temperature), and $J = 96.5$ K kg mol^{-1}. $R(g, t)$, the rate of reaction is defined as the kilogram-moles of SO_3 produced per hour per kilogram of catalyst. The extent of reaction, or the moles of SO_3 formed per unit mass of mixture, is limited by the inlet mass flow of SO_2, which is fixed at 2.5 mol SO_2 per unit mass of mixture. Lee and Aris looked at the maximization of an objective function based on the value of the product stream, catalyst, and preheating costs. However, they assumed adiabatic reactor sections, with cold-shot cooling. In our procedure, we do not enforce the adiabatic restriction, choosing yield maximization as our objective instead.

For this reaction, we observed that for a constraint on residence time of 0.25 s, the maximum reaction extent of 2.42 is obtained in a PFR with the temperature profile shown in Fig. 9a. The resulting optimization problem (555

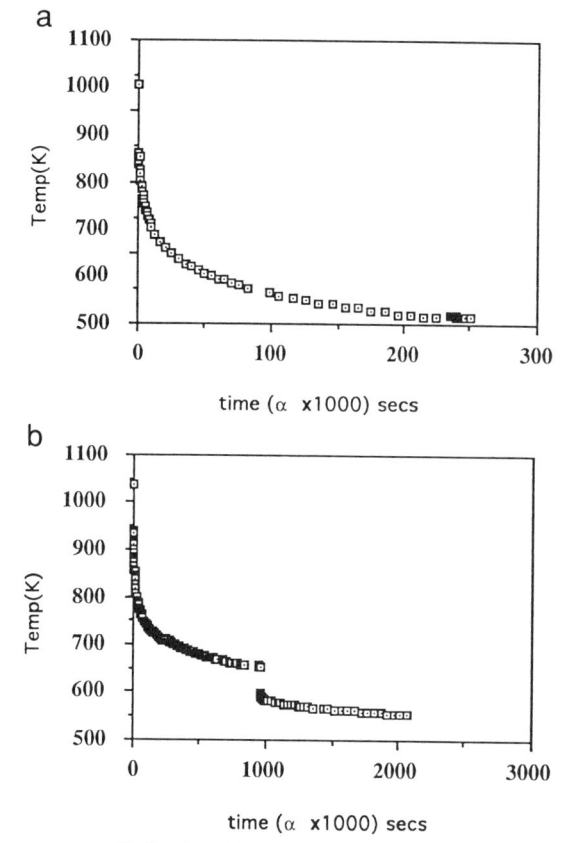

FIG. 9. Temperature profile for Lee–Aris example: (a) Residence time = 0.25 s; (b) large residence times.

equations, 753 variables) took 1503 CPU seconds on a VAX 3200 workstation. However, if the constraint on the residence time is removed, the extent of reaction (as defined by g) asymptotically approaches the upper bound of 2.5 in a PFR with a sufficiently large residence time. Also, for a residence time bound of 2.2 s, the temperature profile as a function of time is shown in Fig. 9b and delivers an extent of reaction of 2.48. Additional nonisothermal examples are considered in Balakrishna and Biegler (1992b).

V. Energy Integration of Reactor Networks

Energy integration involves the matching of heat loads between a set of hot streams and cold streams to minimize the cost of utilities for the network. Algorithms for the "isolated" construction of heat-exchanger networks (HENs) are well known. However, the synergy among process subsystems is a key area for the exploitation of energy integration. Reactor networks, in particular, are associated with significant heat effects and strongly influence the behavior of other subsystems. In this section, we address integration of the heat effects within the reactor with the rest of the process and demonstrate the effectiveness of the optimization formulations of the previous section.

Often, the reactor is the most important unit of the chemical plant because the downstream processing steps and the feed processing steps depend on the selective conversion of raw material into product. Here, the difference between stand-alone reactor network synthesis and reactor-flowsheet integration is that the constraints within the flowsheet allow a variable, but usually bounded, space of inlet concentrations for the reactor. If the space of inlet concentrations is not predetermined, then the problem size increases at each iteration of the constructive approach. This is because the previous reactors (or previous convex hulls) are not fixed, at each iteration, owing to the infinite number of inlet concentrations dictated by the flowsheet constraints. Nevertheless, the optimization formulations developed in the previous two sections can be readily adapted to determine the optimal flowsheet parameters. The development of a model for the energy integration of reactors is presented next. Following this, we apply our formulations to generate reactor extensions and illustrate this approach for the design of energy-integrated, reactor-based flowsheets.

A. Model Formulation for Energy Integration

We consider two approaches for the integration of the reactor and energy network: the *sequential* and the *simultaneous* formulations. In the conventional sequential approach, the reactor and separator schemes appear at a level higher relative to energy integration. In other words, once the "optimal" flowsheet parameters have been determined for the reactor target and the separation system, the reactor network is realized, and the heat-exchanger network is derived in a straightforward manner. However, it is well known that this approach can be suboptimal with respect to the overall flowsheet (Duran and Grossmann, 1986).

For the simultaneous approach, we consider reactor synthesis and energy integration at the same level. This approach is attractive because it considers the strong interaction between the chemical process and the heat-exchanger network. However, this is not a trivial problem because the flowrates and the temperature of the process streams are not known in advance. Moreover, we do not restrict ourselves to adiabatic or isothermal constraints on the temperature profile within the reactor. Therefore, the streams within the reactor cannot be classified as hot or cold streams *a priori*, because the nature of the optimal temperature trajectory within the reactor is unknown. Instead, we discretize the temperature trajectory in the CFR model proposed for nonisothermal reactor synthesis, and introduce the concept of candidate streams within the reactor network. Here, we approximate the optimal temperature trajectory within the reactor by a set of isothermal segments followed by the temperature change between these segments as shown in Fig. 10. The solid curve represents the actual trajectory, while horizontal lines represent isothermal reacting segments and vertical lines represent the temperature changes necessary to follow an optimal trajectory.

The horizontal segments at constant temperature correspond either to hot streams or cold streams, depending on whether the reaction is exothermic or endothermic, while the vertical sections involve heating or cooling; therefore we assume the presence of both heaters and coolers between the reacting segments. Also, we call these hot or cold streams *candidate* streams, because they

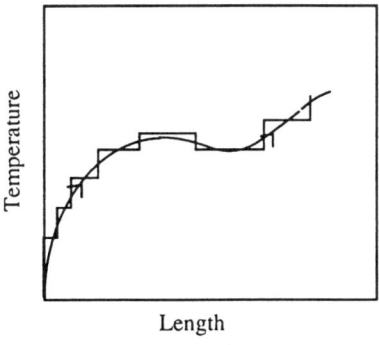

FIG. 10. Piecewise constant approximation of optimal reactor temperature profiles.

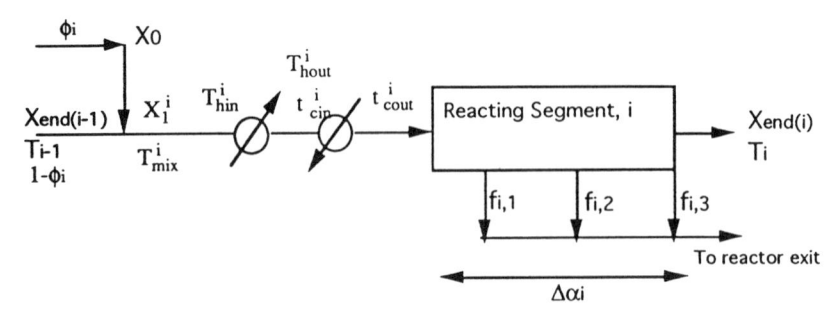

FIG. 11. Reacting segment for heat integration.

may or may not be present in the optimal network. This will depend on the exit flow distribution, which may include only some of the reacting segments and hence the corresponding temperature profiles. The schematic in Fig. 11 shows the reactor representation (of the CFR model for nonisothermal networks) corresponding to the above approximation.

The representation in Fig. 11 is similar to the discretized representation of the cross-flow reactor shown in Fig. 7, except for the additional heat exchangers. The subscript i again represents the ith finite element corresponding to the discretization. T^i_{mix} corresponds to the temperature after mixing the reacting stream with the feed. T^i_{hin}, T^i_{hout} correspond to the temperatures of the streams entering and leaving the heater, and t^i_{cin}, t^i_{cout} correspond to the temperatures of the streams entering and leaving the cooler. In an optimal network, it can be shown that only one of these two heat exchangers will be chosen. $\Delta\alpha_i$ corresponds to the length of the finite element, which may also be a variable in the optimization problem (subject to constraints on error control). Thus, the problem is well posed since we now know the hot and cold streams *a priori*, even if the flow rates and the temperatures are not known. Also, some amount of temperature control can be achieved by mixing, while the remainder can be accomplished by the utilities or the heat flows within the network. Using the framework for reactor targeting from above, we now integrate this within a suitable energy targeting framework. We assume here that utility costs will predominate and will be sufficient for the simultaneous total flowsheet synthesis for preliminary design. However, overall capital cost and area estimates for the heat-exchanger networks (Kravanja and Grossmann, 1990) can also be included here if desired.

Duran and Grossmann (1986) derived analytical expressions for minimum utility consumption as a function of flowrates and temperatures of the heat-exchange streams. They showed that given a set of hot and cold streams, the minimum heating utility consumption is given by $Q_H = \max(z_H^p)$, where

z_H^P is the difference between the heat sources and sinks above the pinch point for pinch candidate p. For hot and cold streams with inlet temperatures given by T_h^{in} and t_c^{in}; and outlet temperatures T_h^{out} and t_c^{out} respectively, $z_H^P(y)$ is given by

$$z_H^P(y) = \sum_{c \in C} w_c[\max\{0; t_c^{out} - \{T^P - \Delta T_m\}\} - \max\{0; t_c^{in} - \{T^P - \Delta T_m\}\}]$$

$$- \sum_{h \in H} W_h[\max\{0; T_h^{in} - T^P\} - \max\{0; T_h^{out} - T^P\}]$$

for $p = 1, ..., N_p$; where N_p is the total number of heat-exchange streams. Here, T^P corresponds to all the candidate pinch points; these are given by the inlet temperatures for all hot streams and the inlet temperature added to ΔT_m for the cold streams. w_c and W_h are the heat capacity flows for the hot and cold streams. Also, y is the set of all variables in the reactor and energy network. The minimum cooling utility is given by a simple energy balance as $Q_C = Q_H + \Omega(y)$; where, $\Omega(y)$ is the difference in the heat content between the hot and the cold process streams, given by:

$$\Omega(y) = \sum_h W_h(T_h^{in} - T_h^{out}) - \sum_c w_c(t_c^{out} - t_c^{in})$$

Based on the above concepts for reactor and energy network synthesis, a unified target for simultaneous reactor-energy synthesis can now be formulated. We first classify the streams within the process into four categories. Let H_R, C_R be the set of hot and cold streams associated with the reactor network and let H_P, C_P be the set of hot and cold streams in the process flowsheet. Also, $h \in H = H_R \cup H_P$, and $c \in C = C_R \cup C_P$. If there are NE isothermal reacting segments and if the reaction is exothermic, then there is a set of NE hot reacting streams from which the heat of reaction is to be removed in order to maintain a desired temperature in each segment. Also, between any two elements, there is a hot stream corresponding to the discretization shown in Fig. 11. Hence, H_R is a set of cardinality 2NE, while C_R is a set of cardinality NE. For an endothermic reaction, C_R and H_R have cardinalities in the reverse order, since the reacting segments now correspond to cold streams. Thus, there are always 3NE candidate streams, some of which may have zero heat content in the reactor network. Let F_h and F_c denote the mass flow rates of the hot and cold streams respectively. $F_{(i)}$ denotes the mass flow at the entry point of reacting segment i, F_0 is the total inlet flow into the reactor, heat capacities are allowed to be temperature-dependent. In addition, ω constitutes the variables in the flowsheet. Based on these assumptions, a unified target for simultaneous reactor-energy synthesis can be obtained by extending (P7) to include the Duran and Grossmann heat integration model for the reactor network and flowsheet. This requires the solution of the following nonlinear programming problem:

$$\text{Max } \Phi(\omega, y, Q_H, Q_C) = J(\omega, y) - c_H Q_H - c_C Q_C$$

$$\text{s.t. } \sum_k X_{ik} L_k'(\alpha_j) - R(X_{ij}, T_{ij}) \Delta\alpha_i = 0 \quad j = 1, K$$

$$X(0) = X_0(\omega, y)$$

$$X_{iend} = \sum_k X_{ik} L_k(t_{end})$$

$$X_{i,0} = \phi_i X_0 + (1 - \phi_i) X_{(i-1)end}$$

$$X_{exit} = \sum_i \sum_j X_{ij} f_{ij}$$

$$\sum_{ij} (f_{ij} - q_{ij}) \alpha_{ij} = \tau$$

$$\sum_{ij} f_{ij} = 1$$

$$F_{(0)} = \phi_0 F_0$$

$$F_{(ij)} = \sum_i \phi_i F_{(i,0)} - \sum_{ij} f_{ij} F_0$$

$$Q_C = Q_H + \sum_{h \in H} W_h [T_h^{in} - T_h^{out}] - \sum_{c \in C} w_c [t_c^{out} - t_c^{in}]$$

$$Q_H \geq z_H^P(y)$$

$$h(\omega, y) = 0$$

$$g(\omega, y) \leq 0$$

(P10)

Here, T^P corresponds to all the inlet temperatures for the hot streams, and $t_c^{in} + \Delta T_m$ denotes all the cold streams. The heats of reaction are directly accounted for by the heat capacity flow rates of the reacting streams as follows. If Q_R is the heat of reaction to be removed (or added, for endothermic reactions) to maintain an isothermal reacting segment, the equivalent $(FC_p)_h$ or W_h for this reacting stream is equated to Q_R, and we assume a 1 K temperature difference for this reacting stream. In addition, the relations involving $h(\omega,y)$ and $g(\omega,y)$ are derived from interactions due to the rest of the flowsheet.

Finally, the max(0, Z) functions, which make up the $z_H^P(y)$ relations, have a nondifferentiability at the origin, which can lead to failure of the NLP solver. In order to provide smooth problem formulations, we approximate max(0, Z) as

$$f(Z) = \max(0, Z) \approx \frac{\sqrt{Z^2 + \epsilon^2}}{2} + \frac{Z}{2}$$

With values of $\epsilon = 0.01$, we were able to obtain a reasonably good approximation to the max function while simultaneously allowing easier solution of problem (P10). The advantage in this representation is that it provides a single function approximation over the entire domain.

B. Extensions from the Targeting Model

The solution of the nonlinear optimization problem (P10) gives us a lower bound on the objective function for the flowsheet. However, the cross-flow model may not be sufficient for the network, and we need to check for reactor extensions that improve our objective function beyond those available from the cross-flow reactor. We have already considered nonisothermal systems in the previous section. However, for simultaneous reactor energy synthesis, the dimensionality of the problem increases with each iteration of the algorithm in Fig. 8 because the heat effects in the reactor affect the heat integration of the process streams. Here, we check for CSTR extensions from the convex hull of the cross-flow reactor model, in much the same spirit as the illustration in Fig. 5, except that all the flowsheet constraints are included in each iteration. A CSTR extension to the convex hull of the cross-flow reactor constitutes the addition of the following terms to (P10) in order to maximize $\Phi(2)$ instead of Φ:

$$\text{Max} \quad \Phi^{(2)}(\omega, y^{(2)}, Q_H, Q_C) = J(\omega, y^{(2)}) - c_H Q_H - c_C Q_C$$

$$\text{s.t.} \quad X_{cstr} = X_{exit} + R(X_{cstr}, T_{cstr}) \tau_{cstr} \quad \text{(P11)}$$

$$\tau \geq 0, \quad X_{cstr} \geq 0$$

Here, X_{cstr} corresponds to the new reactor extension and $y^{(2)}$ is the set of new variables in the reactor and energy network. Besides the variables ω and y in (P10), this includes the variables corresponding to the new CSTR extension, namely, X_{cstr}, T_{cstr}, τ_{cstr}, and three more candidate streams for heat exchange. This is because the two heat exchangers will either cool or heat the feed to the CSTR (only one of these will exist in the optimal network), and one additional exchanger within the CSTR will maintain a desired temperature. If $\Phi^{(2)*} > \Phi^*$, we have a reactor extension that improves the objective function. The next step consists of creating the new convex hull of concentrations and checking for extensions that improve our objective function within the flowsheet constraints. As in Sections III and IV, we continue this procedure until there are no extensions that improve the objective function.

C. Energy-Integration Example

We now provide a small process example to illustrate the simultaneous synthesis of reactor and energy networks. Here, we consider a reaction mechanism in the van de Vusse form, though with kinetic expressions different from those used above. The integrated flowsheet corresponding to the synthesis problem is shown in Fig. 12.

The feed to the plant consists of pure A. This is mixed with the recycle gas stream consisting of almost pure A, and preheated (C1) before entering the reactor. After reaction, the mixture of A, B, C, and D passes through an aftercooler prior to distillation. In the first column, A is recovered and recycled, while in the second column, the desired product B is separated from CD, which is used as fuel. The distillation columns are assumed to operate with a constant temperature difference between reboiler and condenser temperatures (Andrecovich and Westerberg, 1985). The reflux ratios are fixed and the column temperatures are functions of the pressure in the column, which is variable so that efficient heat integration can be attained between the distillation columns and the rest of the process. The reactions involved in this flowsheet are as follows:

$$A \xrightarrow{k_1} B \xrightarrow{k_2} C$$
$$k_3 \downarrow$$
$$D$$

where $k_{10} = 8.86 \times 10^6$ h^{-1}, $k_{20} = 9.7 \times 10^9$ h^{-1}, $k_{30} = 9.83 \times 10^3$ liters mol^{-1} h^{-1}; $E_1 = 15.00$ kcal/g-mol, $E_2 = 22.70$ kcal/g-mol, and $E_3 = 6.920$ kcal/g-mol; and $\Delta H_{A \to B} = -0.4802$ kcal/g-mol, $\Delta H_{B \to C} = -0.918$ kcal/g-mol, and $\Delta H_{A \to D} = -0.792$ kcal/g-mol of A.

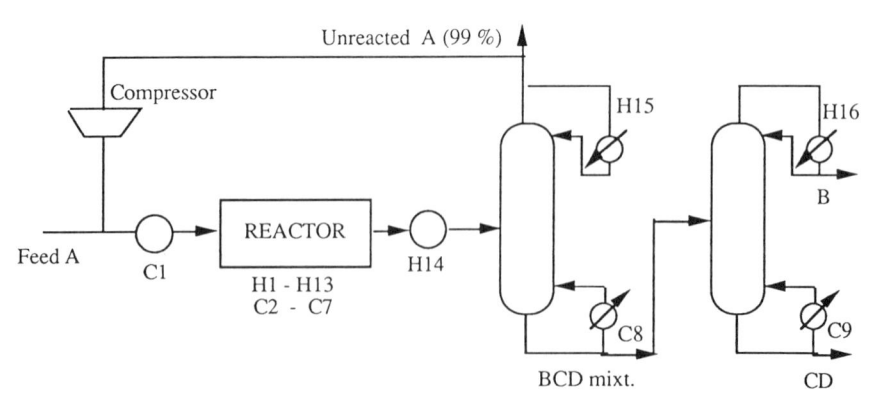

FIG. 12. Flowsheet for reactor–energy network synthesis.

The reactor is represented by the cross-flow discretization shown in Fig. 11. Here, we choose seven reactor segments (NE=7), with uniform segment lengths, $\Delta\alpha_i$. Since the reaction is exothermic, this corresponds to 14 hot streams and 7 cold streams. Thus, the streams in the reactor may be enumerated as hot streams H1–H13 (2NE − 1, since the entry point is fixed to be a preheater), and cold streams C1–C7. The streams H15–H16, and C8–C9 correspond to the condensers and reboilers of the distillation column. As described in (P10), the specific heats are assumed to be linear with the inlet temperatures. The objective function here as the total profit for the plant and is given in simplified form by

$$J = 1.7F_B + 0.8F_{CD} - 6.95 \times 10^{-5}\tau F_0 - 0.4566F_B(1 + 0.01(T_{H15}^{in} - 320))$$
$$- 0.7(F_B + F_{CD}) - 0.2F_{A0} - 0.007Q_C - 0.08Q_H$$

Further details of this formulation can be found in Balakrishna and Biegler (1992b). In this expression, F_B and F_{CD} represent the production rates of B and CD, respectively. F_{A0} is the flow rate of fresh feed. The third term corresponds to the reactor capital cost with τ, the residence time, and F_0, the total reactor feed; the fourth and the fifth terms correspond to the capital cost of the distillation columns. The operating costs of the columns are directly incorporated into the energy network in terms of condenser and reboiler heat loads. We assume that the cost of the reactor can be described by the total residence time and is independent of the type of reactor. The potential error from this assumption can be justified because the capital cost of the reactor itself is usually an order of magnitude or more smaller than the operating costs and the capital costs of the downstream processing steps.

Here, we consider two alternatives. First, we consider the sequential approach, where we optimize the reactor network with an optimal temperature profile, then integrate the maintenance of this optimal profile with the energy flows in the rest of the flowsheet. In the second case, we solve the above problem with the simultaneous formulation proposed in (P10).

The optimization model in the sequential case had 342 equations and 362 variables for the reactor flowsheet optimization (96 CPU seconds on the VAX 6320) and 200 equations and 161 variables for the energy integration (170 CPU seconds). The simultaneous optimization model (542 equations, 523 variables) was solved in 1455 CPU seconds and was initialized with the solution to the sequential model. Table I presents a comparison between the results for sequential and simultaneous modes to synthesis. A target production rate of 40,000 lb/h is assumed for the desired product B.

Clearly, the simultaneous formulation leads to a significant improvement in the overall profit and conversion due to the correct anticipation of the energy costs in the reactor design. Even though the shape of the temperature profiles is not markedly different, the temperatures in the simultaneous case are lower,

TABLE I
COMPARISON BETWEEN SEQUENTIAL AND SIMULTANEOUS FORMULATIONS

	Sequential	Simultaneous
Overall profit	38.98 × 10^5 $/year	74.02 × 10^5 $/year
Overall conversion	49.6%	61.55%
Hot utility load	3.101 × 10^5 BTU/h	2.801 × 10^5 BTU/h
Cold utility load	252.2 × 10^6 BTU/h	168.5 × 10^6 BTU/h
Fresh feed A	8.057 × 10^4 lb/h	6.466 × 10^4 lb/h
Degraded product C	3.112 × 10^4 lb/h	1.44 × 10^4 lb/h
By-product D	0.933 × 10^4 lb/h	1.00 × 10^4 lb/h
Unreacted (recycled) A	1.22 × 10^4 lb/h	1.963 × 10^4 lb/h

as seen in Fig. 13. These lower temperatures lead to a significant reduction in the degradation of product B to by-product C, as seen in Table I. Since the B–C reaction is also the most exothermic, the retardation of this reaction leads to less heat evolution and therefore less cold utility consumption. Furthermore, the more efficient conversion to B results in less consumption of raw material A, leading to higher overall conversion and less total flow within the reactor in the simultaneous case. This results in lower heat capacity flow rates for the reacting hot streams, and hence lower cold utility consumption. However, the optimal reactor in both sequential and simultaneous cases (a nonisothermal PFR) has the same residence time of 0.59 s. Since the temperatures are lower in the simultaneous case, the conversion per pass of A is actually lower in the simultaneous case, leading to higher recycles in the simultaneous case. In either case, of the 20 candidate streams, only 12 streams are active in the optimal network. This is because the strictly falling temperature profiles avoid the use of any cold streams (C2–C7) within the reactor network.

In addition, no mixing was predicted in the solution, so cold-shot cooling was not used at all. However, this decision is directly influenced by the ratio of the raw material to energy costs. For small ratios, even if mixing is not optimal in concentration space, the energy costs may drive the use of cold shots in order to reduce utility consumption. Within the constraints on the residence time ($\tau^{up} =$

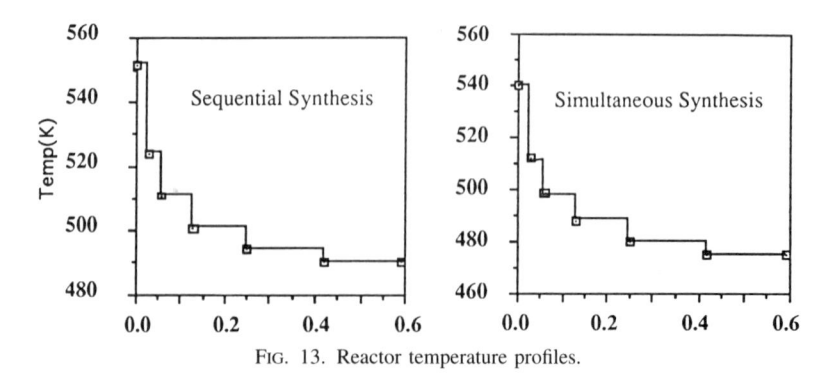

FIG. 13. Reactor temperature profiles.

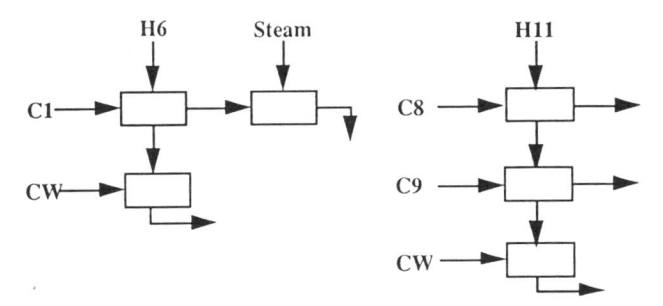

FIG. 14. Heat-exchanger network substructure.

1.00 s), no further extensions that improve the objective function are found from these solutions to either the sequential or the simultaneous targeting model. The pinch points correspond to 546.5 K and 535.1 K for the sequential and simultaneous schemes, respectively. Also, the heat loads for the hot streams are significantly higher than those for the cold streams. Thus, the cooling (or T–Q curves) for the hot streams will be nearly flat. Since the pinch corresponds to the inlet temperature of the hottest hot stream in either case, no part of the T–Q curve for the hot streams will extend beyond the pinch. The matches below the pinch are thus very easily determined. In fact, streams C1, C8, and C9 can be matched with any of the streams from H1 to H11, without any alteration in the utility cost. The network in this case is thus innately flexible; the primary reason for this is the significant disparity in the heat contents due to the high reaction exothermicity.

One feasible network would correspond to the cold streams C1, C8, and C9 diverted to suitable jacketed reactor compartments, as the simple network in Fig. 14 shows. The hot streams not shown in this network are matched directly with cooling water (CW), and the amount of steam used here is very small. Note that this network would require the same minimum utility consumption predicted by the solution of (P10). It can be inferred that the network in Fig. 14 is equally suitable for both the simultaneous and sequential solutions. In fact, Balakrishna and Biegler (1993) showed that, for exothermic systems in which the reactor temperature is the highest process temperature, the pinch point is known *a priori* as the highest reactor temperature (in this case, the feed temperature) and the inequality constraints in (P10), $Q_H \geq z_H^p(y)$, $p \in P$, can be replaced by a simple energy balance constraint. This greatly reduces the computational effort to solve (P10).

VI. Simultaneous Reaction, Separation, and Energy System Synthesis

Previous studies on process integration have generally considered reaction and separation as processes that occur sequentially in a flowsheet. In this section, a unified formalism is presented for the synthesis of reaction–separation systems while ensuring optimal energy management. The synthesis approach is developed in the light of the ideas previously presented on the sequential bounding scheme for reactor targeting. While previous formulations have considered reaction and separation as sequential operations in a flowsheet, our model is developed to consider simultaneous reaction and separation as an option within the network. Such simultaneous events occur, for example, in reactive distillation and membrane reactors.

We first develop a nonisothermal reactor model, which allows for separation as reaction progresses. This is facilitated through a species-dependent residence time distribution. Optimization of this residence time distribution function leads to a separation profile as a function of age. The synthesis model is then formulated as a mixed-integer optimal control problem, where the integer variables account for the fixed costs of separation. The control profiles include the temperature, the separation profile, and some mixing functions defined for the network. Costs for maintaining a separation profile are addressed next by defining a separation index (defined to model the intensity of separation) and a fixed charge for any separation between two components in the reaction mixture. Following this, we consider the amalgamation of this formulation with energy minimization and develop simplifications for systems with highly exothermic reactions. The solution to this model gives us a lower bound on the performance index; we therefore present schemes to successively improve these bounds, based on reactor extensions.

A. COMBINED REACTION–SEPARATION MODEL

Figure 15 gives a schematic of a simultaneous reaction–separation model. To include separation in a reactor targeting model, we postulate a separation function vector (γ) analogous to a residence time distribution function for homogeneous reactors. Here, however, each species has its own residence time distribution function, which is dependent on its separation function γ_c.

If we define $m_c(\alpha)$ as the mass of component c (c − 1, ..., C) at time α, then a mass balance around the network in Fig. 15 leads to

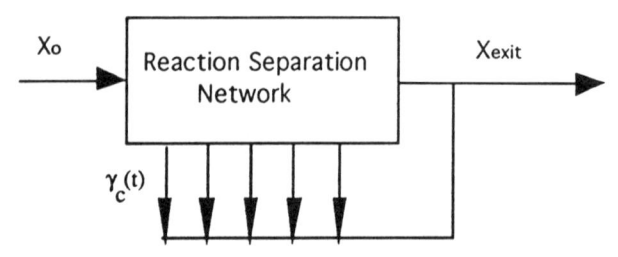

FIG. 15. Flow model for combined reaction–separation targeting.

CHEMICAL REACTOR NETWORK TARGETING AND INTEGRATION 285

$$\frac{d\sum_{c=1}^{C} m}{d\alpha} = -\sum_{c=1}^{C} \gamma_c(\alpha) m_c(\alpha)$$

where $\gamma_c(\alpha)$ is a continuous function of α. For a homogenous system, if ρ is the density of the system, then

$$\frac{dQ}{d\alpha} = -\sum_{c=1}^{C} \frac{\gamma_c(\alpha) m_c(\alpha)}{\rho}$$

We assume constant-density systems for the sake of simplicity, even though variable density could be considered by a straightforward extension of this model.

Now a differential balance around an infinitesimal element of α for component c gives

$$\frac{dX_c}{d\alpha} = R_c(X,T) + X_c(\alpha)\left[\frac{\gamma(\alpha)^T X(\alpha)}{\rho} - \gamma_c(\alpha)\right]$$

With this governing equation, a mathematical model for maximizing the performance index in this reacting environment can be derived as follows:

$$\underset{\gamma,T}{\text{Max}} \quad J_{\text{exit}}(m_{c(\text{exit})}, Q, \tau)$$

$$\frac{dX_c}{d\alpha} = R_c(X,T) + X_c(\alpha)\left[\frac{\gamma(\alpha)^T X(\alpha)}{\rho} - \gamma_c(\alpha)\right]; \quad c = 1, C$$

$$\frac{dQ}{d\alpha} = -\sum_{c=1}^{C} \frac{\gamma_c(\alpha) m_c(\alpha)}{\rho}$$

$$m(\alpha) = X(\alpha) Q(\alpha)$$

$$m_c(0) = m_{c0} \tag{P12}$$

$$\sum_{c=1}^{C} m_c(0) = \sum_{c=1}^{C} m_c(\text{exit})$$

$$m_c(\text{exit}) = \int_0^\infty \gamma_c(\alpha) m_c(\alpha) \, d\alpha$$

$$\tau = \int_0^\infty \alpha f(\alpha) \, d\alpha$$

$$g(\gamma, X, \mu) \leq 0$$

$$h(\gamma, X, \mu) = 0$$

Here, J is an objective function specified by the designer, $X(\alpha)$ is the mass concentration vector of molecules of age α, m_{c0} is the mass flow of each species at the entrance to the reactor, and m_c(exit) is the mass flow of each species at the reactor exit given as the integral of outlet flows at different points within the reacting system. The residence time, τ, is determined from the RTD function $f(\alpha)$. Also, g and h represent the inequality and equality constraints imposed by the process variables (μ) on the reaction system.

Clearly, formulation (P12) is an optimal control problem with differential equation constraints, where the γ_c's, and the temperature are the control profiles. The solution to this model will give us the optimal separation profile along the reactor. It is clear that $\gamma(\alpha)$ models the effect of separation within the reactor network. If all the elements of the vector $\gamma(\alpha)$ are the same (which implies that there is no relative separation between the species in the reactor), the second term for the governing differential equation vanishes, since

$$\gamma(\alpha)^T X(\alpha) = \sum_{c=1}^{C} \gamma_c X_c = \gamma_c \sum_{c=1}^{C} X_c = \gamma_c \rho$$

therefore

$$\left[\frac{\gamma(\alpha)^T X(\alpha)}{\rho} - \gamma_c(\alpha) \right] = 0; \quad \forall \, c \in C$$

Thus, the governing equation to this reactor scheme reduces to that of segregated flow, and the formulation reduces to the segregated flow optimization problem in Section III. Furthermore, the γ_c's can now be directly related to the RTD function (Balakrishna and Biegler, 1993) through the following relation:

$$\gamma_c(\alpha) = \frac{f(\alpha)}{1 - F(\alpha)}$$

where $f(\alpha)$ is the true residence time distribution of the molecules within the reactor network, and $F(\alpha)$ is the cumulative RTD $= \int_0^\alpha f(t) \, dt$.

However, if the γ_c's are not the same for all components—i.e, there exists a separation profile—then the actual RTD for this system is given by:

$$f(\alpha) = \sum_c \frac{\gamma_c(\alpha) q_c(\alpha)}{Q_0}$$

where $q_c(\alpha) = m_c(\alpha)/\rho$.

The solution to (P12) gives us the optimal separation profile as a function of age within the reactor. However, except in the case of reactive phase equilibrium, the assumption of a continuous separation profile is not really required. Furthermore, a continuous separation profile may not be implementable in practice. To address this, we take advantage of the structure of a discretization procedure for the differential equation system. In this case, we choose orthogonal collocation on finite elements to discretize the above model. This results

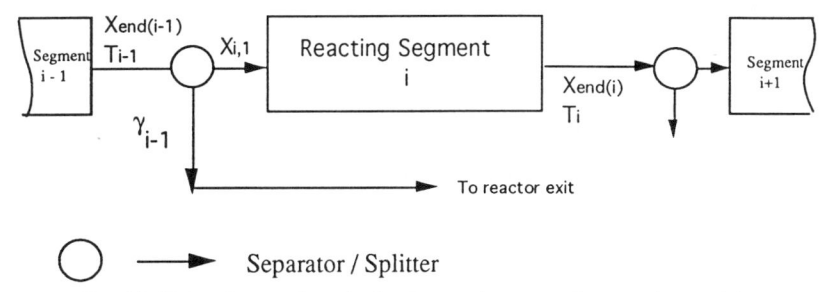

Xend(i-1)
Segment i-1 | T_{i-1} | $X_{i,1}$ → Reacting Segment i → Xend(i), T_i → Segment i+1

γ_{i-1}

→ To reactor exit

○ → Separator / Splitter

FIG. 16. Finite-element discretization for reaction–separation targeting model.

in a reactor structure as shown in Fig. 16, where we restrict separation only to the ends of each finite element. Also, the differential equations are now converted to algebraic equations through collocation, and the optimal control problem is reduced to the nonlinear program shown in (P13). Furthermore, it can be shown that as $\Delta\alpha$ tends to zero, this discretized model is equivalent to the original optimal control problem. [A proof for this equivalence can be found in Balakrishna and Biegler (1993).] The $\gamma_c(\alpha)$ in the original model now reduces to a mass split fraction vector of each species at the end of each element ($\gamma_{c,i}$). Finally, the control profiles, temperature (T), separation fractions (γ_c), and the RTD are assumed to be piecewise constant over each element.

$$\underset{\gamma,T}{\text{Max}} \quad J_{\text{exit}}(m_{\text{exit}}, Q, \tau)$$

$$\sum_k X_{ik} L_k'(t_j)/\Delta\alpha_i - R(X_{ij}, T_{ij}) = 0 \quad @ \ t_j \ \text{s.t.} \ j \neq 0$$

$$X(0) = X_0$$

$$X_{i\text{end}} = \sum_k X_{ik} L_k(t_{\text{end}})$$

$$m_{c,i+1} = [X_{c,i\text{end}}] Q_i [1 - \gamma_{c,i}]$$

$$X_{i1} Q_{(i)} = m_i \tag{P13}$$

$$Q_{(i+1)} = Q_i \left[1 - \sum_{c=1}^{C} X_{c,i\text{end}} \gamma_{c,i}/\rho\right]$$

$$m_{c,\text{exit}} = \sum_i X_{c,i\text{end}} \gamma_{c,i} Q_i$$

$$f(i) = \sum_{c=1}^{C} \gamma_{c,i} m_{c,i}/Q_0$$

$$\tau = \sum_i f(i) t(i) \leq t_{\text{max}}$$

$$\sum_i \Delta\alpha_i = t_{\text{max}}$$

where

X_{ij} = Mass concentration vector at collocation point j in finite element i (point $[i,j]$)
$L_k'(t_j)$ = Derivative of Lagrange interpolation polynomial at $[i,j]$
γ_i = Mass split fraction vector at the end of finite element i (array of $\gamma_{c,i}$)
$f(i)$ = Actual RTD for the system at element i
T_{ij} = Temperature at $[i,j]$
X_{iend} = Mass concentration vector at the end of element i.
m_i = Species mass flow vector entering element i
Q_i = Total volumetric flow rate entering element i

The first three constraints represent orthogonal collocation applied to the differential equations at the collocation points. The next three equations represent mass balances at the separation point. The discretized RTD function and the expression for the mean residence time are given in the final constraints. As $\Delta\alpha \to 0$, this model is equivalent to the original reaction–separation model (P12). The main difference is that we allow separation only at the end of each element; within each element no separation occurs. Although the model appears nonlinear, the nonlinearities are actually reduced when one considers the rates in terms of the mass fractions. The solution to this model then gives us the optimal separation split fractions as a function of time along the reactor.

One important issue that still needs attention is the objective function. It is intuitively obvious that if a separation cost is not associated with it, we will usually end up getting near-complete separations of products, and hence complete conversions to the extent possible within stoichiometric constraints. Thus the AR in concentration space can easily be the entire stoichiometric space. Unfortunately, it is difficult to get an accurate representation for the separation cost, especially when sharp splits are not enforced. Here, we present a simple cost model by assuming that the variable cost of separation is determined by two factors, namely, the difficulty of separation and the mass flow rate through the separator.

We first consider an example for modeling the separation costs. As shown in the schematic in Fig. 17, a stream with components A,B,C and mass flow rates F_A, F_B, F_C undergoes a separation operation into two output streams, with mass flow rates F_{A1}, F_{B1}, F_{C1} and mass flows F_{A2}, F_{B2}, F_{C2}, respectively. The streams A, B, and C are arranged in a sequential order of separability; for example, in the case of distillation, we may assume that A, B, and C are in decreasing order of volatility. The mass fractions γ_A, γ_B, and γ_C are then defined as

$$\gamma_A = F_{A1}/F_A, \quad \gamma_B = F_{B1}/F_B, \quad \gamma_C = F_{C1}/F_C$$

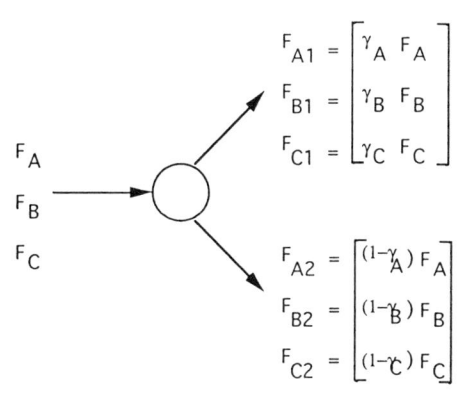

CHEMICAL REACTOR NETWORK TARGETING AND INTEGRATION 289

$$\begin{aligned}F_{A1} &= \begin{bmatrix}\gamma_A F_A\\ \gamma_B F_B\\ \gamma_C F_C\end{bmatrix}\\ F_{B1} &\\ F_{C1} &\end{aligned}$$

$$\begin{aligned}F_{A2} &= \begin{bmatrix}(1-\gamma_A) F_A\\ (1-\gamma_B) F_B\\ (1-\gamma_C) F_C\end{bmatrix}\\ F_{B2} &\\ F_{C2} &\end{aligned}$$

FIG. 17. Separation of mass fractions.

If the split fractions $\gamma_A = \gamma_B = \gamma_C$, we have only a splitting operation without any separation. Otherwise, there is a relative separation between two adjacent components in the mixture and we define $|\gamma_A - \gamma_B|$ as a measure of the intensity of separation between these two components. When $\gamma_A - \gamma_B = \gamma_B - \gamma_C = 0$, we have only a splitting operation among these components, and the cost of separation is identically zero; however, if $\gamma_A - \gamma_B = \pm 1$, we have a sharp split between components A and B. Any intermediate degree of separation could then be modeled by complete sharp split separation followed by mixing in order to achieve the desired composition.

We can generalize this to formulate the separation costs. Let $M = \{m\}$ denote the set of all components in the reacting system and let these be arranged in some monotonic order of relative separability, such as volatilities. If Q is the mass flow rate handled by the separator, the cost of separation may be described by

$$C_{sep} = C_{capital} + C_{operating}$$
$$C_{capital} = C_{fixed(mn)} y_{mn} + \sum_{m;n=m+1} p_{mn} |\Delta\gamma_{mn}| Q$$

Here, y_{mn} is the binary variable associated with the separation of components m and n, such that if $y_{mn} = 0$, then $\Delta\gamma_{mn} = 0$; and if $y_{mn} = 1$, then $\Delta\gamma_{mn} \leq 1$. The second term models the intensity of separation, where the cost coefficient p_{mn} for unit separation between two adjacent components m and n reflects the difficulty of separation between m and n. Q is the net flow through the separation network. The above formulation gives us an exact representation when we have sharp splits between adjacent components. As we mentioned earlier, nonsharp splits can be modeled by sharp splits followed by mixing, and an upper bound on the separation costs can be derived by enforcing $|\Delta\gamma_{mn}| = 1$ whenever $y_{mn} = 1$ (i.e., by assuming sharp splits) while a lower bound on the separation cost is given by the expression above. The operating cost (reboiler and condenser duties in distillation, for example) can be directly incorporated into the energy minimization framework presented next.

The presence of $|\Delta\gamma_{mn}|$ in the cost function makes the objective function in (P13) nondifferentiable, so it is reformulated by adding the following terms within (P13):

$$\text{Max} \quad J_{\text{exit}} = J_{\text{product}}(m_{\text{c(exit)}}, Q, \tau) - C_{\text{fixed}(mn)} y_{mn} - \sum_{m;n=m+1} p_{mn} \Delta_{mn} Q - C_{\text{operating}}$$

$$\Delta_{mn} \geq \gamma_m - \gamma_n$$

$$\Delta_{mn} \geq \gamma_n - \gamma_m$$

Since Δ_{mn} is to be minimized in the objective, it is easy to show that this reformulation would result in $\Delta_{mn} = |\Delta\gamma_{mn}|$, as desired, at the optimal solution.

This approach was applied to the Williams–Otto process (Balakrishna and Biegler, 1993). In previous studies, this process was optimized with a CSTR reactor followed by waste and product separators and a recycle stream. The application of (P12) to this problem led to a significantly improved process, particularly when the separation costs (p_{mn}) were low enough to allow coupled reaction and separation. Without separation, the optimal network is a single PFR with twice the return on investment of previous studies. Allowing for separation leads to a tubular reactor with sidestream separators to remove product and waste as they are created. The resulting process objective has a further fivefold improvement.

B. Unified Formulation for Optimal Energy Utilization

The combined reaction–separation model has advantages because it allows us to consider both reaction and separation within one framework. We now extend this formulation further to include energy minimization by using the concepts of pinch technology. Thus heat effects within the reactor and separator are integrated optimally with the energy flows in the flowsheet. The energy minimization scheme for this network closely follows the development in Section V, where our reactor targeting model was integrated with an energy targeting framework based on minimum utility consumption (Duran and Grossmann, 1986). The schematic in Fig. 18 shows one finite element of the discretized reactor–separator representation of Fig. 16 along with the candidate heat-exchange streams. Based on the development in Section V, a unified reactor–separator–energy target can be derived from the solution to the following mixed-integer nonlinear programming (MINLP) problem:

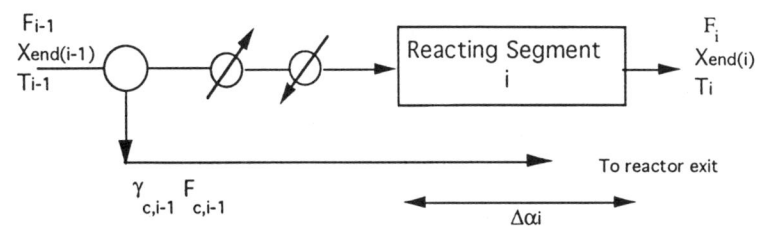

CHEMICAL REACTOR NETWORK TARGETING AND INTEGRATION 291

FIG. 18. Discretized model for energy minimization.

Max $\Gamma(\omega,\psi,Q_H,Q_C) = J(\omega,\psi) - c_H Q_H - c_C Q_C - C_{sep}$

s.t. $\sum_k X_{ik} L_k'(\alpha_j) - R(X_{ij}, T_{ij})\Delta\alpha_i = 0 \quad j = 1, K$

$X(0) = X_0$

$X_{iend} = \sum_k X_{ik} L_k(t_{end})$

$X_{c,i,0} F_i = (1 - \gamma_{c,i-1}) X_{c,(i-1)end} F_{(i-1)}$

$F_{c,exit} = \sum_i \gamma_{c,i} X_{c,(i)end} F_i$

$\Delta_{mn} \leq y_{mn}$

$0 \leq \gamma_i \leq 1$ \hfill (P14)

$\Delta_{mn} \geq \gamma_m - \gamma_n$

$\Delta_{mn} \geq \gamma_n - \gamma_m$

$Q_C = Q_H + \sum_{h \in H} W_h[T_h^{in} - T_h^{out}] - \sum_{c \in C} w_c[t_c^{out} - t_c^{in}]$

$z_H^P(\psi) = \sum_{c \in C} w_c[\max\{0; t_c^{out} - \{T^P - \Delta T_m\}\} - \max\{0; t_c^{in} - \{T^P - \Delta T_m\}\}]$
$\quad - \sum_{h \in H} W_h[\max\{0; T_h^{in} - T^P\} - \max\{0; T_h^{out} - T^P\}]$

$Q_H \geq z_H^P(\psi)$

$g(\omega,\psi,y) \leq 0$

$h(\omega,\psi,y) = 0$

Here, the variables are defined as follows:

ψ = Set of variables in the reaction–separation–energy network
ω = Set of external flowsheet parameters
Q_H, Q_C = Heating and cooling utility loads

c_H, c_C = Cost coefficient for utility loads
w_H, w_C = Heat capacity flow rates for hot and cold streams, respectively
C_{sep} = Total separation cost
T_h^{in}, T_h^{out} = Inlet and outlet temperatures respectively for hot stream h
t_c^{in}, t_c^{out} = Inlet and outlet temperatures respectively for cold stream c
F_i = Total mass flow rate at element i
$F_{c,exit}$ = Mass flow rate of component c at reactor exit
z_H^p = Heating deficit above the pinch

The objective function Γ is a function of the variables within the unified reactor model and the heating and cooling utility loads. The cost model for separation presented in the previous section is incorporated within Γ as C_{sep}. The operating costs for the separation profile (for example, heat loads in the case of distillation) are directly incorporated into the energy minimization formulation. Note that this formulation combines the energy target in (P10) with the separation target in (P12).

The solution to formulation (P14) gives us an optimal network for the reactor flow configuration shown in Fig. 18. However, this flow model may not be sufficient for the synthesis problem, and we need to check for any other reactors that will help us improve the objective function. Using the approach developed in previous sections, we check for the same CSTR extensions from the solution to our unified reactor targeting model. Thus, in addition to the constraints in (P14), we add the CSTR extension constraints in (P11) and define $\Gamma^{(2)}$ as the objective with the same form as that in (P14). Here, $\psi^{(2)}$ is the set of new variables in the reactor energy network; this includes all the variables within ψ, the new variables X_{CSTR}, T_{CSTR}, and τ_{CSTR}, and the corresponding heat-exchange variables for cooling/heating the stream in the CSTR. There are no additional separation variables, since separation is confined to the segregated-flow component of this system. If the optimal solution to this formulation is $\Gamma^{(2)*} \geq \Gamma^*$, we have a reactor extension that improves the objective function. The next step consists of creating the new convex hull of concentrations and checking for any further extensions that improve the objective function within the flowsheet constraints. As before, we continue this procedure until there are no further reactor extensions that improve the objective function.

C. Example Problem

In this section, we reconsider the van de Vusse process to illustrate our synthesis approach. This example also shows the application of the unified reaction–separation–energy integration model. Comparisons are made between sequential and simultaneous modes of synthesis, and the applicability of the simplified model is verified.

CHEMICAL REACTOR NETWORK TARGETING AND INTEGRATION   293

Here, we devise a reaction separation network featuring energy integration for the following system using the proposed targeting scheme.

$$A \xrightarrow{k_1} B \xrightarrow{k_2} C$$
$$\downarrow k_3$$
$$D$$

The feed to the plant consists of pure A. This is mixed with the recycle gas stream, consisting of almost pure A, and preheated (C1) before entering the reactor. The flowsheet in Fig. 12 shows the reaction separation network followed by final separation columns to obtain product streams containing pure B and a C–D mixture. The volatilities of components in the network are given in the following descending order: [A, B, C, D]. The operating costs of the distillation columns (reboiler and condenser duties) are directly incorporated into the energy integration formulation and the columns are modeled in the same way as those in Section V. The reactor here is modeled by the discretized targeting model as shown in Fig. 16, with eight finite elements in the collocation procedure. The discretization procedure results in a total of 18 candidate hot streams and 11 candidate cold streams within the reaction separation network. The objective here is to maximize the total profit given by

$$J = 30F_B - 18F_{CD} - 6.95 \times 10^{-4}\tau F_0 - 4.566 F_B(1 + 0.01(T^{in}_{H15} - 320))$$
$$- 7(F_B + F_{CD}) - 2F_{A0} - C_{sep} - 0.07 Q_C - 0.8 Q_H$$

In this expression, F_B, F_{CD} represent the production rates of B and CD, respectively, and F_{A0} is the amount of fresh feed. The first term corresponds to the product value, while the second term corresponds to the cost of waste treatment for undesired products C and D. The third term corresponds to the reactor capital cost, while the fourth and the fifth terms correspond to the recycling costs. The costs incurred for maintaining a desired separation profile, C_{sep}, is given in (P14).

The operating costs of the columns are directly incorporated into the energy network in terms of condenser and reboiler heat loads. We assume that the cost of the reactor can be described by the total residence time and is independent of the type of reactor. This can be justified on the grounds that the capital cost of the reactor itself is usually orders of magnitude smaller than the operating costs and the capital costs of the downstream processing steps. A target production rate of 960,000 lb/day is assumed for the desired product B.

Here, we consider two alternatives. First, we consider the sequential reaction and separation approach, where we force all the separation fractions to split only fractions. In the second case, we solve the above problem with the formulation proposed in (P14). In this case, the reaction–separation system and the energy network are optimized simultaneously. Table II presents a comparison between

TABLE II
RESULTS FOR SEQUENTIAL AND SIMULTANEOUS FORMULATIONS

	Reaction followed by separation	Simultaneous reaction and separation
Overall profit	53.87 × 10^6 $/year	202.33 × 10^6 $/year
Hot utility load	3.20 × 10^5 BTU/h	2.13 × 10^5 BTU/h
Cold utility load	131.120 × 10^6 BTU/h	126.799 × 10^6 BTU/h

the solutions obtained for simultaneous reaction separation and sequential reaction and separation. The results clearly show that by considering simultaneous reaction and separation as an option within the network, significant increases in overall profit can be obtained for this system.

As shown in Fig. 19, the separation profiles indicate removal of B and CD as reaction progresses, while retaining A for the complete residence time of 0.45 s. The temperature profile is a falling one as long as B and CD remain in the reactor. At every point where B and CD are separated out of the reactor, the temperature rises. This is because, as long as there is only A, a high reaction rate is desired to minimize reactor volume. However, as more B is produced, the temperature profile falls so as to reduce the excessive degradation of B to product C. Thus, the optimal temperature profile in this case is a nonmonotonic one. Also, among the 8 finite elements used in the discretization, only the candidate streams corresponding to 6 elements are active, since at the end of the sixth element, all molecules leave to the reactor exit ($t = 0.45$ s), as shown by the separation profile in Fig. 20. Furthermore, of the 18 candidate hot streams and 11 candidate cold streams, only 12 hot streams and 6 cold streams were active in the optimal network. Also, from the solution of the reactor extension problem, no reactor extensions are observed that improve the objective function for both sequential and simultaneous formulations.

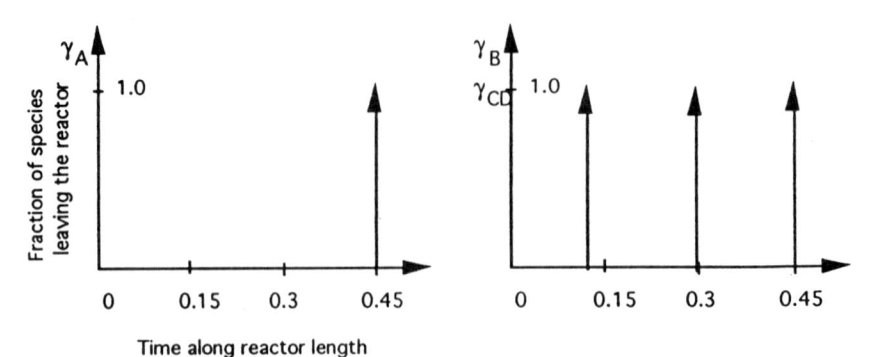

FIG. 19. Separation profiles along reactor length (simultaneous case).

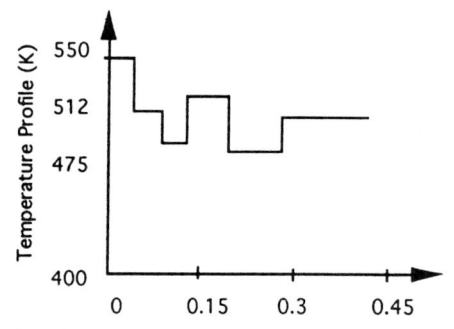

FIG. 20. Temperature profile (simultaneous case).

VII. Summary and Conclusions

In this paper we have addressed the development of mathematical programming strategies for the optimal synthesis and flowsheet integration of chemical reactor networks. Our discussion develops new methodologies for three major aspects of the reactor synthesis problem. These include optimal reactor network synthesis, energy integration of reactor networks, and the development of a combined reaction–separation–energy integration framework for overall synthesis.

In reactor network synthesis approaches based on superstructure optimization, the limitations stem from solutions that may be local or nonunique and that are only as rich as the superstructure chosen. To address these issues, geometric approaches based on attainable region (AR) concepts have recently been developed. Here, a region in concentration space is constructed that cannot be extended with further mixing and/or reaction. Hence, this region includes the concentation trajectories of all possible reactor networks for this system. This geometric approach leads to important insights into the structure of the optimal network. However, as construction of the attainable region is currently based on graphical tools, finding the AR is limited to two- or three-dimensional problem representations.

The synthesis approach proposed in this paper addresses the drawbacks of the superstructure and graphical AR techniques through a constructive, optimization-based targeting methodology. This targeting approach proceeds through the development of simplified reactor models and applies the concept of ARs to verify the sufficiency of these models. Our targeting approach is based on successively generating points of the feasible region for the reaction system with optimization over the convex hull of these points. We started with the segregated-flow limit to this model, which can be solved through a linear programming simplification. The example problems in Section III and in Balakrishna and Biegler (1992a) indicate that, in many cases, the segregated-flow model is sufficient to describe the network. Sufficiency conditions also allow us to evaluate the suitability of the target obtained from the segregated-flow model. When the segregated-flow model is not sufficient, simple nonlinear programs can be solved to enhance the target. Here, each nonlinear program corresponds to a reactor extension from the convex hull of concentrations generated from the previous model, and this procedure is continued until there are no reactor extensions that improve the objective. With this stagewise approach, we therefore solve simpler models and verify their sufficiency in a stagewise manner.

The extension of this approach to nonisothermal systems follows similarly from the above approach. The main difference here is that temperature is an additional control profile and we consider schemes for maintaining a desired temperature profile. We accomplish this by postulating a cross-flow reactor model as the initial targeting model, since this allows for temperature control through feed mixing. From the convex hull of the concentrations available through the cross-flow reactor, we solve small nonlinear programming problems to generate reactor extensions that improve the objective function. In contrast to isothermal synthesis, the variable temperature profile in the initial cross-flow representation itself encompasses a large choice for the AR. In fact, while previous approaches have derived results by placing constraints on the temperature profiles, we show that for suitable temperature profiles, the conversion can asymptotically approach a stoichiometric upper bound for some systems. This is illustrated by the Lee–Aris sulfur dioxide oxidation, where the extent of reaction asymptotically approaches the upper bound, for a temperature profile shown in Section IV.

Reactor network synthesis in isolation, however, fails to address the interaction of the reactor design on the other process subsystems within the flowsheet. For example, reactors are associated with significant heat effects and our targeting approach allows for the integration of the reactor target with synthesis schemes for energy networks. The integration of our reactor targeting formulation with an energy targeting scheme, based on minimum utility costs (Duran and Grossmann, 1986), provides a general formalism for synthesizing energy integrated reactor networks. The results on the example problem indicate that significant increases in profit can be obtained by considering the two subsystems within a unified framework. Also, owing to high reaction exothermicities in the example, the heat-exchanger network is very flexible, as described in Section V.

Finally, we consider a preliminary approach for the optimal synthesis of reactor–separation systems. Here, we formulate a combined reaction–separation model by postulating a species–dependent residence time distribution. The optimization of this distribution function leads to a separation profile as a function of time along the reactor. The costs for maintaining a separation profile are handled through a separation index, which models the intensity of separation, and a fixed charge modeled by binary variables for separation between any two components in the system. The reaction–separation model is then integrated with an energy targeting approach, and is implemented within our targeting framework. Results from a small example illustrate that the flowsheet performance indices can be enhanced significantly by considering reaction–separation systems, in contrast to the more conventional sequential reaction and separation.

While our targeting approach leads to an efficient, systematic strategy for synthesis of integrated reactor networks, there are still many open questions for targeting approaches. These can be summarized in terms of further development of geometric concepts, refinement of optimization strategies, and further extension and application to the design of new and existing processes. These topics are briefly summarized below.

Concepts for the construction of two- and three-dimensional regions have been firmly established by the work of Glasser, Hildebrandt and co-workers. For higher-dimensional systems, Hildebrandt and Feinberg (1992) have established a number of properties that lead to useful insights for processes with reaction and mixing, as well as additional rate processes. However, constructive procedures for higher-dimensional attainable regions still need to be developed. Similarly, optimization formulations still need to be refined in order to exploit the concepts of higher-dimensional ARs. This work presents simple formulations for constructing ARs consisting of PFRs, CSTRs, and some DSRs. However, improved formulations are still needed in order to provide "nonmonotonic" improvements in the network, as observed in Example 3. Moreover, many of the relationships that characterize the existence of DSRs and other connectors in higher-dimensional ARs need to be incorporated within optimization formulations.

Related to the development of AR concepts is the application of more powerful optimization formulations and algorithms. A current limitation to our targeting approach is the application of *local* optimization algorithms to nonconvex problem formulations, even though the resulting AR formulations demand global solutions. Clearly the development and application of efficient global optimization algorithms is still needed. A summary of the state of the art for global optimization in process synthesis is given by Floudas and Grossmann (1994). A related problem is the incorporation of uncertainty into the synthesis of reactor networks. Since reaction rate laws are often uncertain and the process needs to operate under a variety of conditions, it is important that the reactor network be flexible and robust to process changes and uncertain parameters. This problem has yet to be addressed for reactor network synthesis, even though systematic approaches have been developed for other process systems (see Grossmann and Straub, 1991, for a comprehensive survey). For reactor networks, this problem is complicated by the nonlinear character of the process as well as the large problem size.

Finally, further extensions of reactor network targeting include the design of reactive separation processes. Spurred by industrial successes (Agreda et al., 1990), the strong integration of reaction and separation processes can lead to significant improvements and savings in the design of new processes. Section VI presents a preliminary approach for identifying the potential for coupling these two processes, but detailed phenomena have not been included in this approach. Again, the nonlinearity and complexity of the reaction and phase equilibrium models make this problem very difficult. Nevertheless, as with the generation of ARs, geometric insights (e.g., by Barbosa and Doherty, 1988) should lead to simplification of the synthesis procedure as well as refinement of the optimization formulation. In addition to reaction/separation systems, reactor network targeting can be applied to a number of design problems, particularly for the synthesis of waste-minimizing flowsheets. Here, the approaches described in this paper can be applied directly to these problems simply by considering waste generation as part of the design objective. Lakshmanan and Biegler (1994) recently considered this problem and established trade-offs in reactor targeting between process profitability and waste generation.

In conclusion, further development and application of reactor network targeting concepts will change the nature of current chemical process design and focus more attention on the integration of the reaction system and other process subsystems. This will lead to a better quantitative understanding of these systems and their trade-offs and ultimately will lead to processes that are more environmentally benign, more profitable, and less wasteful of raw materials, capital, and energy.

Acknowledgments

This work was supported by the Engineering Design Research Center, an NSF sponsored Engineering Research Center at Carnegie Mellon University, and by the Department of Energy.

References

Achenie, L. E. K., and Biegler, L. T. "Algorithmic Synthesis of Chemical Reactor Networks Using Mathematical Programming," *Ind. Eng. Chem. Fundam.* **25,** 621 (1986).
Achenie, L. E. K., and Biegler, L. T. "Developing Targets for the Performance Index of a Chemical Reactor Network," *Ind. Eng. Chem. Res.* **27,** 1811 (1988).
Achenie, L. E. K., and Biegler, L. T. "A Superstructure Based Approach to Chemical Reactor Network Synthesis," *Comput. Chem. Eng.* **14**(1), 23 (1990).
Agreda, V. H., Partin, L. R., and Heise, W. H. "High Purity Methyl Acetate via Reactive Distillation," *Chem. Eng. Prog.* **86**(2) (1990).

Andrecovich, M. J., and Westerberg, A. W. "An MILP Formulation for Heat Integrated Distillation Sequence Synthesis," *AIChE J.* **31,** 363 (1985).

Aris, R. "The Optimal Design of Chemical Reactors." Academic Press, New York, 1961.

Balakrishna, S. Ph.D. Thesis, Carnegie-Mellon University, Pittsburgh, PA (1992).

Balakrishna, S., and Biegler, L. T. "A Constructive Targeting Approach for the Synthesis of Isothermal Reactor Networks," *Ind. Eng. Chem. Res.* **31,** 300 (1992a).

Balakrishna, S., and Biegler, L. T. "Targeting Strategies for the Synthesis and Heat Integration of Nonisothermal Reactor Networks," *Ind. Eng. Chem. Res.* **31,** 2152 (1992b).

Balakrishna, S., and Biegler, L. T. "A Unified Approach for the Simultaneous Synthesis of Reaction, Energy and Separation Systems," *Ind. Eng. Chem. Res.* **32,** 1372 (1993).

Barbosa, D., and Doherty, M. F. "The Simple Distillation of Homogeneous Reactive Mixtures," *Chem. Eng. Sci.* **43,** 541 (1988).

Chitra, S. P., and Govind, R. "Synthesis of Optimal Serial Reactor Structure for Homogenous Reactions. Part II: Nonisothermal Reactors," *AIChE J.* **31**(2), 185 (1985).

Cuthrell, J. E., and Biegler, L. T. "On the Optimization of Differential Algebraic Process Systems," *AIChE J.* **33,** 1257 (1987).

Douglas, J. M. "A Hierarchical Decision Procedure for Process Synthesis," *AIChE J.* **31,** 353 (1985).

Douglas, J. M. "Conceptual Design of Chemical Processes." McGraw-Hill, New York, 1988.

Duran, M. A., and Grossmann, I. E. "Simultaneous Optimization and Heat Integration of Chemical Processes," *AIChE J.* **32,** 123 (1986).

Dyson, D. C., and Horn, F. J. M. "Optimum Distributed Feed Reactors for Exothermic Reversible Reactions," *J. Optim. Theory Appl.* **1,** 1 (1967).

Fjeld, M., Asbjornsen, O. A., and Aström, K. J. "Reaction Invariants and the Importance of in the Analysis of Eigenvectors, Stability and Controllability of CSTRs," *Chem. Eng. Sci.* **30,** 1917 (1974).

Floudas, C. A., and Grossmann, I. E. "Global Optimization for Process Synthesis and Design," in "Foundationns of Computer-Aided Process Design (FOCAPD '94), Snowmass, CO" (L. T. Biegler and M. F. Doherty, eds.), p. 198 (1994).

Fogler, H. S. "Elements of Chemical Reaction Engineering." Prentice-Hall, Englewood Cliffs, NJ, 1992.

Froment, G. F., and Bischoff, K. B. "Chemical Reactor Analysis and Design." Wiley, New York, 1979.

Glasser, B., Hildebrandt, D., and Glasser, D. "Optimal Mixing for Exothermic Reversible Reactions," Paper 70g, *Annual AIChE Meeting,* Chicago (1990).

Glasser, B., Hildebrandt, D., and Glasser, D. "Optimal Mixing for Exothermic Reversible Reactions," *Ind. Eng. Chem. Res.* **31**(6), 1541 (1992).

Glasser, D., Crowe, C. M., and Jackson, R. "Zwietering's Maximum-Mixed Reactor Model and the Existence of Multiple Steady States," *Chem. Eng. Commun.* **40,** 41 (1986).

Glasser, D., Crowe, C. M., and Hildebrandt, D. "A Geometric Approach to Steady Flow Reactors: The Attainable Region and Optimization in Concentration Space," *Ind. Eng. Chem. Res.* **26**(9), 1803 (1987).

Godorr, S., Hildebrandt, D., and Glasser, D. "The Attainable Region for Mixing and Multiple Rate Processes," *Chem. Eng. J.* (1994).

Grossmann, I. E., and Straub, D. A. "Recent Developments in Evaluation and Optimization of Flexible Chemical Processes," *Proc. COPE '91,* p. 41 (1991).

Hartmann, K., and Kaplick, K. "Analysis and Synthesis of Chemical Process Systems." Elsevier, Amsterdam, 1990.

Hildebrandt, D. Ph.D. Thesis, University of Witwatersrand, Johannesburg, South Africa (1989).

Hildebrandt, D., and Biegler, L. T. "Synthesis of Chemical Reactor Networks," in "Foundations of Computer-Aided Process Design (FOCAPD '94), Snowmass, CO" (L. T. Biegler and M. F. Doherty, eds.), p. 52 (1994).

Hildebrandt, D., and Feinberg, M. "Optimal Reactor Design from a Geometric Viewpoint," Paper 142c, AIChE Annual Meeting, Miami Beach, FL (1992).
Hildebrandt, D., and Glasser, D. "The Attainable Region and Optimal Reactor Structures," *Proc. ISCRE Meet.*, Toronto (1990).
Hildebrandt, D., Glasser, D., and Crowe, C. "The Geometry of the Attainable Region Generated by Reaction and Mixing: With and Without Constraints," *Ind. Eng. Chem. Res.* **29**(1), 49 (1990).
Horn, F. "Attainable Regions in Chemical Reaction Technique," in "Third European Symposium on Chemical Reaction Engineering." Pergamon, London, 1964.
Horn, F. J. M., and Tsai, M. J. "The Use of Adjoint Variables in the Development of Improvement Criteria for Chemical Reactors," *J. Optim. Theory Appl.* **1**(2), 131 (1967).
Jackson, R. "Optimization of Chemical Reactors with Respect to Flow Configuration," *J. Optim. Theory Appl.* **2** (4), 240 (1968).
Kokossis, A. C., and Floudas, C. A. "Synthesis of Isothermal Reactor-Separator-Recycle Systems," *Annual AIChE Meeting*, San Francisco (1989).
Kokossis, A. C., and Floudas, C. A. "Optimization of Complex Reactor Networks. I. Isothermal Operation," *Chem. Eng. Sci.* **45**(3), 595 (1990).
Kokossis, A. C., and Floudas, C. A. "Synthesis of Non-isothermal Reactor Networks," Annual AIChE Meeting, Los Angeles, CA (1991).
Kramers, H., and Westerterp, K. R. "Elements of Chemical Reactor Design and Operation." Academic Press, New York, 1963.
Kravanja, Z., and Grossmann, I. E. "Prosyn—An MINLP Synthesizer," *Comput. Chem. Eng.* **14**(12), 1363 (1990).
Lakshmanan, A., and Biegler, L. T. "Reactor Network Targeting for Waste Minimization," presented at National AIChE Meeting, Atlanta, GA (1994).
Lee, K. Y., and Aris, R. "Optimal Adiabatic Bed Reactors for Sulphur Dioxide with Cold Shot Cooling," *Ind. Eng. Chem. Process Des. Dev.* **2**, 300 (1963).
Levenspiel, O. "Chemical Reaction Engineering." Wiley, New York, 1962.
Linnhoff, B., and Hindmarsh, E. "The Pinch Design Method for Heat Exchanger Networks," *Chem. Eng. Sci.* **38**, 745 (1983).
Narasimhan, G. "Optimization of Adiabatic Reactor Sequence with Heat Exchanger Cooling," *Br. Chem. Eng.* **14**, 1402 (1969).
Nishida, N., Stephanopoulos, G., and Westerberg, A. W. "Review of Process Synthesis," *AIChE J.* **27**, 321 (1981).
Omtveit, T., and Lien, K. "Graphical Targeting Procedures for Reactor Systems," *Proc. Eur. Symp. Comput-Aided Eng. (ESCAPE-3)*, Graz, Austria, *1993* (1993).
Paynter, J. D., and Haskins, D. E. "Determination of Optimal Reactor Type," *Chem. Eng. Sci.* **25**, 1415 (1970).
Pibouleau, L., Floquet, P., and Domenech, S. "Optimal Synthesis of Reactor Separator Systems by Nonlinear Programming Method," *AIChE J.* **34**, 163 (1988).
Ravimohan, A. "Optimization of Chemical Reactor Networks with Respect to Flow Configuration," *J. Optim. Theory Appl.* **8**(3), 204 (1971).
Trambouze, P. J., and Piret, E. L. "Continuous Stirred Tank Reactors: Designs for Maximum Conversions of Raw Material to Desired Product," *AIChE J.* **5**, 384 (1959).
van de Vusse, J. G. "Plug Flow Type Reactor vs. Tank Reactor," *Chem. Eng. Sci.* **19**, 994 (1964).
Viswanathan, J. V., and Grossmann, I. E. "A Combined Penalty Function and Outer-Approximation Method for MINLP Optimization," *Comput. Chem. Eng.* **14**, 769–782 (1990).
Waghmere, R. S., and Lim, H. C. "Optimal Operation of Isothermal Reactors," *Ind. Eng. Chem. Fundam.* **20**, 361 (1981).

OPERABILITY AND CONTROL IN PROCESS SYNTHESIS AND DESIGN

Steve Walsh and John Perkins

Centre for Process Systems Engineering
Imperial College of Science, Technology and Medicine
London, England

I. Introduction	302
II. General Techniques	306
A. Design with Uncertainty	306
B. Screening Tools for Disturbance Rejection	321
C. Optimization of Dynamic Systems	333
III. Neutralization of Waste Water	342
A. Making Discrete Design Decisions	342
B. Design Procedure for Neutralization of Waste Water	344
IV. Modeling of Waste Water Neutralization Systems	353
A. Steady-State pH Characteristics	354
B. Characteristics of Calcium Hydroxide Reagent	355
C. Measurement Response	357
D. Mixing	358
E. Summary	360
V. Design Examples	360
A. Exploration of a Generic Neutralization Control Problem	361
B. Preliminary Design of a Central Effluent Plant	371
C. pH Control of Several Strong Acid Streams with NaOH/HCl	374
D. $Ca(OH)_2$ Neutralization with Highly Variable Titration Characteristic (Central Effluent Plant)	379
E. $Ca(OH)_2$ Neutralization of a Single Effluent Stream at High Intensity	385
VI. Conclusions	391
Notation	393
Appendix A: Worst-Case Design Algorithm	396
Appendix B: Example of Progress of Algorithm	398
References	401

This chapter presents an integrated approach to process design and control in which operability and controller design are considered at the same time as process design. The aim of the approach presented is to assist designers in achieving a good trade-off between cost and risk, uncertainties and disturbances notwithstanding. We address the following questions:

- *Can a particular design or structure be made to satisfy its operating constraints despite uncertainty and disturbances?*
- *What is the impact of dealing with disturbances and uncertainty on the economics of the processing system considered?*

We propose a range of tests that can be applied in a hierarchical manner to give increasingly accurate performance estimates as the number of options being considered is reduced and the design approaches its final form.

The key design tools discussed are:

- *a robust dynamic optimization package*
- *a screening method for evaluating the effect of delays on achievable performance*
- *an optimization method for dealing with parametric uncertainty.*

The application of this approach to chemical waste water treatment is also discussed extensively.

I. Introduction

Operability and control tend to be considered near the end of process design and synthesis. A process may be designed by one group of people, then another design group is handed the process flowsheet and required to add on a control scheme to ensure that it operates successfully. Changes in the process design at this stage are usually expensive both in resources and in time. This approach has long been recognized as far from ideal, but the problem of how to do things better has not yet been resolved satisfactorily.

Integration of operability and control considerations into early-stage process design can be facilitated by organizational measures. Process control experts can be brought into the early-stage design team to identify and investigate possible operational difficulties. Process engineers can receive additional training to enhance their understanding of operational and control issues. Such organizational measures need to be supported by appropriate design procedures and tools if they are to deliver their full potential benefit.

Operability encompasses the ability of the system to cope with uncertainty and disturbances and also with issues of reliability and maintenance. In this contribution, we restrict our discussion to methods of dealing with uncertainty and disturbances. We address the following questions:

- Can a particular design or structure be made to satisfy its operating constraints despite uncertainty and disturbances?
- What is the impact of dealing with disturbances and uncertainty on the economics of the processing system considered?

It should be noted that these questions do not fit comfortably within any of the conventional categories of operability analysis, such as controllability (regulatory capability about a steady state), switchability (ability to switch between operating modes), flexibility (ability to accommodate uncertainty at steady state), and robust control (ability to maintain stability and performance of a control scheme despite perturbations in the characteristics of the controlled system). Rather, we attempt to develop an integrated design approach that allows us to consider operability issues on a par with economic issues, thus permitting design and synthesis decisions to be made within a common framework.

An important strand of previous work addressing the issues of dynamics and uncertainty in process and control system design is the development of design guidelines based on a mixture of experience, theory, and intuition. Among the notable contributors to such guidelines are P. Buckley, F. G. Shinskey, and W. L. Luyben. We value these contributions and believe they repay careful study. However, because such qualitative guidelines do not place operability issues on a par with process design issues, they do not allow integrated consideration of process and control decisions. Our approach exploits advances in optimization methods and computing capabilities in order to achieve this.

Engineers are required to produce designs that will operate safely and legally at all times while providing an adequate return on investment. They must meet these requirements despite incomplete knowledge of the process and equipment characteristics and the operating demands on the plant. Success or failure will depend on how well the designers manage the trade-off between risk and return on investment and how well they anticipate potential operational problems. Integrated design approaches aim to provide tools, procedures, and environments to aid engineers in tackling these issues in a systematic, well-coordinated manner.

The design problem described above is a multifaceted, multiobjective, incompletely defined problem and as such requires the application of creative intelligence. As no computer yet exhibits such creativity, any integrated design approach should be built around a human project team. Computer-based techniques should be designed to assist such a team and should not attempt to carry out tasks that humans are better able to do. The challenge in integrated design is to find a good "division of labor" between human and computer and an appropriate way of interfacing between the two.

With these considerations in mind, we have chosen the following approximation to the integrated design problem. The integrated design objective is to maximize return on investment while ensuring that performance constraints are satisfied for all possible plant parameters and disturbances.

This approximation involves a number of nontrivial assumptions.

- The performance requirements can be represented either by constraints or by a contribution to the operating costs. In some cases, this is straightforward—equipment and material and energy inputs can be included in an economic objective; constraints for which violation leads to plant shutdown are well approximated by a hard constraint. In other cases, the choice of representation is more difficult—excessive quality variations may lead to complete loss of the market for a product (constraint) or reduced price for the product (objective).
- Adequate models must be available to allow evaluation of the effect of design decisions on performance. This can be very difficult to achieve, particularly if we desire to characterize a wide range of options in depth at the outset of a design. We assume that the project team acts to refine and augment the models required for performance prediction as the design progresses. The models can take any form, from correlation curves through differential-algebraic equations.
- Uncertainty and disturbances can be described in terms of mathematical constraints defining a finite set of *bounded* regions for the allowable values of the uncertain parameters of the model and the parameters defining the disturbances. If uncertainty or disturbances were unbounded, it would not make sense to try to ensure satisfaction of performance requirements for all possible plant parameters and disturbances. If the uncertainty cannot be related mathematically to model parameters, the model cannot adequately predict the effect of uncertainty on performance. The simplest form of description arises when the model is developed so that the uncertainty and disturbances can be mapped to independent, bounded variations on model parameters. This last stage is not essential to the method, but it does fit many process engineering problems and allows particularly efficient optimization methods to be deployed. Some parameter variations are naturally bounded; e.g., feed properties and measurement errors should be bounded by the quality specification of the supplier. Other parameter variations require a mixture of judgment and experiment to define, e.g., kinetic parameters.

Matching the above approximation to the original design problem requires the intervention of the project team to define

- objectives and constraints to approximate performance requirements
- models to link these to the choice of design variables
- bounds on uncertain variables to manage the trade-off between cost and risk

Problem definition and solution will usually be an iterative process. Each solution of the approximate problem can be used to provide estimates of cost and to indicate the critical values of the uncertain parameters. This information allows the project team to trade off risk and cost at a very high level. In working with industrial project teams, we have found them to be quite comfortable with defining problems in this way and with exploring the interactions between problem definition and the cost of the resulting design.

If good information is available on the probability distribution of the uncertain parameters, it becomes possible in principle to design to a specified probability of infeasibility (Straub and Grossmann, 1990). This would provide an alternative method of trading off cost and risk. The algorithms to deal with this case are significantly more complex than those for dealing with bounded uncertainty as it is necessary to identify the complete boundaries of the feasible region in order to evaluate the probability of the uncertain parameters lying within the feasible region. Because of the difficulty of solving this problem for general nonlinear systems and the difficulty of obtaining reliable statistical data on the variation of the uncertain parameters, we have not pursued this approach.

In considering the solution of the approximate problem, a number of points are evident. It is highly desirable to tackle the problem with a hierarchy of methods and models. The design process can be likened to a mixture of broad and deep searches over the "space" of design options. The search has two main mechanisms:

- narrowing the search domain by eliminating sets of options
- deepening the search by picking likely "winners"

Eliminating options is assisted by simple criteria or tests that can identify a large number of options as unviable. Deepening the search can be assisted by more heuristic measures as the aim is usually to establish a viable base-case design rapidly, without necessarily discarding other options. The appropriate criteria to use and the balance between search methods is clearly problem-dependent.

There are a number of general techniques suggested by the problem formulation. At the most detailed level of design, the design parameters need to be optimized in relation to performance criteria based on a nonlinear dynamic model. This points to a need for effective tools for dynamic optimization. At a more preliminary level in a hierarchy of techniques, it might be useful to evaluate steady-state performance or to carry out tests on achievable dynamic performance to eliminate infeasible options. Appropriate screening techniques are therefore needed. All these methods can use nominal models for initial analysis, but a full analysis should be based on design with uncertainty.

The rest of this chapter is structured as follows. The next section considers general techniques for use in the integrated design approach proposed: design with uncertainty, screening tools for disturbance rejection, and dynamic optimization. Then a particular problem domain—chemical waste water neutralization—is introduced. A procedure is presented for coordinating the techniques in an integrated design framework for this problem area. Key modeling issues for this problem domain are discussed. A generic example and four industrial case studies are then presented, illustrating different aspects of the use of this procedure. We conclude the chapter by reviewing our experience on the problem domain discussed and highlighting some directions for further extension of the proposed integrated design approach.

II. General Techniques

In the previous section, an approach to synthesis and design taking account of uncertainty and the impact of process disturbances on performance was outlined. In this section, a set of tools to support the approach will be presented. First, we examine techniques for the design of plant for which uncertainties in design data are explicitly represented. These techniques are computationally demanding; currently, they are most suitable at the later stages of design when the most promising designs are being verified and finalized. In the early stages, where rapid screening of a large number of options is required, less computationally demanding tools are more appropriate. A technique based on the ability to reject process disturbances is presented for the rapid evaluation of design alternatives. Finally, since the aim is to evaluate the dynamic behavior of the process, and since optimization is a key component of our approach, algorithms for solving optimal control problems (i.e., optimization problems subject to differential equation constraints) are discussed.

A. Design with Uncertainty

Some approaches to design with uncertainty are reviewed in order to identify the appropriate basic approach and relevant techniques from previous work. A new algorithm is then presented which builds on previous work and attempts to exploit the characteristics of the design problems.

1. Review

As noted by Grossmann *et al.* (1983), the problem of design with uncertainty is not well-defined and many different approaches exist. As discussed above, we are interested in solving problems in which the uncertainty is assumed to be defined by bounds on model characteristics or parameters and in which it is assumed that the constraints must be satisfied at all times. Attention will therefore be restricted to techniques to ensure feasibility for all possible realizations of the uncertainty (worst-case design). This review considers only optimization-based methods that are potentially applicable to general nonlinear systems. Algorithms designed for linear or bilinear systems or involving manipulation of explicit objectives and constraints are therefore not covered.

Two approaches to worst-case design that have been applied successfully to engineering design problems with uncertainty are those of Grossmann et al. (1983; Swaneyand Grossmann, 1985a, b; Grossmann and Floudas, 1987) and Polak et al. (Polak, 1982; Tits, 1985; Polak and Stimler, 1988; Mayne et al., 1990). The problem tackled by both approaches is how to deal with an effectively infinite number of constraints (i.e., constraints must be satisfied for the infinite number of parameter realizations that must be considered), which makes the design optimizations *semiinfinite*. In both cases the approach adopted is to approximate the continuous uncertain parameter space, $v \in V$, by a discrete set, $v \in V_i$ and to update this set until it gives a design for which no constraint violation can be found. This is known as an outer-approximation algorithm, as the constraints associated with $v \in V_i$ define a feasible region which contains (i.e., is an outer approximation to) the feasible region associated with $v \in V$. This is illustrated in Fig. 1, which shows how three scenarios of the uncertain parameters give an approximation to the true feasible region.

Grossmann et al. consider the set V to be defined explicitly by bounds on the uncertain parameters. Polak et al. allow for general constraints on the values of V subject to the requirement that V is made up of the union of a finite number

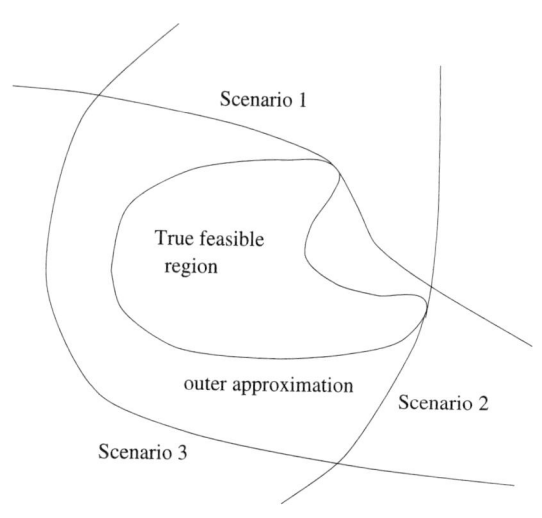

FIG. 1. Effect of outer approximation.

of compact subsets. While many engineering problems fit the simpler case of independent bounded parameter variation, given appropriate models, this approach is not universally appropriate. For example, a feed may have a quality specification on total impurity level as well as specifications on levels of individual impurities. This would give rise to an uncertainty specification of a limit on the sum of the uncertain parameters defining the impurities. We focus below on the simpler case, but comment on extensions necessary to handle the more general case.

The general structure of an outer-approximation algorithm for worst-case design is as follows

1: Choose an initial set $V_i(i = 0)$ to approximate V

2: Carry out a design so that the constraints are satisfied for all $v \in V_i$.
If this stage fails then the problem is infeasible.

3: Find the maximum constraint violation over the set V, c^*, and the corresponding value of v, v^*
If $c^* \leq 0$ then solution found
Else $V_{i+1} = V_i \cup v^*$, $i = i + 1$, goto step 2

The methods differ mainly in step 3.

Grossmann et al. (1983) formulate the worst-case design problem as

$$\min_{p \in P} E \{\min_{v \in V} J(p, v, o) | c(p, v, o) \leq 0\} \tag{1}$$

$$\text{s.t.} \quad \forall v \in V \{\exists o \in O \langle \forall k \in K[c_k(p, v, o) \leq 0] \rangle\}$$

where J is the objective function to be minimized, p is a vector of design variables, v is a vector of uncertain variables, c_k is the kth element of the constraint vector c, and o is a vector of operating variables that may be adjusted to reduce costs and to maintain feasibility in the light of the value of v. The set V is assumed to be a polyhedron defined by simple bounds on elements of v. The infinite constraint may be reformulated as

$$\max_{v \in V} \min_{o \in O} \max_{k \in K}[c_k(p, v, o)] \leq 0 \tag{2}$$

The design variables are chosen to allow constraint satisfaction for *all* the uncertain variable values while the operating variables are adjusted for *each* value of the uncertain variables. If operating variables are eliminated, the problem simplifies to

$$\min_{p \in P} E \{J(p, v)\} \tag{3}$$

$$\text{s.t.} \quad \forall v \in V \{\forall k \in K[c_k(p, v) \leq 0]\}$$

The formulation without operating variables is qualitatively easier to solve as one level of optimization is eliminated. More specifically, finding v^* for the optimization problem defined by Eq. (2) is a nondifferentiable global optimization problem that is extremely difficult to solve rigorously in the general case. It is therefore important to consider the pros and cons of using operating variables carefully.

The motivation for including operating variables is that certain variables may be adjusted during plant commissioning or operation to give improved performance in the light of the actual plant behavior. Requiring such variables to be chosen so as to accommodate all possible uncertain variables, i.e., as design variables, introduces an element of conservatism into the design. This is particularly the case if the operating variables include process inputs which would be adjusted by a control scheme to maintain satisfaction of constraints and if the uncertain parameters are fixed but unknown rather than variable. On the other hand, including operating variables in the problem formulation assumes an *ideal* adaptation of the operating variables to *all* the uncertain variables, which may actually vary over time. This problem will generally be optimistic as the actual "operator" will have only partial knowledge of some of the uncertain parameters and will adapt the operation in a nonideal way. A design generated using optimization of operating variables will therefore *usually* be infeasible for some values of $v \in V$. Failure of a design problem with operating variables indicates that no control scheme for adjusting the operating variables can achieve feasible operation for all the uncertain parameters. Success of such a design problem does not imply that an *implementable* control scheme exists which can achieve feasible operation. Operating variables, therefore, should not be used in determining design parameters, although they may be useful in certain screening tests if the resulting problem can be solved efficiently (see Section II.B).

To go beyond the potential conservatism of having all the variables as design variables and the probable optimism of using operating variables, it is necessary to include the adaptation mechanism (control scheme) within the model. If desired, the parameters controlling this adaptation can be made design variables. Grossmann *et al.* (1983) consider that this "would make the problem virtually unmanageable" at the design stage. In the integrated design context, which involves consideration of process *and* control system design, it is certainly appropriate. More generally, including basic control information may be accomplished simply by requiring that certain variables remain at their setpoints, which may be added to the design variables, and eliminating the operating variables, o, using the extra equality constraints. The key formulation for design with uncertainty in relation to the integrated design approach is that without operating variables. Solving general problems with operating variables is not necessary in the context of our approach to integrated design. However, the penalty paid for adopting a particular adaptation scheme is that we obtain no information from the optimization as to what improvements could be made by a different scheme.

In replacing operating variables by a control scheme with associated design variables, the designer has moved from a test for which "no" means *no* and "yes" means *maybe* to a test for which "yes" means *yes* and "no" means *maybe*. The two approaches are clearly complementary. The operating variable approach is best suited to screening the potential feasibility of a design (we want "no" to mean *no*) or setting targets for design, while it is only appropriate for direct use in design if the designer is confident that approximately ideal adaptation can be achieved in practice. The embedded control scheme approach is best suited to use in design, particularly when it is necessary or desirable to design the control scheme itself, while it is ideally suited for screening only if one can parameterize and optimize the best controller one might conceivably implement. In the new algorithm and examples presented below, we do not make use of operating variables.

It is possible to solve the general formulation (including operating variables) rigorously for certain special cases. The algorithms developed for this purpose are of interest as they include techniques that are useful for developing a general algorithm.

An algorithm for solving the general problem under the assumption that the worst-case uncertain parameters lie at vertices of the parameter space, V, is given by Grossmann *et al.* (1983). This is presented below following the general outer-approximation algorithm structure given above.

1: $V_i(i = 0)$ is selected based on the sign of gradients of individual constraints with respect to v. Positive sign indicates that the parameter would maximize that particular constraint at its upper limit if the constraint is monotonic.

2: Solve for new design parameters, p^*, which minimize the expected cost subject to the constraints associated with each $v_j \in V_i$

$$\min_{p, o_j, j=1, \ldots, N_i} \sum_{j=1}^{N_i} w_j J(p, o_j, v_j)$$

s.t. $c(p, o_j, v_j) \leq 0, \quad j = 1, \ldots, N_i$

(The weighted sum is used by Grossmann *et al.* to approximate the expectation, N_i is the dimension of V_i.)

3: For each vertex, v^v, in V choose o to minimize the maximum constraint violation for the new design parameters.

$$\min_{o \in O} \{c_{\max} | c_{\max} \geq c_k(p, v_j, o), \forall k\}$$

(It is not necessary to carry out the minimization exhaustively as the minimization for a particular vertex can stop once $c_{\max} < 0$.)

Choose v^* as the vertex giving the largest value of c_{max}

If $c_{max}(v^*) \leq 0$ then solution found
Else $V_{i+1} = V_i \cup v^*$, $i = i + 1$, goto step 2

The fundamental limitation of this approach is that it *assumes* the worst-case parameters are always at a vertex. This requires convexity properties that will not be satisfied in all the problems of interest. A *sufficient* condition for a vertex solution (Swaney and Grossmann, 1985a) is that all $c_k(p, v, o)$ are jointly quasi-convex in o and one-dimensional quasi-convex in v. One-dimensional quasi-convexity implies that

$$\max(f(x_1), f(x_2)) \geq f(\alpha x_1 + (1 - \alpha)x_2) \quad (4)$$

$$\forall \alpha \in [0, 1], \forall \alpha_2 \in \mathcal{R}, \forall x_1, x_2 = x_1 + \alpha_2 e_i$$

where e_i is a vector with ith element 1 and all other elements 0. A nonvertex solution requires a maximum in v, for some $\min_o c_k$, which is not at a bound of V. This is precluded by the above conditions. Nonvertex solutions may occur for the design problems of interest. Two examples illustrating this are given below.

1. If the time between two step disturbances is an element of v, then the worst case is not necessarily at either bound of this variable.
2. A sinusoidal disturbance with uncertain frequency will tend to have a worst case near the resonant frequency of the control system design which will not, in general, lie at a limit of the uncertainty range.

However, many elements of v can be expected to have a worst case at a bound, e.g., measurement bias, flow, concentration, buffering, and reaction rates. For the general case, the vertex assumption cannot be relied upon but provides a basis for useful heuristics in the search for a solution.

Even with the assumption of vertex constraint maximizers, vertex enumerations with a large number of parameters can be very time-consuming. Algorithms for efficient exploration of the vertices and a proposed "flexibility index" are presented in Grossmann et al. (1983) and developed further in later papers (Swaney and Grossmann, 1985a, b). These are discussed below.

The flexibility index, F is defined by

$$F = \max \delta$$

$$\text{s.t.} \max_{v \in V_\delta} \min_{o \in O} \max_{k \in K} c_k(p, v, o) \leq 0 \quad (5)$$

$$V_\delta = \{v | (v_0 - \delta \Delta v^-) \leq v \leq (v_0 + \delta \Delta v^+)\}$$

where v_0 is the nominal value of v. Therefore, F is the factor by which a polyhedron representing nominal variability or uncertainty of the parameters can be expanded without leading to constraint violation for any parameter contained within it. An equivalent representation more useful for solution is

$$F = \min_{\tilde{v} \in \tilde{V}} \max_{\bar{\delta}, o} \bar{\delta}$$

$$\text{s.t. } c(p, v, o) \leq 0 \quad \forall \delta \in [0, \bar{\delta}] \quad (6)$$

$$v = v_0 + \delta \tilde{v}$$

$$\tilde{V} = \{\tilde{v} | -\Delta v^- \leq \tilde{v} \leq \Delta v^+\}$$

The condition $\forall \delta \in [0, \bar{\delta}]$ is *assumed* to be satisfied in the solution methods presented if $\bar{\delta}$ gives a feasible point giving the simplified problem

$$F = \min_{\tilde{v} \in \tilde{V}} \max_{\delta, o} \delta$$

$$\text{s.t. } c(p, v, o) \leq 0$$

$$v = v_0 + \delta \tilde{v} \quad (7)$$

$$\tilde{V} = \{\tilde{v} | -\Delta v^- \leq \tilde{v} \leq \Delta v^+\}$$

This index could be used for trading off risk and cost but such indices oversimplify a complex tradeoff. The mapping between F and risk is not direct. Some parameters may never violate their bounds, others may have a significant probability of doing so. The use of F treats all parameters as having a uniform likelihood of violating their bounds, which is not generally appropriate.

The calculation methods for the flexibility index discussed below can be applied to step 3 of the worst-case design optimization procedure to maximize constraint violation instead of minimizing δ.

Under the vertex solution assumption, the flexibility index can be calculated by evaluating the maximum δ along each vertex direction and taking the minimum of the results. This approach becomes computationally impractical for more than about 15–20 parameters, so two procedures are presented which give upper bounds on F more efficiently (a vertex search method and a branch-and-bound method). Of the two methods, the vertex search method is found to be more efficient on the examples considered and therefore seems the best candidate for use to get an approximate solution for the worst-case vertex where appropriate. The vertex search procedure is given below.

1: Set ρ^{max} (see step 3) and choose an initial value of v

2: Update v using $\text{sign}(\tilde{v}_i) = -\text{sign}(\partial \delta / \partial v_i)$
until either δ fails to decrease as predicted or
the method predicts a vertex already examined

3: For each $c_k < 0$ compute the maximum increase Δc_k in the constraint, based on the constraint gradients, and identify the corresponding vertex
Compute the projected fractional change $\rho = -\Delta c_k/c_k$
If the maximum value of ρ over all the constraints is less than ρ^{max} or no new vertices are identified then STOP
Else select the new vertex value of v giving the maximum value of ρ and goto 2:

ρ^{max} governs the thoroughness of the local search. $\rho^{max} = 1$ corresponds to examining a vertex if the linearized constraints at the present vertex indicate that some currently inactive constraint might become active. $\rho^{max} = 0.5$ corresponds to examining a vertex if the linearized constraints at the present vertex indicate that the constraint will move halfway to becoming active.

Swaney and Grossmann (1985b) note that the assumption of vertex solutions can be relaxed somewhat by carrying out a local search from the solution vertex if a descent direction exists and give one example where this approach identified a nonvertex worst point.

Grossmann and Floudas (1987) present a complementary approach to evaluating the flexibility index, F. This approach introduces integer conditions which for linear constraints allow prediction of the subset of constraints that will be active at the solution, based on the gradients of the constraints with respect to the operating variables, o. These conditions can be directly incorporated into a mixed-integer linear program (MILP). For nonlinear constraints, a modified approach is presented to decompose the problem into a series of smaller nonlinear programs (NLPs) corresponding to the predicted active constraint sets. In the absence of operating variables, this approach reduces to solving the same number of NLPs as the number of inequality constraints, with the inequality constraints replaced in each case by a single equality constraint. For general problems the method requires that the constraints be monotonic in o for all v. Global solutions to the NLP subproblems are guaranteed if the active constraints are jointly quasi-concave in o and v and strictly quasi-convex in o for fixed v. This approach seems to have little merit for the problems of interest for which the required mathematical properties are unlikely to be met.

Polak (1982) gives a different algorithm for solving the general worst-case design problem that attempts to avoid reliance on special convexity or concavity properties. The allowable values of v are given by a set of inequalities defining a compact subset rather than by simple bounds. Tits (1985) notes an error in this algorithm and this method appears to have been abandoned. Tits suggests using a vertex assumption or using local searches from all or a subset of the vertices at each iteration to improve the likelihood of a global maximum. This is not far from the algorithm of Grossmann *et al*.

In a more recent paper Polak and Stimler (1988) note that

> To date, the use of semiinfinite methods in worst-case control system design with parametric uncertainty has been extremely limited because of the above

mentioned computational problem of evaluating the maximum of the constraints.

The paper also states that

> The most general semiinfinite optimisation problems that are solvable by existing algorithms are of the form
>
> $$\min_{p}\{J(p)|c(p) \leq 0; \max_{v^i \in V^i} c^i(p, v^i) \leq 0, i \in \{1, ..., I\}\}$$

This confirms the difficulty of solving problems with operating variables, o, in a rigorous manner. Note that V^i represents the ith subset of the uncertain parameter space and not an outer approximation set (V_i).

The algorithms used for solving this problem are based on the application of general global optimization methods, such as grid enumeration, for solving the constraint maximization problem without operating variables, o. When applying general global optimization methods to the constraint maximization problem, it does not matter greatly whether the set V is defined in terms of simple bounds, linear inequalities, or nonlinear inequalities. So long as checking whether or not an element v is within V is computationally cheap, the effort to deal with the different cases will not vary much. However, as the definition of V becomes more complex, it becomes more difficult to find efficient approximations to the global optimization problem or special cases for which more efficient methods can be applied. For the case of linear inequalities, the vertices of the linear constraints provide a natural extension of exploring the vertices defined by the bounds. Linear equality constraints on elements of v can be used to eliminate uncertain variables and are therefore beneficial rather than detrimental.

Polak and Stimler (1988) develop approximations to some worst-case controller design problems which are more amenable to solution. This is accomplished by replacing the constraint

$$c(p, v) \leq 0 \quad \forall v \in V \qquad (8)$$

by

$$\tilde{c}(p, v) \leq 0 \quad \forall v \in V \qquad (9)$$

where

$$c(p, v) \leq \tilde{c}(p, v) \quad \forall v \in V \qquad (10)$$

This operation is referred to as majorization and allows simpler constraints to be used, facilitating solution at the expense of reducing the feasible space for p. This method is not readily applicable to nonlinear dynamic optimization, where there is no formula for the constraints; but trading conservatism in design for ease of solution can be achieved for the problems of interest by a different strategy discussed in Section II.A.2.

A more recent paper (Mayne *et al.*, 1990) suggests a technique for reducing the number of global maximizations required in outer approximation methods by tracking local minima previously identified as global minima until they cease to give constraint violations. They recommend gradually increasing the effort deployed on constraint maximization at each iteration, provided a constraint violation is obtained.

None of the methods reviewed provides a complete solution to the type of problems of interest. They suffer either from making assumptions that are not likely to be satisfied or from relying on a computationally unrealistic degree of brute force. This review has, however, highlighted a number of methods that could usefully be exploited in an algorithm for worst-case design.

1. Exploiting the heuristic that many local minimizers of constraints lie at vertices of the parameter space through vertex searches and local searches from promising vertices.
2. Constructing suitable approximations to the original problem to make it easier to solve while still giving meaningful solutions.
3. Increasing the effort employed on constraint maximization as the optimization progresses.

These ideas are used below in developing a new algorithm.

2. A New Algorithm for Worst-Case Design

This section presents a new algorithm for design with uncertainty. This algorithm does not represent a radical departure from previous approaches, but rather represents a new *variant* of the basic outer approximation algorithm discussed above. The key feature is the use of approximate global optimization to give reduced reliance on convexity properties compared to the algorithm of Grossmann *et al.* discussed above, while not relying on the rigorous but expensive global optimization methods favored by Polak *et al.* The algorithm seeks to reduce computational cost by relying on local search methods in initial iterations and by introducing a systematic overdesign factor to the design iterations.

The discussion below is supplemented by a pseudocode listing in Appendix A.

a. General Observations. No operating variables are included in the optimization formulation handled by this algorithm. Such adaptation of system variables as is possible is assumed to be embedded in the model via the control scheme so that no operating variables are necessary in the optimization problem. This gives the optimization formulation

$$\min_{p}\{J(p, V) | \max_{v \in V} \max_{k \in K}[c_k(p, v)] \leq 0\} \tag{11}$$

where $J(p, V)$ represents an objective function that depends on the design variables and the set of uncertain variables. This encompasses nominal cost, expected cost, and maximum cost. Maximum cost

$$J(p, V) = \epsilon | J(p, v) \le \epsilon \quad \forall v \in V \quad (12)$$

or a weighted cost

$$J(p, V) = \sum w_j J(p, v_j), \quad v_j \in V_i \quad (13)$$

provide two readily implementable forms of $J(p, V)$. The weights are defined so that one particular model (usually a nominal model) can be given a fixed weight between 0 and 1 and the other active models are given equal weight to make $\sum w_j$ equal to 1.

The algorithm presented could be adapted to the case with operating variables. The main problems for extension to this case are as follows:

1. The local search procedures are more likely to suffer problems due to discontinuities either from changes in the identified local minima with respect to o (discontinuity in value or first derivative) or from changes in the active set of constraints at a minimum with respect to o (discontinuity in first derivative). This is likely to cause failure to converge to a local maximum with respect to v in at least some cases.
2. The computation time increases as an optimization replaces a function evaluation.
3. The design problem is of much larger dimension than p if the dimensions of o and the set V_i are large.

The uncertain and variable parameters are assumed to lie within a polyhedron defined by simple bounds on the variables ($v \in V$). The advantage of this description is that the boundaries of the uncertain parameter space, V, are easily identified and many constraint maximizers can be expected to lie at the vertices of the uncertain parameter space. Similar advantages could be obtained for more general uncertainty descriptions such as linear inequalities, although the implementation would be more complex. The extension to uncertain parameters lying in the union of multiple bounded sets can be made straightforwardly but has not been implemented. For more general representations of V, the heuristic of exploring vertices has no clear meaning and would have to be dropped, with correspondingly greater emphasis on general global optimization techniques.

Vertex constraint maximizers are not *assumed,* although they are expected to be common. Both local and global searches into the interior of V may be carried out as part of the search procedure, the effort expended in finding a new constraint maximum being bounded by user-defined variables.

In order to obtain a solution in reasonable time, it is extremely important that the design algorithm not take too many iterations to identify an outer approximation adequate to force a feasible design. One method of tackling this is to use the method of Mayne *et al.* (1990), in which previous local maximizers are tracked following the update of design variables and both the original and updated maxima are included in the design set. This is not particularly well suited to the case in which many maximizers are expected to be at vertices of V, as is typical of engineering problems. It is, however, possible to develop an alternative, heuristic method to trade off number of iterations against accuracy of solution.

The method used involves building a projection factor, ϵ_p, into the model. Setting $\epsilon_p = 0$ corresponds to the actual problem definition. $\epsilon_p > 0$ corresponds to a performance specification that is more demanding than the actual specification, with the difficulty of the problem increasing with increasing ϵ_p. For each set of uncertain parameters (scenario) that is to be added to the scenarios considered for design, setting ϵ_p to a value greater than zero should make it more difficult to find a satisfactory design. The resulting design, which is able to accommodate the selected scenarios with the chosen, nonzero ϵ_p, should then be more likely to be feasible for all $v \in V$ than a design based on the same scenarios and $\epsilon_p = 0$. This characteristic may allow convergence to a feasible solution of the actual problem using fewer iterations of the outer approximation algorithm than if ϵ_p were set to zero in the design problems. Faster convergence will, however, be at the expense of a degree of overdesign. The final design will give an objective function value lying somewhere between the cost of a design that gives guaranteed performance for the original specification ($\epsilon_p = 0$ is used in constraint maximization to check feasibility) and the cost of a design that gives guaranteed performance for the chosen, nonzero ϵ_p (used in the design optimization).

The appropriateness of this approach depends on being able to find a sensible definition of ϵ_p. In engineering problems, performance is typically limited by the maximum disturbance magnitude or the maximum or minimum plant throughput. The design problems might therefore be solved with throughput variations or disturbances scaled up by a factor of $1 + \epsilon_p$ compared to the actual scenarios being considered. Feasibility would then be checked by maximizing constraint violation with the actual throughput or disturbance conditions. The projection factor ϵ_p may be large on initial evaluation of designs (where speed is at a premium) and be reduced for a more precise optimization of the final design (where accuracy is at a premium).

For this heuristic strategy to be appropriate, the key conditions are as follows:

1. There should be a sensible way of defining ϵ_p, e.g., a fractional increase of disturbance amplitude or throughput.
2. Adaptation of the design variables to increased ϵ_p should improve performance over most or all of the uncertain parameter set, V, e.g., through increased steady-state offset from the active constraints or increased process capacity.

The first condition is straightforward and usually trivial. The second condition is necessary to exclude cases for which increased ϵ_p pushes the design in a direction which expands the feasible region locally but creates new constraint violations elsewhere in V, and hence gives no net benefit. Although this condition cannot be verified *a priori*, the heuristic was found to work effectively in the design examples (Section V), indicating that this condition is adequately satisfied in the problems examined.

The new worst-case design algorithm is discussed further below following the general outer-approximation algorithm structure; constraint maximization, initialization, and multi-model design.

b. Constraint Maximization. The constraint maximization algorithm makes use of the heuristic that the maximizing values of the uncertain variables will often lie on the bounds. It also uses the idea of progressively increasing the amount of effort applied to the constraint maximization as the search proceeds. This progressive increase of effort is implemented by stopping the search if a constraint violation is found by the time the previously applied depth of search has been completed. If not, the search proceeds further, subject to a specified maximum depth of search, until a constraint violation is found and the effort required is recorded to set the minimum effort in the next constraint maximization.

1. In the early iterations of the design, local vertex searches as used by Grossmann and Swaney provide an efficient way to generate new maximizers and push the design toward the required robustness. Local searches from a subset of the vertices may be used to try to identify nonvertex maximizers. $nloc1_{max}$ is set to limit the number of local searches used to supplement the local vertex search. ρ_{min}^{max} defines the maximum depth of the local vertex search. See Appendix A for further details.

2. If the local search methods are unsuccessful on any iteration, global search procedures are activated for use in all subsequent iterations. In *global* constraint maximization, vertex enumeration may be used (if the designer has selected this by setting *ivert* = 1). A multi-start random search may be used in which the random search is biased toward the vertices by specifying an *a priori* probability for each variable lying on a bound at the worst case (*pvert*). When an increased value for the maximum constraint is found by the random search, a new local search is initiated. The number of random points examined is limited to $nrand_{max}$. The number of local searches allowed during the global phase of the search is limited to $nloc2_{max}$. Local searches are initiated whenever a global search procedure identifies an increased value of c_{max} for which $\partial c_{max}/\partial v$ indicates an ascent direction within V. The use of *pvert* was found to be essential for an effective random search as nonvertex maximizers usually have only

OPERABILITY AND CONTROL IN PROCESS SYNTHESIS AND DESIGN 319

one or two values away from their bounds and these can often be identified by *a priori* physical argument.

3. Both local and global searches stop either when the computational effort employed exceeds the maximum used in previous iterations and a constraint violation has been identified or when the specified maximum effort has been employed unsuccessfully. This ensures monotonically increasing search effort as the outer approximation improves and the design converges.

This approach provides the flexibility to trade off confidence in final solution against the constraints of finite computational power, making extensive use of appropriate heuristics to improve efficiency of solution. The philosophy in applying the heuristics is to use cheaply obtainable constraint maximizers to push the design toward its final form before applying the more expensive procedures.

In carrying out local searches for

$$\max_{v \in V} \max_{k \in K} c_k(p, v) \qquad (14)$$

it is desirable to find the solution in a single NLP problem while avoiding the nondifferentiability of the optimization problem above. A differentiable approximation to $\max_{k \in K} c_k(p, v)$ of the form

$$c_{\max} = \epsilon \left| \log \left(\sum_{k=1}^{N} \exp \left(\frac{c_k(p, v) - \epsilon}{c_{\text{sig}k}} \right) \right) \right| = 0 \qquad (15)$$

can be used. This approach is preferred because

$$\max_{v \in V} c_{\max} \qquad (16)$$

requires the solution of fewer NLP problems for a local search than the more obvious approach to avoid nondifferentiability

$$\max_{k \in K} \max_{v \in V} c_k(v, x) \qquad (17)$$

When $\epsilon_k > 0$, the algorithm must decide whether to add the original model or the projected model to the outer approximation set V_k. If applying ϵ_p gives an increased value of $c_{\max}$, the projected model is added; otherwise, the unprojected model is added.

c. Initialization. Initialization includes defining the sets V and P. Set V is defined in terms of simple bounds on the design parameters; P may be defined by a mixture of simple bounds and more general constraints on p as this does not imply any increase in complexity. These constraints are simply added to the outer approximation constraints for each design iteration.

The initialization must also define the parameters controlling the maximum effort to be applied to the optimization. At one extreme, the search could be limited to a crude vertex search by excluding vertex enumeration ($ivert = 0$) and random searches ($nrand_{max} = 0$) and setting $\rho_{min}^{max} \geq 1$ to limit the depth of the vertex search. At the other extreme, vertices could be exhaustively enumerated ($ivert = 1$) and extensive random searches carried out ($nrand_{max} > 2^{n_v}$, where n_v is the number of uncertain parameters). The initial vertex exploration can be supplemented by up to $nloc1_{max}$ local searches and the global searches can be supplemented by up to $nloc2_{max}$ local searches. Setting $nloc1_{max}$ and $nloc2_{max}$ to about 10 effectively removes any constraint on the use of local searches.

pvert sets the probability of the worst-case parameter value being at an upper or lower bound. Unless there is a physical reason to expect a worst case that is *not* at an extreme, this should be set in the range .8–.999.

ϵ_p must be set to control the degree of precision used in solving the problem. Values of .05–.2 are typical.

Initialization must provide an initial outer approximation, V_0, on which to base the design. The selection of the *nmods* parameter set(s) making up V_0 may be based on prior judgment or analysis (first *nset* elements of V_0) or it may be based on a simplified constraint maximization algorithm (the remaining elements). Grossmann et al. (1983) use the gradients of the constraints at a *nominal point* to make a prediction of the set of vertices maximizing the individual constraints. An initial application of a local vertex search algorithm (see Appendix A) to each constraint seems more appropriate. Only distinct vertex maximizers corresponding to violated constraints are included in the initial set, as there is a high computational cost for including unnecessary points. If $\epsilon_p > 0$, then *either* the projected *or* the original point identified by the constraint maximization is included in V_0, depending on which gives the larger violation of the constraint. Including both points will usually give no benefit while doubling the computational effort in the first design iteration.

d. Multimodel Design. In the multimodel design stage, it is desirable to carry out the design with the minimal set of models that define the optimum while not dropping and reidentifying models that do affect the solution. The approach adopted is to predict the "active set" of models, design with this active set, and check the solution against all models identified; then if any additional model indicates constraint violations, add it to the predicted active set and rerun the optimization. The active set used is formed from all models that have been active for at least one of the previous three design iterations or have previously had to be added back at the end of a design optimization. Although quite heuristic, this strategy appears reasonable.

The use of the "ϵ_p" heuristic can potentially slow convergence if the unprojected constraint maximizers included in the multimodel design become active again as the design progresses. This behavior would not be expected in typical problems—nor has it actually been observed—but it is guarded against by checking the unprojected models at the same time as other inactive models and adding them to the design set if necessary.

e. Some Practical Considerations. It is important to emphasize that, despite the measures introduced to make the worst-case design algorithm more efficient, it is not practical to deploy the global optimization capabilities using more than about 10 uncertain parameters for dynamic problems or about 15 for steady-state problems. This means that the engineer must exercise judgment as to the key parameters to be given the most rigorous treatment. Engineering judgment may identify some variables that are expected to make all performance measures worst at one extreme of their range or groups of parameters that have an essentially equivalent effect. This may allow elimination of many uncertain parameters on a fairly rigorous basis. Other parameters whose influence is unclear but judged to be minor may be frozen at nominal values if necessary to get a problem of reasonable size. Uncertain cost parameters, which appear only in the objective and not in the model equations or constraints, may be conveniently dealt with by evaluating the expected objective over these parameters and eliminating them from the optimization level.

B. Screening Tools for Disturbance Rejection

It is often important to get a rough solution to a design problem quickly. This may be critical in the early stages of a project, when many alternatives are being considered and when the problem definition is still changing rapidly. Under these circumstances, detailed modeling and analysis may be too slow and too costly to be effective, but key design decisions still have to be made. Screening methods are required which allow rapid assessment of alternative designs. This section reviews some screening methods proposed in the literature and presents an improved method for analyzing the effect of delays on achievable disturbance rejection.

Within an integrated design procedure, steady-state modeling and optimization may be used as a screening tool within the overall process of developing a design that can meet the performance requirements at all times and over all uncertain parameters. In the second Shell Process Control Workshop (Prett *et al.*, 1990), Campo *et al.* use steady-state design with uncertainty as a screening tool for linear systems. They use the method of Grossmann *et al.* discussed above. Our modifications of this method are used for this purpose in the design procedure presented below.

Here, we focus on screening tools, addressing dynamic performance in response to disturbances and uncertainty.

1. Review

Methods relevant to assessing the effect of dynamics may be divided into two groups: controllability indicators and design validation tests. Controllability indicators may be used to direct designers away from options that are likely to be difficult to control but have an inherent ambiguity when weighed against economic considerations. Design validation tests should determine whether any design within a group of possible designs might achieve the required performance, without actually having to select a working design. If such tests can be carried out efficiently, they are particularly useful as they are unambiguous.

a. Controllability Indicators. Controllability indicators provide a *qualitative* indication of how difficult a plant will be to control; they cannot, however, easily be related to the economics and operating constraints for a particular problem. This means that the controllability indicators cannot be weighed on the same scales as the process economics. Within this framework, a choice between a plant with good steady-state economics but poor controllability indicators and one with inferior steady-state economics but better controllability indicators cannot be made without a *judgment* of how much the better controllability indicators are worth. In addition, there are multiple controllability indicators which are likely to give contradictory messages for any given plant. Finally, controllability indicators are usually based on linear models of the system to be controlled, leaving the effect of the nonlinearity of the true plant and process unexamined. Controllability indicators can play a useful role in highlighting when difficulties might arise, but they cannot be used to rigorously eliminate design options. Their main use in this context is to provide the designer with information that may be of assistance in finding a base-case solution quickly. A review of controllability indicators is presented in Morari (1992).

b. Design Validation. The design validation problem is to determine whether performance requirements can be met without actually carrying out a full design.

The controller validation problem of determining whether any controller exists which will allow a particular plant design to meet its performance requirements under process disturbances and uncertainty has been given particular attention. The *Fundamental Process Control* methodology (Prett and Garcia, 1988) addresses this validation problem as an integral part of the approach to control system design advocated. Their approach is of general interest as it represents an attempt in the control area to deal with uncertainty and constraints systematically. It is emphasized that a problem specification for a control system design, using their approach, should contain quantitatively defined objectives and constraints together with a process model and an uncertainty description. The design problem is divided into two stages—validation and analysis. The validation stage attempts (1) to determine whether performance requirements can be satisfied irrespective of controller structure and type, and (2) to allow the performance indices to be refined if necessary before moving on to the analysis of the worst-case performance of particular control system designs. Such an objective is clearly desirable, so it is important to consider both some specific problems with their formulation and some general problems in separating validation from analysis.

Prett and Garcia (1988) pose the validation problem as a discrete time linear optimal control problem under uncertainty. The uncertainty is defined by simple bounds, giving a polyhedral set of uncertain parameters V. For this problem, certain forms of uncertainty, e.g., in gains only, together with a quadratic performance index can be shown to satisfy the convexity requirements for the worst-case parameters to lie at vertices of V. This allows the algorithm of Grossmann et al., based on examination only of vertices of V, to be applied (see Section II.A.1). The mathematical formulation is

$$\min_{\Delta u(k)} \max_{v \in V} \{ \|y(k+1) - y_s(k+1)\|^2_{W_y^T W_y} + \|\Delta u(k)\|^2_{W_u^T W_u} \} \qquad (18)$$

subject to bounds on the outputs y, the control moves $\Delta u(k)$, and the control values $u(k)$. Here, y_s is a vector of target values for the outputs, and W_y and W_u are weighting matrices for deviations from target values and control moves, respectively. This formulation implies that the controller has no knowledge of the uncertain parameters, i.e., that the controller operates completely blindly. This assumption can cause this validation problem to give very pessimistic results.

Consider, for example, a typical stable system, $y(s) = (k1 e^{-k2s}/(1 + k3s)) u(s)$, which is known exactly except for an uncertainty in $k1$. It is required that the control bring the system from a given initial condition to target steady state at some arbitrary time in the future in the absence of disturbances. Each possible input sequence, $u(k)$, achieves this for one and only one value of $k1$. The validation problem formulation above will therefore indicate that no controller exists satisfying this problem specification. In fact, any PI controller giving a stable closed-loop system will achieve the performance specified.

The problem of performing the validation irrespective of the control system is confirmed by Garcia in the *Second Shell Process Control Workshop* (Prett et al., 1990):

> In a nutshell, since the optimization problem . . . does not consider any controller explicitly, it searches for a *fixed* sequence of moves into the future that meets all performance criteria for all plants in the uncertainty description. This can be shown to be impossible for trivial cases. . . . Our current thinking is leaning towards solving the *design analysis* problem . . . for the most comprehensive controller that the designer can implement (e.g. DMC) and use this controller for validating design decisions.

The analysis problem for DMC itself is as yet unsolved. Garcia also notes the limitations of the use of linear models:

> Most likely, if a controllability problem arises it is probably due to nonlinearities. In such cases a linear process representation may not be sufficient to allow the solution of the problem requiring the use of a more detailed or sophisticated model. Therefore, the procedure should include the option of finding a nonlinear description and solving the nonlinear control problem under uncertainties.

Leaving aside the emphasis on DMC and the limitation of the design to control, this is consistent with the design approach used here (see Section II.A).

A possible method of salvaging the validation problem is to solve *both* a minmax (Eq. 18) and a maxmin optimal control problem:

$$\max_{v \in V} \min_{\Delta u(k)} \{\|y(k+1) - y_s(k+1)\|^2_{W_y^T W_y} + \|\Delta u(k)\|^2_{W_u^T W_u}\} \tag{19}$$

If the maxmin analysis indicates the problem to be infeasible, it is indeed infeasible as there exists some v for which no input sequence can be found to satisfy the performance specification. If the minmax problem indicates feasibility, there is some *fixed* input sequence which satisfies the constraints for all v.

For an optimal control problem, the complementary maxmin problem is equivalent to the controller having perfect knowledge of the plant parameters and disturbances, including *future* disturbances. This formulation could be termed the "crystal ball" approach to control. For many problems, the crystal ball control will be successful, but this says very little about whether any *realizable* control system exists that can meet the performance specification.

The most likely outcome of solving the complementary maxmin/minmax problems is that the maxmin problem is feasible (for each v there exists an input sequence that can satisfy the constraints) and the minmax is infeasible (there is no single input sequence that can satisfy the constraints for all v). This is unfortunate, as this outcome is the least informative as to whether a realizable controller exists.

For example, optimizing control moves for disturbance rejection in a blending system subject to step disturbances of variable magnitude would give perfect disturbance rejection for a maxmin/perfect-knowledge formulation and a poor (possibly infeasible) disturbance rejection for a minmax/no-knowledge formulation. This arises as disturbances can be canceled at the system inlet by adjusting flows to maintain the correct blend; thus the blending system is essentially a mechanism for compensating for model uncertainty, particularly uncertainty regarding the disturbances. The merit of a particular blending system lies in how effectively it compensates for lack of knowledge and facilitates extraction of knowledge. The mathematically convenient extremes of no knowledge and perfect knowledge do not provide good bounds on performance for this type of problem. This means that useful results are unlikely unless the mechanism for knowledge extraction is embedded in the optimization by implementing a control system in the model used to evaluate performance. Unfortunately, this destroys the desirable separation between design validation and design analysis.

An additional point with regard to the use of combined maxmin/minmax optimal control formulations is that they give rise to problems of comparable or greater difficulty to carrying out a worst-case design (Section II.A) with a specified control system. Given that the bounds they produce for performance are generally loose as discussed above, they do not seem to be an efficient use of resources in a design procedure.

2. A Screening Test for Disturbance Rejection in Nonlinear Processes Subject to Time Delays

Time delays are a feature of many process systems. Where present, they imply a fundamental limitation on achievable performance that no feedback controller, however sophisticated, may overcome. In this section, a screening test for analyzing achievable disturbance rejection in dynamic systems involving time delays is presented. The test is based on calculating the minimum time before a feedback control system can begin to counteract a disturbance and testing whether the open-loop response violates a constraint before this time has elapsed. Passing this test is a necessary condition for the existence of a controller that can meet the specified disturbance rejection requirements. A variation of this test can be used to calculate the fraction of the disturbances that can be rejected. This provides a measure of controllability that can be used to judge whether an implementable control scheme is likely to be successful. The method includes consideration of process and disturbance dynamics and can be applied to uncertain multivariable nonlinear systems.

Previous process controllability work on analyzing the effect of delays on achievable control performance has centered on defining and computing the effective delay(s) in multivariable systems. This work has generated some useful controllability measures, which are reviewed below, but does not in itself provide an answer to the key question of whether a particular disturbance can potentially be rejected by control action so as to avoid constraint violations. This section presents a method for extending the earlier controllability measures to answer this question directly and unambiguously.

a. Review of Previous Work. Previous work on controllability analysis of systems with delays has centered on calculating the effective delay associated with multivariable systems. Holt and Morari (1985) present several measures of the effective delay. The first measure is the set of minimum delays from any input to each output. This provides an obvious lower bound on the time for control action to reach each output. In this paper they note that this bound may not always be achievable with a stable causal controller. The minimum delays to make a change in each output without disturbing any other output are computed to provide an achievable upper bound on the minimum delay associated with each output. Perkins and Wong (1985) provide an alternative approach based on functional controllability analysis, computing the minimum delay before an arbitrary trajectory can be imposed on each output independently. This corresponds to the maximum of the upper bound delays computed by Holt and Morari and has the virtue of providing a single measure of the effective delay. Holt and Morari demonstrate that increasing delays in the process dynamics can improve the achievable *decoupled* response subject to delays.

These measures of delay all provide useful indicators of the effective delay, but they do not in themselves indicate whether the effect of the delay prevents a disturbance from being rejected before causing constraint violations. In Holt and Morari's analysis, the disturbances are assumed to appear as steps on the outputs, making the question of whether the disturbance causes constraint violation trivial. In practice, disturbances are often well approximated by steps, but the effect of the step usually propagates dynamically through part of the process before affecting the outputs. This means that the effect of the process dynamics in attenuating the disturbance should be included in order to assess disturbance rejection.

It should also be noted that the analyses of delay times discussed above do not distinguish between feedback and feedforward control. In practice, disturbances may often be measured with a smaller delay than the delay between the disturbance and any of the constrained outputs. This allows feedforward control to make a contribution to achievable control performance which should be considered explicitly.

A common performance estimation method in "classical" single-loop feedback controller design is to check the open-loop disturbance rejection of a system up to the point in time at which the controller action is assumed to take effect. If constraints are violated during this time, the controller cannot prevent the violation. The use of this test can be traced back to Velguth and Anderson (1954).

They consider a single disturbance, a fixed linear process model (series lags), and a heuristic estimate of the controller response time based on the sum of delays and minor lags in the control loop. This estimate of the delay is based on trying to approximate the peak in the response of a proportional-integral-derivative (PID) controller, and a number of variants (including computing the response time by using the Ziegler–Nichols rules for computing effective delay) have appeared, e.g., the work of McMillan (1983). This analysis directly addresses achievable disturbance rejection, but does so in a heuristic manner which does not provide a rigorous bound on performance, even for the special case of PID controllers.

The analysis presented below draws on both the areas of previous work discussed above to develop general and direct methods of determining whether the effect of delays prevents control action from meeting constraints.

b. Analysis of Delay Effects on Disturbance Rejection. The basic technique in the methods discussed below is to compute the open-loop response to the disturbance over the interval from disturbance onset to the minimum delay for effective controller response elapsing. Any constraint violations up to this time indicate that even "ideal delay-limited control" is not adequate. Success with ideal delay-limited control is a necessary, but not sufficient condition for success with an implementable feedback controller such as PI (see Fig. 2).

The response of the ideal delay-limited controller is not specified, but is assumed to be able to prevent constraint violations after the minimum delay has elapsed.

This test is particularly useful for disturbances such as steps and pulses because their full impact is felt immediately. However, the test is not restricted to these disturbances and may also be used to consider the initial effect of sinusoids, ramps, or other disturbances.

It should be noted that if the open-loop response is evaluated by numerical integration, there is no requirement for linearity of the process model. Convenient analytical solutions can be found for some linear systems, such as series stirred tanks with instantaneous reaction and constant flow.

The discussion below covers

1. the definition of the minimum delay
2. the formulation of feasibility tests for disturbance rejection
3. the formulation and interpretation of a controllability measure
4. the simplified analysis for series stirred tanks.

c. Definition of the Minimum Delay. For a single-input/single-output system with a single disturbance entry point, the time between a disturbance occurring and the controller response *beginning* to take effect, t_d, is made up of the time

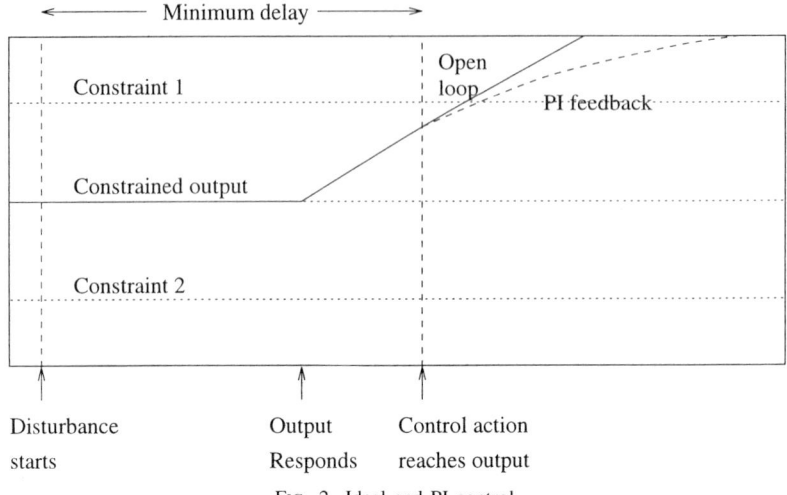

FIG. 2. Ideal and PI control.

delay between the disturbance and the measured value ($t_{d_{d,y}}$) and the delay between the manipulated variable and the measured value ($t_{d_{u,y}}$):

$$t_d = t_{d_{d,y}} + t_{d_{u,y}} \qquad (20)$$

The main complication for multivariable systems with a single disturbance entry point or multiple independent disturbance entry points is generalizing t_d for each constrained output. For a general multivariable controller with inputs u_i and outputs y_j the generalization, for output $y_{\bar{j}}$, is

$$t_{d_{\bar{j}}} = \min_j \{t_{d_{d,y_j}}\} + \min_i \{t_{d_{u_i,y_{\bar{j}}}}\} \qquad (21)$$

This expression for the minimum delay from control action to the constrained output y_j, $\min_i\{t_{d_{u_i,y_j}}\}$, corresponds to Holt and Morari's lower bound on the minimum delay. Use of their upper-bound values is not appropriate in general, as they are based on a requirement for decoupling which may not be necessary or even desirable for disturbance rejection. If decoupling is assumed to be an additional performance requirement, the upper-bound delays could be substituted for those used. It should be noted that, in the absence of a decoupling requirement, t_d and hence the predicted disturbance rejection, can never be improved by increasing any process delay.

Situations will arise in which paths with minimal delay are technically present but are not realistically useful. For example, pressure variations in a distillation column may accompany disturbances affecting composition, but do not readily provide a means of countering such a disturbance. In general, only certain manipulated variables, $u_i | i \in I_j^s$, will have a strong enough effect on y_j to be able to counter the disturbance. Similarly, the disturbance can usefully only be detected by using a subset of the measured variables, $y_j | j \in J_d$. These restrictions combine to eliminate weak connections which would form no practical control feedback paths and prevent unrealistic solutions. In assessing the effect of control structure, it is necessary to impose a particular control structure in which only certain inputs, $u_i | i \in I_j^c$, are connected to certain measurements. Combining these restrictions gives

$$t_{d_j} = \min_{j \in J_d} \{t_{dd, y_j} + \min_{i \in I_j^s \cap I_j^c} \{t_{du_i, y_j}\}\} \tag{22}$$

The difference between feedforward and feedback control can be conveniently represented by using an additional *unconstrained* measurement with the appropriate delay in relation to the disturbance to represent the feedforward measurement. The effective delay will usually be less for feedforward than for feedback control, giving an improved bound on disturbance rejection.

d. Feasibility Tests. For a nominal process model, the test for feasibility of disturbance rejection is

$$\exists p \in P \quad \text{s.t.} \quad \max_{k \in K} \max_{t \in [0, t_{d_k}]} \{c_k(p, t)\} \leq 0 \tag{23}$$

where t_{d_k} is the minimum delay associated with the kth constraint c_k, and p is a vector of design parameters lying in the set P.

In general, there will be uncertain or variable properties to consider and it will be required that the constraints be satisfied for all combinations of the corresponding uncertain parameters. This uncertainty is assumed to be parameterized by a parameter vector v, the values of which lie within a polyhedron V. v may include both parameters defining the possible disturbances and model parameters affecting the initial open-loop response, such as biases on measurements used for control.

Assuming that the effect of control in steady state is defined by a set of controller setpoints and fixed values for a subset of uncontrolled inputs, it is appropriate to use a worst-case design formulation in which a single choice of p is required to accommodate all possible values of the uncertain parameters. In other words, the only steady-state adaptation to uncertainty is that implicit in the choice of controller setpoints. This is potentially conservative; however, it is a fair approximation to industrial practice, where minimizing the need for operator intervention and control system adjustment is often desirable in normal operation. The feasibility test then becomes

$$\exists p \in P \quad \text{s.t.} \quad \max_{v \in V} \max_{k \in K} \max_{t \in [0, t_{d_k}]} \{c_k(p, v, t)\} \leq 0 \quad (24)$$

$$V = \{v \mid -\Delta v^- \leq v \leq \Delta v^+\}$$

A more general formulation allows certain variables (o), known as operating variables, to be adjusted in response to the variation in the uncertain parameters. As discussed in Section II.A, this formulation is more difficult to analyze and will often give inappropriately optimistic results. The formulation with operating variables is, however, given below for completeness.

$$\exists p \in P \quad \text{s.t.} \quad \max_{v \in V} \max_{o \in O} \max_{k \in K} \max_{t \in [0, t_{d_k}]} \{c_k(p, v, o, t)\} \leq 0 \quad (25)$$

$$V = \{v \mid -\Delta v^- \leq v \leq \Delta v^+\}$$

If desired, an objective function, min $J(p,V)$, could be combined with any of the above constraint formulations to obtain the optimal feasible value of p. For example, p might include capacities of equipment and setpoints for the outputs, y, and the objective $J(p,V)$ might be annualized cost. The optimized cost would then give a rough estimate of the effect of process disturbances and dynamics (and uncertainty) on the optimal process cost.

Optimization problems based on ideal disturbance rejection with delays, as above, are much easier to solve than the corresponding problem involving design of a specific controller for the following reasons:

1. The timespan to be simulated is much less than would be required for controller design.
2. A controller need not be included in the model, thus reducing model complexity and the number of design variables.
3. The number of uncertain parameters required to model uncertainty is reduced as some parameters, such as those associated with measurement dynamics, do not affect the open-loop response.

These problems have the additional advantage that they provide a result that bounds the performance of a set of controllers rather than just a particular controller considered in a full design. The "ideal delay-limited control" analysis has been found in the case studies (Section V) to provide a useful bound on the performance achieved with implementable controllers.

In some problems, the delays will themselves be functions of the parameters p and v. This introduces a potential problem for the optimization in that, even if each delay is a smooth function of the parameters, the path determining the minimum delay may change, introducing a discontinuity. This *might* require the use of nonsmooth optimization methods. As noted by Gill *et al.* (1981), smooth algorithms may often be successful on problems such as those with occasional discontinuities and should be tried first.

The optimization problems above may be used to provide a rigorous bound on the achievable performance if delays are calculated as discussed in Section II.B.2. If it is desired to obtain a heuristic estimate of performance rather than a rigorous bound, the control response delays, $t_{d_{u_i,y_j}}$, within control loops could be replaced by heuristic values such as $\pi/2\omega_n$, where ω_n is the frequency (rad/s) at which the phase shift of the dynamic response between the input and output is π radians. This heuristic corresponds to making $t_d = t_u/4$, where t_u is the natural (ultimate) period of the loop (i.e., the period of oscillation of an open-loop stable system when the feedback gain is just sufficient to sustain an oscillation). The heuristic delay, $t_u/4$, is nearly equivalent to the rigorous method for process dynamics consisting of a small pure delay and a large first-order lag. An example presented in Section V.A.3 confirms this heuristic to be effective in considering PI control of pH in a stirred tank.

e. Controllability Measure. The feasibility test is useful in itself, but a positive or negative result raises additional questions.

1. If the test indicates that control performance may be adequate for several different processes/control schemes, which of these should be pursued in detail first?
2. If the test indicates that feedback control would fail but feedforward control might succeed, what can be said about the likely success of a combined scheme?

One way of addressing these questions is to determine what fraction of the disturbance, δ_f, can be handled by particular schemes assuming ideal delay-limited control. The likelihood of success for alternative schemes can then be estimated from the value of this fraction. Combined feedforward/feedback control is unlikely to be successful unless the relative feedforward error due to measurement and modeling errors is less than the calculated δ_f for feedback control. That is, if δ_f without feedforward control is $x\%$, the feedforward must be accurate to within $\pm x\%$ of value for the combined control to have a chance of success. In considering otherwise comparable design alternatives, the alternative with the larger δ_f should be pursued first. The disturbance fraction therefore represents a useful controllability measure to guide further design effort.

To calculate the disturbance fraction, it is only necessary to include it in the model and apply the optimization formulations discussed above using $J(p, V) = -\delta_f$ as an objective:

$$\min_{p \in P, \delta_f} -\delta_f \quad \text{s.t.} \quad \max_{v \in V} \max_{k \in K} \max_{t \in [0, t_{dk}]} \{c_k(p, v, t, \delta_f)\} \leq 0 \tag{26}$$

$$v = \{v | -\Delta v^- \leq v \leq \Delta v^+\}$$

f. Simplified Analysis for Series CSTRs. Although general problems require optimization of a nonlinear dynamic model as discussed above, the analysis can be greatly simplified for some special cases. The case of particular interest for the problems considered later is that of continuous-flow stirred-tank reactors (CSTRs) in series. In this case, it is desired to add reagent so as to keep variations in the net concentration of effluent and reagent, *cnet*, at the exit of the last tank below a certain level, δ_{cnet}, in the face of step disturbances in the inlet concentration of magnitude Δ_{cnet}. This objective can be expressed as a required disturbance attenuation, δ_c, where

$$\delta_c = \frac{\delta_{cnet}}{\Delta_{cnet}} \tag{27}$$

If the reagent flow is much less than the effluent flow and the disturbance is symmetric (equally severe with respect to the upper and lower constraints), the equation

$$\delta_c = \frac{Fr_h - Fr_l}{2 \, \Delta Fr} \tag{28}$$

can be used, where ΔFr is the change in reagent flow required to cancel the disturbance and Fr_h and Fr_l are the maximum and minimum reagent flows to remain within the exit limits.

It can be shown that the unit step response of n tanks in series with mixing lag t_{cmix} and no delays is given by

$$y(t) = 1 - e^{-t/t_{cmix}} \sum_{i=0}^{n-1} (t/t_{cmix})^i / i! \tag{29}$$

Given the minimum delay t'_d in the control response relative to the disturbance response at the treatment system exit ($t'_d = t_d - t_{dd,y}$, where y is the exit measurement), the best possible disturbance attenuation, δ_a, is given by

$$\delta_a = 1 - e^{-t'_d/t_{cmix}} \sum_{i=0}^{n-1} (t'_d/t_{cmix})^i / i! \approx \frac{1}{n!} \left(\frac{t'_d}{t_{cmix}} \right)^n \tag{30}$$

Tanks of different sizes may be analysed similarly, giving

$$\delta_a = 1 - \sum_{\substack{i=1 \\ j \neq i}}^{n} \frac{e^{-t'_d/t_{c\text{mix}}(i)} t_{c\text{mix}}^{n-1}(i)}{\prod (t_{c\text{mix}}(i) - t_{c\text{mix}}(j))} \approx \frac{1}{n!}\left(\frac{t'^n_d}{\prod_{i=1,n} t_{c\text{mix}}(i)}\right) \quad (31)$$

If the reagent added to each tank is based on concentration measured at the tank exit without any measurement delay, then t'_d will simply be the minimum value of $t_{d\text{mix}}$, where $t_{d\text{mix}}$ is the delay between reagent addition and measurement response due to imperfect mixing.

To take account of the effect of uncompensated minor time lags on the performance of PI controllers, replace t'_d in the above formulas by the minimum value of $t_u/4$, where t_u is the natural period of a control loop around a single tank. Using this effective t'_d makes the calculated performance bound heuristic rather than rigorous. A good default is 10 seconds for t'_d when estimating the best practicable performance, while the heuristic method will give typical t'_d values of about 30 seconds.

Similar expressions can be derived for responses to ramp disturbances. If the disturbance is a decaying exponential, then this is equivalent to an extra lag on a step disturbance, and the response can be evaluated by using the formulas for non–equal–sized tanks. The improvement in achievable attenuation from replacing a step by an exponential disturbance of time constant τ' or a ramp reaching a new steady value after τ' is approximately $t'_d/(n + 1)\tau'$.

C. Optimization of Dynamic Systems

Dynamic optimization is an important subproblem underlying both dynamic worst-case design (Section II.A) and the general version of the disturbance rejection test (Section II.B.2). Here, we address methods for solving such problems and discuss the approach we have taken in our work.

The optimization problem of interest is

$$\min_{u(t)} J(x(t_f), z(t_f), u(t))$$

$$\text{s.t. } f(\dot{x}, x, z, u(t), t) = 0$$

$$g(x, z, u(t), t) = 0$$

$$q(x, z, u(t)) \leq 0 \quad 0 \leq t \leq t_f \quad (32)$$

$$L(x(\phi), z(\phi), u(t), \phi) \leq 0$$

$$M(x(\phi), z(\phi), u(t), \phi) = 0 \quad \forall \phi \in \Phi$$

$$\Phi \subset t$$

where J is the objective of the optimization, $u(t)$ is a vector of optimization parameters that may vary over time, f and g define a model that will in general constitute a mixed system of differential and algebraic equations (a DAE system) with differential variables x and algebraic variables z, q defines path constraints that must be satisfied at all times, and L and M define interior point constraints that must be satisfied for a discrete set of times Φ.

It should be noted that $u(t)$ in formulation (32) can be replaced by time-invariant parameters θ by parameterizing the variation with time variation explicitly, e.g., by using a piecewise linear function of time. This allows standard finite-dimensional optimization methods to be applied and is the approach adopted in our work.

Dynamic optimization is a rapidly maturing field which we will not review in detail. Approaches can be distinguished by

- whether the dynamic equations are solved by conventional simulation techniques (feasible path) or as part of the optimization using approximation techniques to transform the problem to algebraic constraints (infeasible path)
- how constraints on values of variables over time (state or path constraints) are dealt with (see below)
- how gradients are calculated, e.g., adjoint equations or sensitivity equations
- the optimization methods applied, e.g., SQP, iterative dynamic programming.

A general review is given by Biegler (1990).

Our judgment is that feasible path methods in which the solution of the model equations over time is carried out by conventional integration software, which has been extensively developed and refined, are at present more reliable than infeasible path methods. Feasible path optimization methods are also easier to implement as the size of the optimization problem is much smaller. For these reasons, we have pursued feasible path methods despite evidence that infeasible path methods are more efficient on some problems.

We have used sensitivity equation methods (Leis and Kramer, 1985) for gradient evaluation as these are simple and efficient for problems with few parameters and constraints. In general, the balance in efficiency between sensitivity and adjoint methods depends on the type of problem being addressed. Adjoint methods are particularly advantageous for optimal control problems in which the inputs are represented as a large number of piecewise constant input values and few interior point constraints exist. Sensitivity methods are preferable for problems with few parameters and many constraints.

We have chosen to use an SQP method (Chen, 1988) as the basis for local solution of the dynamic optimization problem. This choice is based on the potentially rapid convergence of this method and its computational efficiency on small problems.

We present below some easily implementable methods for improving the robustness and efficiency of feasible path dynamic optimization codes which have proved useful in our work. Here, we cover methods for preventing simulation error from disrupting optimization, representation of path constraints, and handling poor local approximations during the optimization.

1. Noise Control

All numerical integration methods solve problems only approximately and will usually provide some error control tolerances to adjust the accuracy of the solution. Typically, these include an absolute and relative tolerance on the local error during integration (*atol* and *rtol*), an event tolerance for the location of discontinuities (*evtol*), and a tolerance for the precision to which the equations must be satisfied on initialization (*sstol*). Generally, *rtol* and *atol* are combined to give an error control weighting, $w_i = rtol|y_i| + atol$, and are often given the same value. For feasible path dynamic optimization, these tolerances must all be chosen so that they meet the following criteria:

1. The numerical solution of the equations does not in itself introduce a significant modeling error.
2. Noise or precision error does not seriously disrupt progress of the optimization, thus causing slow convergence or failure.
3. The computation time used does not substantially exceed that required to satisfy these requirements.

The first requirement is generally easily met as the error in the equation solution typically becomes small compared to overall modeling error for moderate values of the error control tolerances. The tradeoff between the second and third requirements is more difficult and depends on the particular numerical characteristics of the system equations and the particular values of the optimization parameters. It is desirable to have some means of estimating and adjusting the precision error to optimize this tradeoff. This requires that the precision error be estimated, its effect on the optimization assessed, and the integrator tolerances adjusted appropriately.

a. Estimating Precision Error. When the sensitivity method for evaluating gradients is used (Dunker, 1984; Leis and Kramer, 1985), the system equations are reintegrated during the gradient evaluation following a successful line search by the optimizer. The variable trajectories ($x^s(t)$, $z^s(t)$) generated during the gradient evaluation are not identical to the trajectories generated previously ($x(t)$, $z(t)$). The difference is due to the jacobian of the system equations, which is used in the iterative solution of the equations being updated more frequently when calculating gradients than when just integrating the model equations.

The precision error in the variables can be estimated by using the formula

$$\epsilon(y) = \frac{\max_i |\Delta^k y_i|}{\sqrt{\frac{(2k)!}{(k!)^2}}} \qquad (33)$$

(Gill et al., 1981) based on successive differencing of variable values generated using closely and uniformly spaced parameter values. Numerical trials indicate that

$$\delta(y) = |y^s - y| \qquad (34)$$

is comparable to $\epsilon(y)$. The values often agree to within about 20%, with $\delta(y)$ usually being greater than $\epsilon(y)$. As $\epsilon(y)$ requires about six extra integrations to generate, there is a clear case for using $\delta(y)$, which is available without additional system integrations, as an estimate of the precision error.

This estimate strictly addresses only precision error in the variable trajectories, and not errors in the gradients. However, as noted by Dunker (1984), the gradients given by the sensitivity method are exact gradients of the computed solution trajectory regardless of truncation errors. This suggests that controlling the accuracy of the integration of the DAE system will generally be sufficient to give accurate gradients. Numerical experience indicates that $\epsilon\left(\frac{\partial y}{\partial \theta_i}\right) / \left|\frac{\partial y}{\partial \theta_i}\right|$ is indeed comparable to $\epsilon(y)/|y|$. This experience is all based on sensibly scaled problems with values of y and $\partial y/\partial \theta_i$ within a few orders of magnitude of 1, but does support the suggestion that control of the variable accuracy is sufficient to control the gradient accuracy.

If the sensitivity method is not being used to generate the gradients, the precision can be computed using extra integrations to generate $\epsilon(y)$. This estimate could be mapped straightforwardly to precision error in gradients calculated by numerical differencing. Estimating errors in the gradients calculated using adjoints would require additional integrations of the adjoint equations to generate $\epsilon(\partial y/\partial \theta_i)$.

2. Evaluating Effect on Optimizer

The estimated error in objective and constraint values may be mapped through calculations in the optimizer to determine whether the noise is likely to impede progress. The appropriate measure will depend to some extent on the optimization code used. Noise in the objective and constraints may interfere with the evaluation of the optimizer termination conditions or with comparisons between performance with different values of the optimization variables, e.g., during a line search. Noise in the gradients may result in selection of a search direction that is not a descent direction, particularly if the descent direction is chosen using a second-order algorithm with an ill-conditioned Hessian estimate. The Hessian estimate may itself be degraded by noise in the gradients.

A typical SQP termination condition for a constrained optimization problem

$$\min_\theta J(\theta)$$

$$\text{subject to} \quad c_i(\theta) = 0, \quad i = 1, 2, ..., meq \tag{35}$$

$$c_i(\theta) \leq 0, \quad i = meq + 1, ..., m$$

is

$$\max\left[\sum_{i=1}^{meq} |c_i| + \sum_{i=meq+1}^{m} \max(0, c_i), \right.$$

$$\left. \left(|\nabla_x J^T \delta_k| + \sum_{i=1}^{meq} |\lambda_i c_i| + \sum_{i=meq+1}^{m} \lambda_i \max(0, c_i)\right)\Big/(1 + |J|)\right] \leq optacc \tag{36}$$

where λ is the vector of estimated Lagrange multipliers, δ_k is the step estimated from the quadratic program on the kth iteration, and *optacc* is the specified optimization accuracy.

Comparisons between points are usually based on some form of merit function, P_m, e.g.,

$$P_m(\theta) = J(\theta) + \sum_{i=1}^{meq} \mu_i |c_i(\theta)| + \sum_{i=meq+1}^{m} \mu_i \max(0, c_i(\theta)) \tag{37}$$

where the penalty parameters μ_i are closely related to the $|\lambda_i|$.

Controlling the combined precision error

$$error = \frac{\delta(J)}{1 + |J|} + \sum_{i=1}^{meq} \max\left(1, \left|\frac{\lambda_i}{1 + |J|}\right|\right) \delta(|c_i|)$$

$$+ \sum_{i=meq+1}^{m} \max\left(1, \left|\frac{\lambda_i}{1 + |J|}\right|\right) \delta(\max(0, c_i)) \tag{38}$$

approximately controls the noise in the termination condition and merit function. If *error* is kept below *optacc*, noise effects on the termination condition and on the comparisons of values of the merit function are unlikely to disrupt convergence.

Evaluation of the effect of gradient errors would be much more difficult but, as discussed above, has not been found necessary.

3. Adaptation of Integrator Error Tolerances

The value of *error* can be used for feedback control of the integrator tolerances. This feedback is complicated by the fact that in the algorithm used for integration there are four tolerances that can be adjusted to control accuracy. Experimentation indicated that the computational cost of the integration showed little dependence on the initialization tolerance, *sstol*, or on the accuracy to which discontinuities were located in time, *evtol*. These tolerances could, however, have a substantial effect on noise levels. It was therefore decided to set these tolerances to very stringent levels, 10^{-10} being chosen as a default. As noted previously, the remaining tolerances *atol* and *rtol* that govern the acceptance of an integrator step are generally equated. Adopting this practice gives a single parameter *simacc* to adjust, with *rtol* = *atol* = *simacc*.

The mean value of *error* may be expected to increase monotonically as *simacc* increases while the computation time may be expected to decrease monotonically as *simacc* increases. The control objective is therefore to maintain a value of *error* that is *just small enough* to prevent serious noise problems. The precise relationship between *simacc* and *error* is unclear and will vary from system to system and integrator to integrator. As in any feedback scheme with a poorly defined system model, the feedback system must be suitably conservative to avoid instability. The scheme below has been found to be effective in extensive trials.

if (*error* > 2.*optacc/factor*) *simacc* = *simacc*/10

if (*error* < *optacc/factor*) *simacc* = *simacc*.min(1/.5,

$$(optacc/error.factor)^{1/5})$$

where *factor* defines the target margin of safety from levels of noise comparable to the termination threshold *optacc*. A value of 5 was found to be adequate as further increases in *factor* did not reduce the number of iterations required for convergence. The factor of 2 in the above algorithm provides a deadband so as to prevent the adaptation mechanism from responding to insignificant changes in the measured error. If the error seems too large, *simacc* is tightened vigorously to eliminate the high noise promptly. If the response to noise levels below the target is similarly vigorous, the error control is liable to oscillate continuously, giving repeated spikes of excessive noise. Hence a more cautious adaptation to low noise levels is used. The precise detail of the adaptation is not very important. Any scheme incorporating vigorous response to high noise and sufficiently cautious response to low noise so that oscillation does not occur would give essentially the same results. To prevent the error control correction from disrupting line search convergence, the function and constraint values should be recalculated whenever the integration accuracy is tightened.

This method has been found to eliminate the need for application-specific consideration of integrator error control while requiring no significant extra computation and obviating the use of very tight error control on the integrator. The number of iterations required for optimization was found not to differ significantly from that required with fixed small values of *simacc*, while the time for integration was reduced substantially. Failures attributable to noise were eliminated.

The main limitation of the method is that gradient errors are not directly estimated, analyzed, or controlled. If problems occur for this reason, the parameter *factor* could be increased, or fixed values of the integrator tolerances could be chosen by trial and error.

4. Path Constraint Representation

Problems may also arise regarding the representation of constraints of the form

$$q(x, z, \theta, t) \leq 0 \quad \forall t \in [0, t_f] \quad (39)$$

known as path or state constraints. The most common approach to this problem is to use the method of Sargent and Sullivan (1979) in which the original infinite constraint is transformed to

$$\int_0^{t_f} \max(0, q(x, z, \theta, t))^2 dt = 0 \quad (40)$$

This representation is single-valued, differentiable, and precisely equivalent to the original constraint; but it has two undesirable properties:

1. As the constraint violation approaches zero, its gradient approaches zero and the Lagrange multiplier associated with the constraint goes to infinity.
2. A path constraint that is inactive by even a small amount is invisible to the optimizer.

These limitations do not usually prevent convergence in practice, provided noise is controlled as discussed above. However, they may lead either to slow progress due to constraints that flip repeatedly from active to inactive or to the need for stringent noise control as the constraint and its gradients go to zero and the Lagrange multiplier grows.

A simple modification of the path constraint representation (40), is given by

$$\int_0^{t_f} \max(0, q(x, z, \theta, t))^2 dt \leq c_{\text{sig}} \quad (41)$$

This representation makes slightly feasible constraint values visible to the optimizer, and reduces noise problems by avoiding the multiplication of a small constraint by large Lagrange multipliers or penalty factors within the optimizer calculations. The disadvantage is that the modified problem may deviate significantly from the actual problem if c_{sig} is too large. For abstract mathematical problems, this disadvantage is significant, but for engineering problems in which the constraint violation has a meaning it should always be possible to define c_{sig} to be the smallest violation judged to give a significant deviation from the original problem specification.

The scaling of the constraint is also important. The scaling below was found to work well. Within the model

$$\dot{x} = max(0, q(x, z, \theta, t))^2/c_{sig}, \quad x(0) = 0 \tag{42}$$

was used, giving sensible values for the states associated with active constraints.

Many optimizers use the feasibility test $\Sigma |c| \leq optacc$ (cf. Eq. 36). The model generates a path constraint, $x(t_f)$, for which a value of 1 represents an insignificant constraint violation. Using

$$c = (x(t_f) - 1)optacc \leq 0 \tag{43}$$

is enough to make the significance levels consistent, but would allow numerically feasible points with path constraint violations greater than c_{sig}. The constraint

$$c = (2x(t_f) - 1)optacc \leq 0 \tag{44}$$

ensures that any numerically feasible point will have path constraint violations less than c_{sig}. The numerical solution may therefore have active constraints with path constraint violations between $c_{sig}/2$ and c_{sig}, which seems appropriate.

This modified path constraint representation was found to give a reduction of up to a factor of 2 in the number of iterations required for successful optimization.

5. Poor Local Approximation

Local approximations (linear or quadratic) are often particularly poor in dynamic optimization problems. For instance, this situation is found to occur when taking the full step predicted from the local approximation, δ_k, causes a path constraint to become active or the system to become unstable.

The step based on the local approximation may have to be reduced by several orders of magnitude to satisfy the conditions for a new value to be accepted. This may be accomplished using a line search method to select a scaling factor α for the step to satisfy some criteria. If the line search is to be accomplished efficiently, large reductions in step size must be allowed ($\alpha_{j+1} \approx .1\alpha_j$, where j is the line search iteration). If the line search accepts the first point satisfying the search criteria, it may reduce the step magnitude much more than necessary and be unable to improve the local approximation. This can require many full iterations of the optimization to approach the optimum, each involving costly gradient evaluations.

This problem may be tackled by modifying the line search procedure. The line search should be allowed to reduce the step size until a point α_j is found satisfying the line search criteria. This point then defines a minimum step size, α_{min}, while the previous point, α_{j-1}, defines a maximum step size α_{max}. The line search then proceeds as follows:

if $\alpha_{max}/\alpha_{min} > 2$ $i = 1$
do until $i = 0$
    $\alpha_{j+1} = \min(2\alpha_j, (\alpha_j + \alpha_{max})/2)$
    $\theta_{j+1} = \theta_{k-1} + \alpha_{j+1} \delta_k$
    Evaluate J, c for new θ
    If line search criteria satisfied and merit function reduced then
        if $\alpha_{max}/\alpha_{j+1} > 2$
            $j = j + 1$
        else
            $\theta_k = \theta_{j+1}$
            $i = 0$
        end if
    else
        $\theta_k = \theta_j$
        $i = 0$
    end if
end do

This algorithm essentially tries to locate the largest step satisfying the line search criteria, α^*, within a factor of 2, assuming that the criteria are met for all $\alpha \in [0, \alpha^*]$ and not satisfied elsewhere.

This method has been found to reduce the overall computational effort in approaching the optimum by up to 30% compared to that obtained by accepting the first point satisfying an Armijo cone condition on the merit function while allowing step reductions of up to a factor of 10 at each line search iteration and using a quadratic interpolation formula to estimate the step reduction.

A more precise, gradient-based line search such as that discussed by Gill et al. (1981) is unlikely to be beneficial as the increased line search cost would require an implausibly large reduction in the number of line searches to give a net benefit.

6. Discussion

We have developed a code incorporating these characteristics and applied it successfully to more than ten industrial design problems, some of which are discussed later in this chapter. Limited testing on standard problems also supports its effectiveness. This is problem 5.1 in Vasantharajan and Biegler (1990) illustrating that some difficulties with feasible path methods were solved without any difficulty using the code developed.

III. Neutralization of Waste Water

Neutralization of waste water has been chosen to illustrate the integrated design approach as it is a relatively simple system of considerable industrial importance and it is subject to large disturbances and a high degree of uncertainty. We start by explaining the general background to neutralization system design. Following this, a procedure applying the general integrated design approach to this application area is presented.

The design of waste water neutralization systems for an industrial plant or site generally involves an interaction between "end-of-pipe" treatment system design and decisions about the design and operation of other processes. The end-of-pipe neutralization system typically consists of a small number of CSTRs or in-line mixers arranged in series. Reagent is added to each mixer based on measurement of its exit pH. The design of the end-of-pipe treatment system is a compact design problem amenable to general treatment. The more general design decisions are application-dependent and open-ended and do not lend themselves to explicit incorporation within a design method for waste water neutralization systems. A useful design method for waste water neutralization should, however, facilitate interaction with this broader design context as decisions at this level can have a more substantial effect on total treatment cost than the end-of-pipe design itself.

Before presenting our procedure for waste water neutralization system design, we discuss general design methods for the selection of structural or discrete design variables.

A. Making Discrete Design Decisions

Real design problems involve discrete design decisions, such as layout of reactors and choice of control scheme, as well as selection of continuous design parameters, such as reactor sizes and controller tuning. The design tools presented have focused on techniques for analyzing performance of particular designs and for optimizing continuous design variables.

One approach to making discrete design decisions is to pose a mixed-integer nonlinear program (MINLP) and apply a suitable optimization method to the selection of discrete design parameters. We have not adopted this approach for the following reasons.

1. The MINLP approach requires all the objectives and design options (including models) to be fully defined before solution can commence. In typical waste water neutralization system design problems, models and costs are generated and refined for options as the results of the design process indicate a need to do so. The problem definition itself may be modified substantially as the design progresses owing to interaction with design decisions in the broader process context. These characteristics are not conducive to solution methods based on one pass through a single large optimization problem.
2. MINLP algorithms typically capture the results of NLP solutions for particular sets of discrete variables by using some form of linearization at the NLP solution. A design engineer can potentially extract much more information than this and bring judgment and experience to bear in deciding which option for the discrete design parameters to examine next. If the NLP problem can be solved in seconds, a MINLP algorithm will often have the edge over a designer as it can extract limited information about many design options in the time it will take the designer to extract a lot of information about a single option. If the NLP solution time is around an hour or more, as may be the case in dynamic optimization or worst-case design, the designer can be expected to gain the edge over the algorithm.

An alternative to using a MINLP algorithm is to use a heuristic-based, designer-driven search, coupled with a bounding strategy to efficiently eliminate structures. This approach has considerable potential when good rules exist for modifying designs and when bounds on achievable performance can be obtained for candidate structures without too much effort. Examples of this approach are given by Chan and Prince (1988) and Mizsey and Fonyo (1990). Chan and Prince present a heuristic strategy for modifying flowsheets to synthesize a flotation cell separation circuit and show that it gives reasonable results. Mizsey and Fonyo looked at the use of heuristics in design more generally, coining the term "predictor-based bounding strategy" to describe the use of bounds to eliminate structures as candidates selected by a search are evaluated rigorously.

Predictor-based bounding appears appropriate for the integrated design approach proposed. A range of tools can be used for generating bounds on performance:

1. nominal and worst-case ideal delay-limited control evaluation
2. nominal and worst-case steady-state evaluation
3. nominal and worst-case dynamic evaluation

By applying these tools appropriately and using sensible rules to guide the design choices it is possible to develop efficient integrated design procedures.

B. Design Procedure for Neutralization of Waste Water

1. Problem Definition

The basic design problem is defined by which streams are to be treated and the effluent characteristics to be achieved prior to discharge from the treatment system. This definition may be a design decision at a higher level. Therefore a range of problems may need to be considered to provide information on which higher-level design decisions can be taken. This places particular emphasis on efficient preliminary design methods that will allow higher-level options to be considered without costly detailed design analysis or experimental work.

A complete characterization of a neutralization problem for design purposes might require the following:

1. a set of titration curves (pH values tabulated as a function of reagent concentration) or a thermodynamic model covering the range of conditions the treatment system will encounter
2. a definition of the acceptable variation in output characteristics, usually in terms of pH
3. a definition of any additional process constraints (e.g., lower bound on pH within the treatment system to avoid gas formation)
4. a description of the variation over time of effluent flows and titration characteristics/compositions (disturbances)
5. properties of reagents that may be used for neutralization, including reagent particle size distribution and reaction kinetics if the reagent is used in solid form
6. a definition of any equipment outside the scope of the treatment system that provides smoothing of the composition or flow variations in the effluent streams
7. constraints on achievable mixing characteristics, particularly minimum mixing delay and minimum residence time
8. constraints on process equipment selection and sizing, e.g., whether in-line mixers can be used and maximum total volume
9. constraints on measurement type and positioning, e.g., whether flow measurements are available and whether in-line probes are acceptable

10. constraints on reagent addition system type and positioning, e.g., only enough acid reagent available for a final stage pH adjustment
11. maintainable performance levels of equipment, particularly measurements, e.g., maximum lag and maximum bias
12. constraints on controller type, e.g., only direct PI control of reagent addition to be used
13. costs for all design options that have a significant effect on the cost of the final system.

It is impractical to expect to know all these characteristics *exactly* for any problem; moreover, they may be highly variable over time. Thus the specification must be allowed to include sets of possible characteristics rather than just nominal values. Default values and bounds for many of the above characteristics are given in Section IV along with advice on experimental procedures to generate design data. Approximate costs for most options are given in Walsh (1993). It is not necessary to have all the information in order to begin the design.

In defining the disturbances, the need to start the plant up should be considered. A reasonable base-case set of disturbances is startup from zero flow to at least 25% of maximum flow in a single step and a 10% load change at maximum flow. A treatment system that cannot satisfy these requirements is unlikely to be adequately operable. These base-case disturbances may be used in preliminary analysis if actual data are not available.

For preliminary analysis of dynamic response, the information required is

1. the allowable pH range for the treated waste water
2. the minimum time between the disturbance reaching the outlet of the system and the feedback control response reaching the outlet of the system
3. the disturbance condition giving rise to the maximum pH change at the outlet within this time (this may be found by worst-case design if necessary)
4. the relationship between pH and reagent concentration within the allowable pH range under this disturbance condition, i.e., at least a partial titration curve.

This information provides an adequate summary of the overall characteristics for the purpose of evaluating whether an ideal (delay-limited) controller could achieve the required exit pH performance. The worst-case disturbance is often obvious and is therefore included as part of the specification information. It should be noted that if there is more than one tank between the initial disturbance effect and the treatment system exit, the worst-case condition for a given change in effluent load will involve maximum flow. The effect of flow in reducing the concentration change for a given load change (flow^{-1}) will be outweighed by degraded disturbance rejection (flow^{-n}, where n is the number of tanks). If the worst-case condition is not obvious, the methods for worst-case design may be used to identify this condition given a model and uncertainty description.

Steady-state properties of the neutralization system are generally an issue only when neutralizing acid effluents with solid alkali reagents. For evaluating steady-state properties, the following information is required:

1. the allowable pH range for the discharged stream
2. the properties of the reagent to be considered
3. the combination of acid concentration and flow in terms of maximizing the pH change downstream of the treatment system exit (this may be found by worst-case design if necessary)
4. the relationship between pH and reagent concentration within the allowable pH range under this load condition.

The overall design procedure is summarized in Fig. 3. It should be noted that the design techniques are deployed in order of increasing computational load and design data requirements. This allows most design options to be rejected

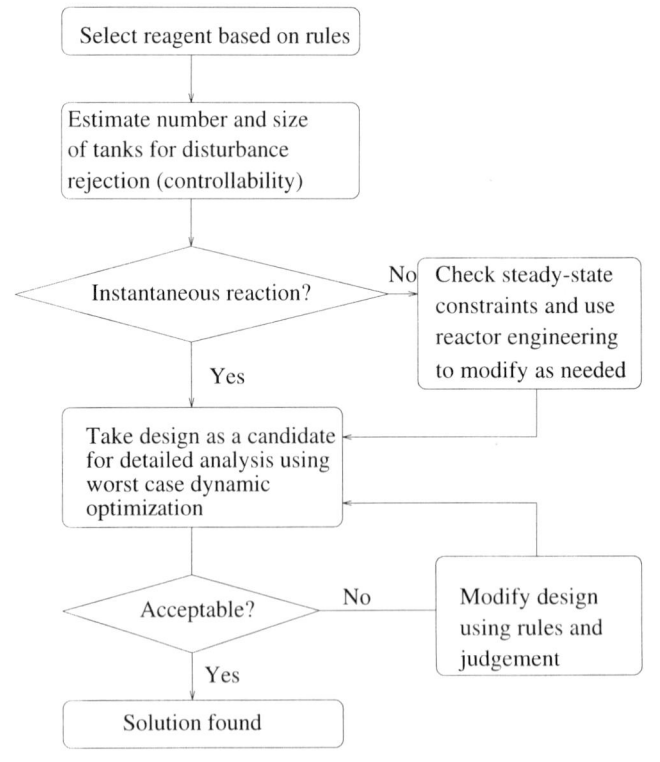

FIG. 3. Integrated design procedure for waste water neutralization.

with little effort. Only a few options will be analyzed using worst-case dynamic optimization.

2. Preliminary Design

The recommended procedure for preliminary design is

1: Select reagent

2: Evaluate design requirements for adequate transient control performance using ideal control analysis or optimal PI control performance estimates

3: If solid alkali reagents are being used then investigate steady-state conversion
  If system chosen by control analysis has acceptable reagent conversion then take as a candidate design for more detailed analysis
  Otherwise modify design based on reaction engineering

4: Take the resulting design as a candidate design for more detailed analysis

The first three steps are discussed further below.

a. Reagent Choice. Many rules exist for guiding reagent choice, scattered over the open literature. These rules range from clear constraints on when certain reagents may feasibly be used to fuzzy preferences and warnings. These rules are summarized in Walsh (1993), along with approximate costs for common reagents.

The following procedure for selecting a reagent is recommended:

1. Identify the set of feasible reagents.
2. Identify relevant design considerations for these reagents.
3. Decide whether any of these considerations override using the cheapest feasible reagent.

The key treatment system properties to consider in carrying out this analysis are (1) whether the effluent is acidic or alkaline, (2) the desired final pH, (3) the presence of ions forming poorly soluble salts with the reagent, and (4) the total acid load.

A typical feasibility criterion is "Carbonate-based reagents cannot be used to give a final pH above 5." A typical design consideration is "Caustic (NaOH) gives reduced sludge formation compared to calcium- or carbonate-based reagents."

This analysis will usually give a clear indication of the best reagent for a given application. The most common choice is calcium hydroxide.

b. *Control Analysis.* The data requirements for the general case are defined in the previous section. In the most difficult cases, preliminary control analysis requires the solution of a worst-case design problem (see Section II.B.2). In most cases, a much simpler approach can be used. The worst-case disturbance condition may often be evident, e.g., a wash operation on the plant or startup of the treatment system. The worst-case disturbance characteristics can usually be bounded by the maximum step change in effluent load, combined with the maximum pH sensitivity to reagent concentration and the maximum flow. If the worst-case condition is clear, it is possible to calculate the required disturbance attenuation δ_c (Section II.B.2). The acceptable δ_{cnet} can be computed from the titration curve (or a thermodyamic model) and the required output pH properties, with due allowance for the effect of pH measurement bias. The inlet concentration variation, Δ_{cnet}, can be computed from the flows and concentrations associated with the worst-case disturbance. The required overall disturbance attenuation is then given by $\delta_{cnet}/\Delta_{cnet}$.

The disturbance attenuation capability, δ_a, of a particular design can be calculated using Eqs. (31) or (52). Equation (31) will give either a rigorous or a heuristic bound, depending on whether the true or effective delay is used for t'_d (Section II.B.2). Equation (52) gives an *estimate* of the optimal performance *with PI control* (see Section V.A.3). The calculated disturbance attenuation can then be compared to the required disturbance attenuation. If the rigorous bound indicates that a design is inadequate, then it can be straightforwardly rejected. If the heuristic bound ($t'_d = t_u/4$) indicates that a design is inadequate, it should be rejected unless some means of compensating for the minor lags in the dynamic response is to be used. If the PI performance estimate indicates a design to be inadequate by a reasonable margin, say 30%, then the design can be rejected if PI controllers are to be used.

It is possible to determine the optimal volume of n CSTRs in series to achieve the required disturbance attenuation by equating the required and estimated attenuation (see Section V.A.4).

This analysis is a good starting point for preliminary design as it can give performance bounds and estimates for all neutralization systems regardless of reagent kinetics.

If on–off control is being considered, the achievable performance can be estimated by considering a limit-cycle at the natural period, t_u. The change in concentration from maximum to minimum reagent addition, Δ_{cnet}, must be attenuated sufficiently so that the outlet concentration variation is less than the acceptable variation, δ_{cnet}. Attributing the output variation to the first harmonic of the square-wave input and assuming reagent flow is negligible compared to effluent flow, we have the following necessary condition for successful on–off control:

$$\delta_{cnet} \geq \frac{4\Delta_{cnet}}{\pi} \prod_{i=1,n} \frac{1}{(1 + (2\pi t_{c_{mix}}(i)/t_u)^2)^{1/2}} \qquad (45)$$

If the worst-case disturbance is not evident, the ideal delay-limited control analysis is better approached by using the worst-case design tools on a dynamic model, as discussed in Section II.B.2. This optimization can be solved in a few minutes (on a SPARC 2) for typical problems.

c. Solid Reagent Conversion. Given a design that is potentially adequate in its dynamic response but uses slowly reacting solid alkali reagents, the next stage is to consider whether the reagent conversion is adequate. Even with adequate pH response at the exit of the treatment system, unreacted reagent at this point may cause downstream pH drift, giving rise to violation of the specified limits. If the design fails this test, it is necessary to modify the design to improve its conversion characteristics. This may be tackled as a reactor engineering problem, bearing in mind that reaction order is greater than 1 for these reagents (Section IV.B). Adding a plug-flow reactor stage early in the mixing scheme is likely to be beneficial, and moving to an increased number of backmixed reactors may be desirable. Alternatively, total volume may be increased.

There are a number of options for carrying out this analysis:

1. Carry out a full experimental investigation of reagent kinetics to generate a model and uncertainty description and apply worst-case optimization.
2. Use the default models from Section IV.B.
3. Apply a shortcut heuristic procedure.

A shortcut procedure should ideally allow simple computation and avoid optimistic results. The procedure outlined below meets these requirements.

The worst-case conditions for reagent carryover must be identified, and the associated reagent conversion requirements δ_l (permissible fractional carryover of reagent from the treatment system exit) computed. It should be noted that the worst-case load condition may be quite different from the worst-case disturbance condition for dynamic analysis. The worst-case load characteristics will normally be bounded by the maximum neutralization load at the maximum flow and the maximum pH sensitivity and may be identified more precisely by worst-case design if desired.

The apparent reaction time constant, τ_r, may be estimated from a batch neutralization experiment based on the pH response as it approaches a steady value in the target pH range. If δ_l is the allowable fractional carryover of reagent, the time from $3\delta_l$ of the reagent remaining unreacted to the time for δ_l of the reagent remaining provides a reasonable estimate of the effective τ_r. Estimated in this way, τ_r may be used as a cautious estimate of an effective first-order kinetic time constant and the residual reagent, δ_{res}, may be estimated as

$$\delta_{res} = \frac{e^{-\tau_{PFR}/\tau_r}}{\prod_{i=1,n} (1 + t_{c_{mix}}(i)/\tau_r)} \tag{46}$$

where τ_{PFR} is the total plug flow reaction time available. δ_{res} should be less than δ_l to ensure that carryover of reagent is not excessive. This method has a sound basis in that it looks at the response associated with the portion of the particle size distribution which will dominate the residual carryover at the critical level, δ_l. It is pessimistic in that it then treats this response as characteristic of the entire particle-size distribution.

If a full steady-state model is used, the reagent conversion may be analyzed by steady-state simulation. All the reagent flows should be specified within the model in some way, e.g., by target pH values within the treatment system. Fixing the pH of each CSTR and analyzing the conversions of reagent added to each tank in turn may be useful to provide insight into the design problem (see Section V.D for an example).

d. Summary. At the conclusion of preliminary design analysis, there should be a clear picture of what the worst-case conditions are likely to be as well as a candidate design that is likely to meet both dynamic response requirements and reagent conversion requirements. The tools used are all quite efficient, so that a large number of options can be explored quickly. This allows the problem specification to be reviewed if necessary and provides a good starting point for more detailed design analysis.

3. Complete Design

Complete design utilizes a full model of the process, control system, and disturbances, including all *significant* uncertainty and variability, to determine whether performance requirements can be met. The design structure is initially based on the preliminary design work, which can also be used to define an initial set of models for worst-case design and initial values for some of the design variables. The controller and process parameters may be adjusted using worst-case optimization. If steady-state issues have not been fully explored in the preliminary design stage, it is sensible to run a steady-state worst-case design prior to the full design. This may identify the need for structural modifications or modifications to the problem formulation in a small fraction of the time it takes to solve a dynamic worst-case design.

Including all design issues within the mathematical formulation may complicate the model excessively, potentially slowing the solution down by an order of magnitude. Measurement noise, controller sampling effects, and detailed reagent delivery characteristics all fall into this category.

In a well-engineered system, measurement noise is unlikely to affect the performance of a PI controller significantly (Section V.A.2), but it may limit the use of derivative control action and will certainly limit the performance of advanced control schemes that attempt to approach ideal control (Section V.A.5). Our approach to PID control has been to optimize PI controllers and leave the possible benefit of derivative action to the commissioning engineers.

The effect of controller sampling time is not as bad as the effect of an increased delay in the system and can if necessary (sampling interval greater than about 2 s) be approximated by a delay equal to the sampling interval. This avoids the introduction of frequent discontinuities in the control trajectory.

Reagent delivery requires careful attention. However, it may be considered after the other design issues have been settled because it is generally amenable to satisfactory solution by detailed design without making a substantial difference to the overall design or to total cost (Walsh, 1993). In initial analysis, deliberately oversized idealized valves can be considered so as not to interfere with the control response. A conservative sizing of the actual valves can be based on evaluating the maximum reagent delivery capacity obtained assuming ideal valves. If this sizing indicates extreme rangeability requirements, then specialized equipment may be needed. If the rangeability requirement is moderately high, say 100:1, the effect of reducing the maximum reagent capacity should be checked, as valves saturating for short periods of time do not necessarily degrade performance significantly. If the rangeability requirements are less than about 50:1, use of standard equipment is acceptable. Equal-percentage valves, characterized to appear linear to the controller and fitted with positioners, are a good default reagent addition element. The effect of valve errors can be considered by including a detailed valve model in the simulation, although this is likely to slow down the integration dramatically. Section V.A.2 presents some evidence that reagent valve errors are not usually critical together with an expression for estimating the exit deviations caused by deadband errors (Eq. 51).

Reliability and maintenance issues may have a significant effect on the total cost, but are unlikely to justify reworking of the design provided some basic issues are accounted for. The main effect of reliability/maintenance analysis on existing plants is that most measurements and valves must be duplicated with appropriate isolating equipment to allow maintenance. Provided this is allowed for approximately in defining any costs considered in optimizing the design, the detail of the implementation of maintenance systems is not likely to affect the design choices. Operation in conditions likely to severely degrade reliability should be avoided by adding constraints or cost penalties to the design as necessary. For example, steady pH values outside the range 2–12 are associated with increased measurement bias, and pH values above 7 encourage fouling of electrodes with carbonate. Therefore, pH operating points between 2 and 7 are preferred for good reliability.

Compliance with the specified pH limits is taken to be required 100% of the time. This is increasingly the accepted design basis and is actually easier to handle than requirements in which pH limits must be satisfied x% of the time. One method of handling "x%" constraints would be by excluding worst-case combinations that are expected to occur less than $(100 - x)$% of the time. This introduces an extra level of iteration to the design problem. Specifications that the pH must be outside certain bounds for less than a certain time can be readily included in analyses based on the full dynamic model.

4. Rules for Modifying Design

The procedures presented so far largely leave the structural decisions to the designer. There are, however, a number of rules that may be useful, particularly if the designer does not have extensive experience with these problems.

If the worst-case disturbance is caused by a flow change as opposed to a concentration change, consider using ratio reagent addition with feedback, adjusting the ratio rather than the reagent itself. This requires gain-scheduling for control around a backmixed tank (dividing controller error by flow would be effective). If the worst-case disturbance is a concentration disturbance, use feedforward if the load can be estimated within about ±20%.

If steady-state reagent conversion of solid alkali is the limiting constraint, consider the following options:

1. Use a plug-flow reactor or a small CSTR as the first stage, using underneutralization to enhance overall reagent conversion. Feedforward control of reagent addition to the PFR, if practicable, avoids the poor feedback control characteristics of the PFR (see Section V.E).
2. Provide acid reagent to downstream stages to compensate for carryover of reagent at intermediate stages in the treatment system when this is the limiting factor.
3. Increase the reactor volumes.

If control is adequate apart from large limit cycles following a major disturbance, apply input conditioning so as to eliminate the limit cycle. This is much preferable to eliminating the limit cycle by reducing controller gain as this would cause an overall degradation in performance. As input conditioning can be trivially implemented in modern control hardware, it is worth implementing in the initial design if there is a well-defined and significant nonlinearity.

On–off control should be considered if titration curve variability is high and disturbances are comparable to the total neutralization load.

If the appropriate control scheme is still not effective, increase the backmixed volume or add another controlled stage, possibly an in-line mixer.

If actuator saturation causes problems, consider increasing valve size, shifting operating points to achieve a better distribution of load between stages, or speeding up response of upstream control stages by reducing mixing delay and using fast response injector probe assemblies so as to reduce transient loading of downstream stages.

If actuator hysteresis effects cause excessive pH fluctuation, consider reducing valve size, using parallel valves, or using specialized reagent addition equipment. It is not likely that such effects will require an increase in backmixed volume as the 1% hysteresis associated with valves using positioners is unlikely to be the worst disturbance to exit pH with sensibly sized equipment (see Section V.A.2).

IV. Modeling of Waste Water Neutralization Systems

> We must remember that the most elegant and high-powered mathematical analysis based on a model which does not match reality is worthless for the engineer who must make design predictions (Levenspiel, 1972).

This section discusses the development of the models used in the case studies presented in Section V, providing an illustration of the challenges involved in developing models and uncertainty descriptions to support integrated design.

All design methods must include a model (at least implicitly) to be effective. As models are rarely precise, it is important to associate an uncertainty description with each model to allow the appropriate degree of robustness to be incorporated into designs utilizing that model. An inadequate model or uncertainty description may lead to an unworkable design or to unnecessary and expensive overdesign.

In choosing between several models giving equally good predictions, we prefer the model requiring least process-specific experimental input. For each model selected for use in design, the sources of uncertainty and variability are discussed, and guidelines for experimental work to reduce the uncertainty are given where appropriate.

The following modeling issues are considered:

Steady-state relationships between pH and reagent addition (Section IV.A)

Reaction kinetics of hydrated lime ($Ca(OH)_2$) (Section IV.B)

Dynamic response and bias errors of measurement (Section IV.C)

Effect of mixing on system characteristics (Section IV.D)

A. STEADY-STATE pH CHARACTERISTICS

The complex nonlinear relationship between pH and neutralizing reagent concentration, often expressed graphically as a titration curve, is a key characteristic of waste waster neutralization systems. It determines the sensitivity of pH to concentration disturbances and hence the required disturbance attenuation. Its nonlinearity may lead to large-amplitude limit cycles in response to disturbances if not adequately compensated for, and may force controller detuning due to the variability in the sensitivity of pH to reagent concentration. Steady-state pH characteristics are therefore an appropriate starting point in considering the modeling of waste water neutralization systems. Modeling the relationship between pH and reagent addition requires a steady-state mapping from compositions to the pH of the solution. The main approaches to constructing this mapping are the use of thermodynamic equilibrium relationships and the direct use of titration curves.

Titration curves have the virtue of simplicity, while thermodynamic models have, at least in principle, greater capacity to be adapted to changing process conditions. In practice, the amount of effort required to develop a thermodynamic model based on real chemical species and physical properties is usually prohibitive. Semiempirical models based on notional species and concentration equilibria can be developed quite readily (Gustafsson and Waller, 1983). However, these have extrapolation properties identical to the original titration curves. In both cases the pH of a mixture of components can be predicted accurately, subject to the assumptions that mixing two solutions of equal pH results in a solution with the same pH and the pH measurement is ideal (Gustafsson, 1982; Luyben, 1990). The validity of this assumption is discussed in Walsh (1993).

Titration curves are therefore used as the standard model representation in the case studies discussed in Section V. The titration curves are represented as tables of pH values and the reagent concentration changes required to move between consecutive pH values. This representation facilitates specifying uncertainty in the sensitivity of pH to reagent concentration within particular pH ranges. The concentration of reagent is given in molar units (M) with concentration taken by convention as zero at a reference pH and negative below that pH. The table function is interpolated using monotonic cubic splines.

In generating titration curves, care should be taken that any reactions are allowed to go to completion (check for pH stabilization over about ten minutes if slow reactions are observed). The material to be titrated should be at the same temperature as the effluent to be treated, and care should be taken to avoid the loss of volatile components from samples to the atmosphere. The reagent used should be the main reagent to be used for neutralization. If problems are experienced in metering small quantities of lime-based reagents, caustic (NaOH) can be used near neutral to get sufficient resolution of the titration curve—at least one point per unit pH change should be generated.

In characterizing uncertainty in titration curves (variability of slope), it is important to ensure that a representative set of samples are titrated. In particular, consideration should be given to titration characteristics during shutdowns, as the shutdown of a plant that provides a lot of weak acid or base wastes may lead to dramatically increased sensitivity of pH to disturbances and reagent adjustment. If backmixed buffering tanks are to be introduced as part of the design, it is useful to look at flow-averaged samples (samples are collected in proportion to total flow over a period of time and the aggregate sample is titrated) as well as grab samples (instantaneous samples). An averaging period comparable to or less than the residence time of the buffer tank will give a picture of the titration characteristics downstream of the buffer tank with less effort and error than combining results from many grab samples.

Titration curves may be misleading when the reagent and effluent reach an equilibrium pH on a time-scale comparable to or longer than the residence time in the treatment system. Unless the reactions involved are modeled, predictions based on the titration curve(s) may be quite inaccurate. The most common instance of this involves the use of calcium hydroxide reagents (slaked or hydrate lime), which is discussed below. If other slow reactions take place, considerable modeling effort may be needed, or very careful interpretation of results.

B. Characteristics of Calcium Hydroxide Reagent

The most common reagents for neutralizing acidic waste water are solid lime-based alkalis, particularly $Ca(OH)_2$. The kinetics of these reagents are not well defined in the literature on waste water treatment (Shinskey, 1973) and may be the dominant factor in equipment sizing in many applications. It is therefore important to develop and validate improved models for these reactions.

Experimental results by Haslam *et al.* (1926) and Yagi *et al.* (1984) support a mass transfer limited mechanism for the dissolution of CaO and $Ca(OH)_2$ in water. This mechanism was verified to be consistent with observed experimental data, using the correlation of Levins and Glastonbury (1972) to predict mass transfer coefficients for suspended particles in a tank. Available correlations for mass transfer were consistently accurate in predicting the observed rate of shrinkage of spherical $Ca(OH)_2$ particles with a radii of 100 microns (μ).

There is some controversy as to how the mass transfer rate varies with particle size, with the transition mechanism between boundary layer transfer and molecular diffusion appearing to depend on particle density. Levins and Glastonbury simply sum the molecular diffusion and boundary layer terms, giving a smooth transition between the two regimes. Brucato *et al.* (1990) present experimental results with dense particles ($\rho_p > 2$ kg/m^3), showing little effect of particle size down to a 15-μ particle radius (the Levins and Glastonbury correlation predicts an increase in dissolution rate of about 66% compared to large particles). This observation is consistent with some early experimental work on lime reagents by Haslam et al. (1926), which showed less than 5% variation in mass transfer coefficient between a 45-μ and 500-μ particle radius. Most of Levins and Glastonbury's work was with low-density particles. However, they did have some data which indicated that high-density particles showed less variation of mass transfer coefficient with size than lower-density particles. Asai et al. (1988) indicate that the transition between turbulent boundary layer mass transfer and molecular diffusion mass transfer is sharper than indicated by Levins and Glastonbury and is better approximated by the maximum of the two contributions rather than by their sum.

This supports the use of a simplified correlation, for dense particles, in which the mass transfer is taken as the maximum of the mass transfer at 100-μ radius (determined experimentally or using Levins and Glastonbury's correlation) and the mass transfer rate due to molecular diffusion, which typically becomes dominant around 10-μ diameter in a well mixed solution (mixing power density $\approx$ .2 W/kg). Including terms for the diffusion of acids to the particle surface (Walsh, 1993) we obtain the model below.

$$\frac{\partial r_p}{\partial t} = -k_{\text{mix}} \cdot \max\left(1, \frac{10}{r_p \cdot k_{\text{mix}}}\right)(k_1 + {}_2[H^+] + k_3 cweak) \; \mu/s \quad (47)$$

where r_p is the particle radius in microns, $[H^+]$ is the concentration of free hydrogen ions, and $cweak$ is the concentration of weak acids that will dissociate below a reference pH (pH$_{\text{ref}}$) of 7. k_{mix} is used to adjust the degree of turbulent mixing and should be about 1 for a well mixed vessel. k_1 was estimated to be about 0.1, k_2 about 7.5, and k_3 about 1.5. It should be noted that the concentration of Ca(OH)$_2$ in the bulk solution is assumed to be negligible.

The above model assumes complete suspension of the particles. Fluid velocities in pipes of 1 m/s are ample for suspension of typical solid alkali reagents. Typical agitation intensities of around .2 W/kg are adequate for suspension of typical calcium hydroxide preparations in tanks.

To complete the model of lime dissolution, a discretized particle size distribution is defined, with $[x_i]$ being the molar concentration of size fraction i with radius r_{p_i}. For an ideal continuous stirred tank reactor (CSTR) with residence time τ the equation for updating the concentrations of the solid lime size fractions is

$$\frac{\partial [x_i]}{\partial t} = 3\frac{\partial r_{p_i}}{\partial t}[x_i]/r_{p_i} - \frac{\partial p_{p_{i+1}}}{\partial t}[x_{i+1}]/(r_{p_{i+1}} - r_{p_i})$$

$$+ \frac{\partial r_{p_i}}{\partial t}[x_i]/(r_{p_i} - r_{p_{i-1}}) + ([xin_i] - [x_i])/\tau \quad (48)$$

where $r_{p_{i+1}} > r_{p_i}$ and $[xin_i]$ is the inlet concentration.

This model was validated against industrial experimental data as discussed in Walsh (1993). Between 6 and 12 particle size fractions were found to be needed to characterize the particle size distributions, which resembled truncated log-normal distributions.

The main sources of uncertainty in this model are discussed below.

1. Errors due to variations in process chemistry: This will mainly affect k_3, which is strongly influenced by the diffusivity of the weak acid species and has about a 50% uncertainty band. Reaction inhibiting impurities may have a very major effect, which should be excluded by experimentation as it cannot readily be compensated for.
2. The variation of the reagent particle size distribution: This depends very much on the quality of the reagent source and should be tested by sampling of the supply.
3. The error due to variation in mixing conditions and uncertainty about their effect: This error is expected to be small for stirred tank reactors using typical power levels of around .2 W/kg ($k_{mix} \approx 1$). There is significant uncertainty in the effect of plug-flow mixing; but so long as the particles are properly suspended, $k_{mix} \in [.5, .9]$ can be expected.

To characterize the reagent properties for particular applications, batch (dynamic) titrations should be carried out with a range of final pH values across the operating region. The batch should be mixed using an agitator rather than a bead stirrer to get realistic mixing conditions. The suggested sampling rates for monitoring the pH response are every 10 seconds for the first minute after addition of lime, every 30 seconds for the next 4 minutes, and every 2 minutes thereafter. The particle size distribution should be measured either by *wet-screening* or by use of a laser counter such as a Coulter counter. If the time between preparing the suspension and completing the measurement is more than a few seconds, water should not be used in making up the suspension unless the fraction of $Ca(OH)_2$ required to saturate the water is small. The particle size distribution can then be used with the default model parameters given to check whether the standard model is appropriate. If deviations are substantial, the model parameters should be adjusted empirically. The most likely need for adjustment lies in the alkali region where reaction inhibition or back-reactions may occur.

C. MEASUREMENT RESPONSE

The response of electrodes used to measure pH is commonly modeled as a first-order lag ranging from one second to several minutes. The actual behavior of electrodes is known to be very complex and a number of researchers have attempted to develop improved models (McAvoy, 1979; Hershkovitch et al., 1978; Johansson and Norberg, 1968). These models are more complex than the first-order lag model and require much more data to define (compositions, diffusion coefficients, dissociation constants). The structural and parametric mismatch to the actual response has not been demonstrated to be reduced relative to the first-order lag approximation, so the use of these models does not seem justified. As a first approximation, the work on more complex models seems to support a first-order lag as being a good match to the form of the electrode response, although the appropriate lag coefficient is expected to vary with operating conditions.

pH probe dynamic response in CSTRs will be taken, based on industrial experience, as a first-order lag with a default maximum value of 30 seconds and minimum value of 5 seconds. A bias error of ±.25 pH will also be taken as a default and it will be assumed that the pH setpoint lies between 2 and 12 to avoid the potentially large errors outside this region (Walsh, 1993). These defaults can, of course, be modified; e.g., for a well maintained multiple probe injector assembly, a faster response and smaller bias error would be appropriate (McMillan, 1984).

Attempts to reduce the uncertainty by experimentation should be approached cautiously as many design parameters affect the apparent response, including fluid velocity near the probe, operating pH, effluent composition, and probe aging and fouling over time.

D. MIXING

The most commonly used model of a mixed vessel is the "fractional tubularity" (delay-lag) model in which some part of the reactor is taken as exhibiting plug-flow conditions and contributing a delay ($t_{d_{mix}}$) and the rest of the reactor is taken as perfectly mixed (uniform concentrations) contributing a first-order lag ($t_{c_{mix}}$). The delay and lag in series are taken as describing the reactor residence time distribution (RTD). The delay-lag representation was validated using both CFD analysis and experimental residence time distributions (Walsh, 1993).

The delay for this model can be estimated for stirred tanks using the correlation of Hoyle (1976)

$$t_{d_{mix}} = kV_t^{0.85}/(Fagit) \qquad (49)$$

For a tank with reagent added at the top of the tank and removed near the bottom at the opposite side $k = .9$, and $k = 1.8$ for a tank with the reverse flow pattern. V_t is the tank volume in cubic meters and $Fagit$ is the agitator pumping rate in cubic meters per hour. It is observed that it is difficult to reduce $t_{d_{mix}}$ below about 9 seconds without causing excessive splashing and air entrainment.

This observation is supported by published correlations for pumping rate and air entrainment which indicate a minimum delay of about 7 seconds (see Walsh, 1993).

The dynamic response of in-line mixers generally approximates a pure delay (Chemineer, 1988). In using in-line mixers, it is important to ensure that injection velocities are high compared to stream velocities and to recognize that about 7 seconds elapses before reagent added to the line reaches the measuring device. These measures ensure smooth reagent addition and adequate micromixing, thereby avoiding excessive variation on the measured pH.

To avoid the need for special procedures and modification of the integration algorithm, delays may be modeled by using rational approximations, e.g., Padé functions or multiple first-order lags in series. Experimentation suggests that 10 series lags is adequate for most applications, so this is used as a default. The approximation should be checked by comparison with a more detailed model where it is believed to be particularly significant.

The above discussion considers mixing only in terms of time response associated with mixing and does not consider the interaction between mixing and reaction kinetics. For fast neutralization reactions and reactions with first-order kinetics, this emphasis is correct; but the time response or residence time distribution (RTD) is insufficient to characterize the effect of mixing on more general reactions. This is because the RTD does not uniquely determine the degree of micromixing of particles; e.g., a set of parallel plug-flow reactors can match the RTD of an ideal CSTR to any required precision, but the time at which particles of different ages first become mixed is very different from that of an ideal CSTR.

When RTDs are well represented by fractional tubularity models, the extremes of micromixing can be achieved by placing the CSTR before the plug-flow section (maximum mixing) or by placing the CSTR after the plug-flow section (maximum segregation). For predominantly second-order reactions, maximum segregation will improve conversion. In the absence of further information, placing the CSTR first will generally give the worst yield for a given RTD and provide a conservative approximation to the micromixing effects for premixed reactants. If the reactants are not premixed, it is possible that the reactants will not mix at all in the plug-flow stage. In this case, the conservative approach is to model the effect of delay on the control but to neglect its effect on reagent conversion. This may be implemented very efficiently by delaying only the reagent flow and not delaying any concentrations, a method that gives a general-purpose conservative model. This model is used in the case-studies on solid reagents (Section V).

The delay associated with mixing in a tank is assumed to be greater than 7 seconds and can be estimated from correlations. A sensible default value for the delay in a stirred tank is 10 seconds, as it will usually be desirable to approach this lower bound. No uncertainty is associated with the delay, on the understanding that the value used in the model is the maximum expected value and that reduced delay will simply improve controller stability and/or performance. The effect of delay on reagent conversion may depend on the detailed mixing pattern, and conservative approximations are used.

One source of uncertainty that has not been tackled is the deviation of the age distribution of solid particles from the age distribution of the liquid. This is not very well understood and varies with particle size. This effect can be minimized by ensuring a good margin from the conditions under which particles are just suspended.

E. Summary

Models for the steady-state relationship between pH and reagent concentration have been explored. In general, titration curve models are more convenient and more readily determined, and these are used in the case-study work presented below. Guidelines for generating titration curves are presented.

A model for reaction of the solid alkali $Ca(OH)_2$ has been developed using general mass-transfer analysis principles. The model has been validated against experimental data and results in the literature. Guidelines for experimental validation and tuning of the model are given.

Measurement characteristics have been reviewed and a default model including an uncertainty description presented.

Mixing properties have been reviewed and a series delay-lag model chosen for general use. Methods for predicting the value of the delay and a lower bound on the achievable delay are discussed and a default value given.

Once the basic modeling foundations have been laid as above for a given application area, the modeling effort for any particular problem is greatly reduced.

Modeling issues in waste water neutralization are discussed in more detail in Walsh (1993), which also discusses redox and precipitation reactions and the reaction of carbonate reagents.

V. Design Examples

The design examples presented in this section cover a wide range of waste water neutralization problems. The examples studied confirm the power of the design methods developed. We consider that the success of the integrated design approach on this challenging and industrially important problem area establishes

OPERABILITY AND CONTROL IN PROCESS SYNTHESIS AND DESIGN 361

its industrial relevance and motivates further work to explore the use of the method on more general process problems. The examples examined cover the following:

Exploration of a generic neutralization problem (Section V.A)
Preliminary design of a central effluent plant (Section V.B)
pH control of a strong-acid/strong-base system (Section V.C)
Neutralization and precipitation with $Ca(OH)_2$ in a central effluent plant with highly variable effluent characteristics (Section V.D)
Neutralization of a well-defined, highly concentrated acid stream with $Ca(OH)_2$ (Section V.E).

The last four examples are based on industrial case studies.

A. Exploration of a Generic Neutralization Control Problem

Before presenting the industrial case studies, it is helpful to use the tools developed to explore a generic problem, allowing some general results and insights to be developed.

1. Problem Definition

To qualify as a generic problem, the disturbance conditions, treatment system structure, and performance targets should be typical of real problems while avoiding unnecessary complexity. These aspects of the generic problem are defined and discussed below.

a. Disturbance Conditions. Disturbances involve both flow and concentration changes. Concentration changes are the most challenging, as it is always possible to minimize the direct effect of flow by use of flow measurements, e.g., by using pH feedback to the *ratio* between reagent and effluent flow. The disturbance was therefore taken as a disturbance in concentration only, for simplicity. The most difficult disturbances are those that occur as step changes, so the disturbance is taken as a pulse change in concentration at the inlet to the treatment system. The duration of the pulse is taken as 30 minutes, allowing recovery to steady-state between the rising and falling edge of the pulse. The precise duration of the pulse is not important.

b. Treatment System. The treatment system was assumed to be made up of continuous-flow stirred-tank reactors (CSTRs) in series as this is the standard industrial system. Most of the examples presented below use equal-sized tanks with $t_{d_{mix}} \approx .05\ V_t/F_T$, as this is typical of industrial systems. Variations from this typical system were made to check the generality of conclusions, where appropriate. It should be noted that only the *ratio* of time constants and delays will affect performance, as an overall scaling of the model time constants simply scales the time axis of the response obtained. Measurement lags were varied to illustrate the effect of minor lags on performance. Actuator dynamics and reagent kinetics are neglected for simplicity, although the associated lags have an effect on control performance similar to measurement lags if they are all small compared to the residence time. Both acid and alkali reagent addition is assumed to be available. For most of the analysis, the reagent addition is assumed to have infinite rangeability and precision, as adequate rangeability and precision can normally be provided by appropriate detailed design. A deadband error effect is considered where an approximation to a practical "sticky" valve is required (Section V.A.2). Where controllers are implemented, they are assumed to be PI controllers from the exit pH of a CSTR to reagent addition at its inlet, as this is the standard industrial system. The final part of this section considers the changes required to move from typical PI performance to the ideal delay-limited control bound.

c. Performance Requirements and Setpoints. The target pH is assumed to lie in the range pH 5–9 (typical discharge consents). The titration characteristic is assumed to be strong-acid/strong-base

$$\text{pH} = -\log_{10}\left(\frac{-cnet + \sqrt{cnet^2 + 4.10^{-14}}}{2}\right), \quad cnet < 10^{-4}$$

$$\text{pH} = -\log_{10}\left(\frac{10^{-14}}{cnet}\right), \quad cnet \geq 10^{-4}$$

(50)

as this gives the strongest steady-state nonlinearity in the target pH range. This means that the acceptable concentration range is $\pm 10^{-5}$ N. As the disturbance is approximated by two equal but opposite steps in concentration and hence produces equally severe acid and alkali deviations, the optimal operating point in relation to ideal delay-limited control or a linear feedback controller (such as a PI controller acting on concentration) is 7 pH (0.0 N concentration, midway between the upper and lower concentration bounds). As the disturbance is in concentration only and perfect actuators are assumed, setpoints other than that of the final stage do not affect the dynamic response for ideal or linear control. Setpoints therefore need to be optimized only when PI controllers based on uncompensated pH measurements are considered. A requirement for the control response to recover to near a steady-state is included by requiring that the concentration variations one hour after the disturbance be less than 10% of the concentration variation that would cause violation of the 5–9 pH bounds. In terms of pH, this corresponds to bounds of 6–8.

The overall design problem is to select number of tanks, sizes of tanks, number of controllers and controller gains, and integral action times to achieve a required disturbance rejection at minimum cost. The generic problem is represented in Fig. 4.

2. Why pH Control Is Difficult

A useful starting point is to examine the characteristics of the pH control problem that make it difficult. The most fundamental reason for the difficulty is the stringency of the performance requirements, which may make it necessary to add reagent to within less than .1% of the ideal amount just to stay within legal limits on the pH of the treated stream.

Pure delays in dynamic response impose fundamental bounds on achievable disturbance rejection, as discussed in Section II.B.2. Uncertainty imposes additional limitations. For example, if the measurement lag were accurately known, it could be canceled and would not limit control performance. But the measurement response is more complex than this and it is variable over time; thus it cannot readily be canceled and combines with the delay to limit controller performance (see Sections V.A.3 and V.A.5). Limits on the reagent addition rate prevent the ideal delay-limited control bound from being achieved in some cases, even in the absence of uncertainty, as infinite actuator range is required (see Section V.A.5). However, these limits are not observed to have a major effect on PI control performance as the degree of transient overshoot is moderate compared to the steady-state control output change required.

Three effects that have a less clear-cut impact on achievable performance are the pH nonlinearity (if uncanceled), the precision limits on the reagent addition, and measurement noise. These are investigated below, in the context of the generic problem of Section V.A.1, to complete the picture of why pH control is difficult.

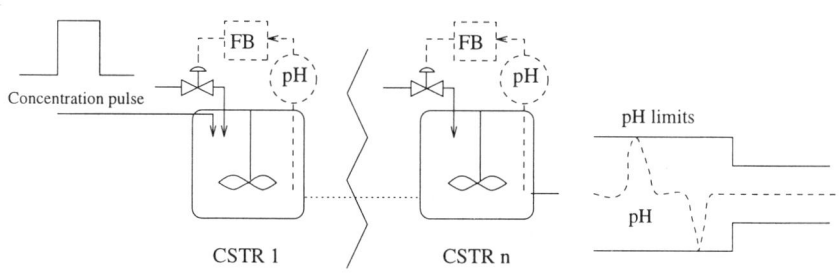

FIG. 4. Generic pH control problem.

a. Effect of Nonlinearity. Using a tank having a residence time (V_T/F) of 3 min, $t_{d_{\mathrm{mix}}} = 9$ s, and $t_{c_{\mathrm{mix}}} = 171$ s, the achievable disturbance rejection was found by optimizing a PI controller based on pH and a PI controller based on concentration. Secondary lags were neglected.

Delay-limited feedback analysis using Eq. (30)

$$\delta_a = 1 - e^{-t'_d/t_{c_{\mathrm{mix}}}} \sum_{i=0}^{n-1} (t'_d/t_{c_{\mathrm{mix}}})^i/i! \approx \frac{1}{n!}\left(\frac{t'_d}{t_{c_{\mathrm{mix}}}}\right)^n$$

indicates that the maximum concentration pulse disturbance level is 1.95×10^{-4} N (calculated as $10^{-5}/\delta_a$ using $t'_d = t_{d_{\mathrm{mix}}}$). A PI controller based on pH measurement was found by optimization to be able to handle 1×10^{-4} N disturbance level. A PI controller based on concentration measurement can handle a 1.6×10^{-4} N concentration pulse. The nonlinearity therefore has a significant quantitative effect on the achievable performance, reducing the maximum disturbance from 80% of the ideal control bound to 50% of the bound. At least for this problem, the dominant limitation is given by the process dynamics, as the PI controllers in both cases approach the theoretical bound on feedback controller performance imposed by the dynamics. This indicates that the effect of the nonlinearity on controller response does not necessarily have a strong influence on the design, even with the extreme strong-acid/strong-base nonlinearity.

Input conditioning can often be applied to minimize the effect of nonlinearity and chemical buffering usually reduces the nonlinearity compared to the strong-acid/strong-base case considered above. Given that the example above shows that even the extreme strong-acid/strong-base nonlinearity does not necessarily degrade performance qualitatively, the rest of the analysis is carried out assuming the titration nonlinearity to be sufficiently compensated as to be negligible.

b. Effect of Precision Error in Reagent Addition. McMillan (1984) notes that reagent control valve precision error is dominated by deadband error, where "[d]eadband error is the change in signal required to start the stroke from a stationary position or to change the stroke direction upon a change in signal direction." The deadband in valve stem position error is typically about 1% of range when a positioner is used. The discussion below assumes a linear valve characteristic (relation between flow and stem position for a fixed pressure drop) for simplicity.

This effect was explored by using a PI controller based on concentration around a tank with $t_{d_{\mathrm{mix}}} = 9$ s and $t_{c_{\mathrm{mix}}} = 171$ s as above. Again, there is no reason to expect the conclusions to be sensitive to the particular mixing conditions. The definition of deadband given leaves some ambiguity as to the valve behavior. Two models were considered for a unit deadband error:

If $u_d > u + 1$ then $u = u_d - 1$

If $u_d < u - 1$ then $u = u_d + 1$

and

If $abs(u_d - u) > 1$ then $u = u_d$

The first model corresponds to the valve position tracking the demanded output, with an offset of 1 unit (valve sticking and then gliding). The second model corresponds to the valve position not responding to discrepancies, with the demanded output smaller than 1 unit and moving promptly to the demanded value when the difference exceeds this (valve sticking and lurching). The "sticking-and-gliding" model is appropriate if the friction resisting movement is independent of the rate of change of valve position. The "sticking-and-lurching" model is appropriate if friction becomes negligible once motion commences. Sticking and gliding is probably closer to the real behavior, but sticking and lurching is also considered as it is not ruled out by the definition and is likely to give worse performance.

Numerical experiments (Walsh, 1993) indicate that the peak-to-peak deviation of the exit concentration can be bounded, for fairly tightly tuned controllers, by calculating the response of the exit concentration to a reagent valve exhibiting a square wave oscillation with peak-to-peak amplitude equal to the deadband error. The exit concentration variation can therefore be estimated as

$$\frac{4\Delta_{cnet}}{\pi} \prod_{i=1,n} \frac{1}{(1 + (\pi t_{c_{mix}}(i)/t_u)^2)^{1/2}} \tag{51}$$

where Δ_{cnet} is the inlet concentration change resulting from a change in valve position equal to the deadband. This error would not be the limiting factor on transient performance unless the disturbances were much smaller than the total load or the deadband error was much higher than the value of 1% of range achievable using a valve positioner.

c. Effect of High-Frequency Measurement Noise. In a well-designed pH measurement system, a moderate amount of high-frequency noise is usually present (about ±.1 pH). This noise has not to our knowledge been analyzed in the detail required to carry out a rigorous evaluation of its likely effect. In our experience, it does not present any significant difficulty when using PI control. It will combine with other forms of process uncertainty to limit the performance achievable with more advanced control strategies (see Section V.A.5).

d. Summary. pH nonlinearity, valve precision, and high-frequency measurement noise do not have a major impact on the design requirements when using series CSTRs and PI control. The key factors limiting design performance are *delays and uncertainty*.

3. Examination of Bounds from Delay-Limited Control Analysis

The performance bound from delay-limited control analysis (Section II.B.2) may be computed for a number of series CSTR systems and compared to the performance obtained by optimizing PI controllers. For simplicity and generality, this section assumes the pH nonlinearity to be canceled at the controllers.

Extensive numerical optimization results (Walsh, 1993) allow a number of observations to be made.

1. The optimal PI performance was within 30% of the correlation

$$\delta_a = \frac{3 t_d'^n}{2 \prod_{i=1,n} t_{c_{\text{mix}}}(i)} \tag{52}$$

   This represents a degradation of performance by a factor of 1.5n! compared to ideal delay-limited control.
2. If one tank in a multitank chain was uncontrolled, the disturbance attenuation was further increased by about 50%.

For preliminary design purposes, the estimate from Eq. (52) is an adequate predictor of achievable PI performance when all tanks are tightly controlled. If control on one tank of a multiple CSTR system is rendered ineffective—due to uncertainty, high delay compared to the minimum delay, or simply the absence of a controller—the predicted disturbance attenuation should be increased by 50%. An exception to this is the case of variation in the sensitivity of pH to concentration on the *final* CSTR. In this case, no degradation of performance from that obtained with the minimum buffering (maximum titration curve slope) will occur as the performance required in terms of concentration deviations relaxes along with the controller performance.

Section II.B.2 suggested that when minor lags are present in addition to pure delays, a heuristic "effective" delay could be computed based on the natural period of the control loop divided by 4. Numerical optimization suggested that Eq. (52) remained approximately valid, although the optimized performance was slightly improved compared to the pure delay case.

Summary. Estimating t_d' based on physical delays and using Eqs. (30) and (31) as appropriate gives a rigorous bound on the achievable performance with any controller. Estimating t_d' as $t_u/4$ and using these equations gives a sensible heuristic performance bound for PI feedback control (and probably for any controller that does not cancel the measurement dynamics and hence is subject to the same bandwidth limitations). Estimating t'_d as $t_u/4$ and using Eq. (52) provides a realistic, albeit heuristic, estimate of achievable performance with PI controllers, neglecting uncertainty. Loss of effective control on a single CSTR in a multiple CSTR system does not have a dramatic effect on the achieved disturbance rejection with PI control.

4. Optimal Sizing of Series CSTRs for pH Control

A striking discrepancy between theory and practice in pH control systems using individually controlled series CSTR systems is the choice of the relative size of the tanks. There is a strong consensus in the literature that equal-sized CSTRs are undesirable for control purposes (Shinskey, 1973). However, equal-sized CSTRs are the industrial norm. Using the delay-limited control performance bound (Eq. 31)

$$\delta_a = 1 - \sum_{i=1}^{n} \frac{e^{-t'_d/t_{c\text{mix}}(i)} t_{c\text{mix}}^{n-1}(i)}{\prod_{\substack{j \neq i}} (t_{c\text{mix}}(i) - t_{c\text{mix}}(j))} \approx \frac{1}{n!} \left(\frac{t'^n_d}{\prod_{i=1,n} t_{c\text{mix}}(i)} \right)$$

and a typical cost function of the form $k_1 + k_2 V_T^x$, it is possible to evaluate the desirability of using tanks of different sizes. The requirement to provide a disturbance attenuation, δ_c, can be approximated well by

$$\delta_c \geq \frac{1}{n!} \left(\frac{(F_T t'_d)^n}{\prod_{i=1,n} V_T(i)} \right) \tag{53}$$

where F_T is the total flow through the system and $t'_d \ll t_{c\text{mix}}$. The magnitude of the disturbance that can be rejected is approximately proportional to the product of the tank volumes, regardless of the number of tanks, if the minimum delay, t'_d, is assumed to be independent of the tank volumes. The economic selection of number and size of tanks to give a required disturbance attenuation, δ_c, at a flow F_T therefore takes the form

$$\min_{n, V_T(i), i=1,n} \left\{ k1.n + k2 \sum_{i=1,n} V_T(i)^x \right\} \quad \text{s.t.} \quad \prod_{i=1,n} V_T(i) = (1/n!)(F_t t'_d)^n/\delta_c \tag{54}$$

where $x \neq 0$. For fixed n the problem becomes

$$\min_{V_T(i), i=1,n} \left\{ \sum_{i=1,n} V_T(i)^x \right\} \quad \text{s.t.} \quad \prod_{i=1,n} V_T(i) = (1/n!)(F_t t'_d)^n/\delta_c \tag{55}$$

Cost for a given number of tanks and disturbance rejection capability is therefore always minimized by equal-sized tanks and the overall problem simplifies to

$$\min_{n} n(k1 + k2V_{T_{\text{opt}}}^x) \qquad (56)$$

with the optimal volume, $V_{T_{\text{opt}}}$, given by $V_T t'_d(n!\delta_c)^{-1/n}$. A similar analysis can be carried out using the PI performance estimate Eq. (52), as the predicted performance is again approximately proportional to the product of the tank volumes for a given problem. The optimal volume for this case is given by $F_T t'_d (1.5/\delta_c)^{1/n}$. The difference between $V_{T_{\text{opt}}}$ with n tanks using the PI performance estimate and using the delay-bound equation with the same value of t'_d corresponds to a factor of $(1.5\, n!)^{1/n}$, that is, a factor of 1.5 for one tank rising to a factor of 2 for three tanks.

The use of this analysis is illustrated below. It might be required to estimate the optimal series CSTR system to achieve a disturbance attenuation of .001 with a flowrate of .03 m³/s and $t'_d = 10$ s, using PI control. Using the cost function $20{,}000 + 2000V_t^{0.7}$ (based on Cushnie, 1984) and Eq. (52) gives a minimum cost CSTR system with two tanks having a volume of 11.6 m³ each and a cost of £62,000. For comparison, a single-tank system would cost £164,000 and a three-tank system would cost £75,000. Increasing t'_d or F_T proportionately increases the required volume for a given n. An increase of a factor of 4 in F_T would shift the economic optimum to three tanks with a cost of £100,000 and a tank volume of 13.6 m³. It should be noted that the optimal volume can be expected to lie below $(k1/k2)^{1/x}$ (27 m³ in this example) as the optimization of n will tend to increase the number of tanks until $k1$ becomes the dominant part of the tank cost. Optimal volumes can be expected to lie between 10 and 30 m³ if the cost expression used is appropriate.

For typical values of F_T of 30–300 m³/hr, the predicted optimal volumes of 10–30 m³ correspond to residence times of between 2 and 60 min. It should be economically attractive to achieve mixing near the achievable bound (see Section IV.D) with tanks of this size, so that $t_{d_{\text{mix}}} \approx 10$ s independent of tank volume. The main secondary lag in typical applications is the measurement lag. The value of this lag varies with installation conditions but is not dependent on tank volume and may be generally taken as less than 30 s. This gives an effective $t'_d \approx 25\text{--}30$ s and a rigorous $t'_d \approx 10$ s, with both values independent of tank volume. This shows that the two main assumptions of the above analysis ($t_{d_{\text{mix}}} \ll t_{c_{\text{mix}}}$ and $t'_d \neq f(V_T(i),\ i = 1,\ n)$) are roughly satisfied at the optimum for typical problems. This confirms the validity of the analysis for typical treatment systems. If flowrates were much above 300 m³/h or the CSTR cost function was quite different, these assumptions might not hold.

The above analysis does not consider other known economic benefits that arise from choosing equal-sized tanks—buying in bulk, simplified construction, and reduction in number of spares required. These factors all reinforce the economic desirability of equal-sized tanks.

Summary. It is possible to rapidly estimate the economically optimum number and size of CSTRs in series required to achieve a given disturbance attenuation. The design obtained from this analysis will always have equal-sized tanks due to the dependence of the constraints on the product of the tank residence times. The tanks will typically be between 10 and 30 m^3 in size with residence times between 2 minutes and 1 hour. This is consistent with industrial practice and contradicts the recommendation in the literature, based on frequency response arguments, that tank sizes should be split in a ratio of about 1:4 or greater (Shinskey, 1973; Moore, 1978; McMillan, 1984). This provides a simple illustration of the ability of the integrated approach to balance operability and economic considerations in a way that qualitative argument cannot.

5. Toward Ideal Control

In the light of the analysis in the rest of this section, let us review what the cost of using PI control is in comparison to achieving the ideal delay-limited control performance bound, and let us examine some measures that may reduce this gap.

In Section V.A.4, we showed that optimal volume is proportional to t'_d and that there is a factor of about $(1.5n!)^{1/n}$ between the optimal volume with PI control (without uncertainty) and the volume required, assuming the performance bound is reached. The PI performance is governed by the effective delay ($t'_d = t_u/4$) rather than the actual pure delay. The maximum effective delay, assuming a mixing delay $t_{d\text{mix}}$ of 10 s, a probe lag of up to 30 s, and a CSTR with $t_{c\text{mix}} \gg t_{d\text{mix}}$, is 28.7 s. The measurement lag may therefore imply almost a threefold increase in the required volume, compared to that required with an instantaneous measurement response.

The volume factor between PI and ideal control for a given t'_d, $(1.5n!)^{1/n}$, may be ascribed to the fact that the PI control response differs from the ideal control response. It is necessary to characterize the ideal control response in order to identify the characteristics preventing its attainment. The initial control action taken by the ideal controller must be sufficient to reverse the direction of change of concentration at the exit of the *n*th tank of a *n* tank treatment system. For one tank, this is achieved by any control action that turns a reagent excess into a deficit or vice versa. For more than one tank, infinite control action at the treatment system inlet is required to reverse the direction of change because the initial process gain to a pulse input is zero. Finite reagent delivery capacity therefore prevents the ideal control bound from being reached for more than one tank with control on the first tank only. In order not to make the disturbance worse, the ideal delay-limited control must not take a control action more than twice that required to cancel the disturbance—at least if the constraints are symmetric about the initial value. This means that the ideal controller must identify the disturbance magnitude and type quite accurately from the initial rate of change of pH observed. This will be obstructed by noise and errors in the model relating pH changes to concentration changes. PI control forms an estimate of the disturbance magnitude without using derivative information and is therefore slower and more cautious than an ideal controller.

The most significant factor that would allow closer approach to the ideal bound is reducing or canceling the pH measurement lag. Cancelation of the lag by lead-lag filtering would be a very effective method of improving control performance in the absence of uncertainty. As discussed in Section IV.C, the probe response is very complex (although the effect on the control performance can be approximated by an uncertain first-order lag). A greatly improved understanding of probe behavior would be needed to allow effective cancelation of the probe dynamics within a control system. High-frequency measurement noise will also limit the degree of lead-lag filtering that is practicable.

Probe lag is a function of installation conditions, in particular decreasing with increasing fluid velocity past the probe. This velocity is relatively low in a stirred tank compared to a properly designed sampling system or injector probe assembly (McMillan, 1984); thus the use of alternatives to mounting the probe directly in the CSTR should be given careful consideration. The other key factor in probe response time is fouling and probe maintenance. The long-term solution to slow response due to fouling may come from advances in measurement technology; but until such time, the potential benefits of careful maintenance and operating in conditions that minimize fouling (underneutralization rather than overneutralization) should be noted.

At the design stage the choice of the maximum probe lag to be considered has a major impact on the final design if transient performance is the limiting factor. In most of the case studies presented in this section, 30 seconds has been used based on discussion with experienced engineers familiar with systems using probes mounted in stirred tanks. This figure can probably be reduced for other types of installations, as in the case study in Section V.C. Improved understanding of measurement response would be helpful in choosing the appropriate value.

If it is desired to evaluate advanced control schemes, the models used in this work should be regarded as generating an upper bound on practically achievable performance; further work is required to generate models suitable for this purpose. At a minimum, the performance of advanced control algorithms should be checked in the presence of correlated noise and actuator deadband error, and an improved model of the probe response is likely to be needed.

Summary. The greatest scope for moving performance toward the ideal control bound lies in reducing measurement lag and improving understanding of measurement response. Constraints on reagent addition rate prevent ideal control from being achieved for more than one tank with control on the first tank only. Finally, model errors and measurement noise obstruct the achievement of ideal control.

6. Summary

The exploration of the generic problem has allowed a number of issues to be clarified. The key conclusions are summarized below.

1. pH control is difficult because of delays and uncertainty. The pH nonlinearity, reagent addition precision errors, and measurement noise play a relatively minor role.
2. The bounds on disturbance rejection based on the ideal delay-limited control analysis are reasonably tight. The achievable performance with PI control (in the absence of uncertainty) may be estimated quite accurately using Eq. (52).
3. The optimal series CSTR configuration to minimize cost for a given disturbance attenuation requirement and instantly reacting reagent comprises equal-sized tanks for typical treatment systems.
4. The greatest potential for moving control performance toward the ideal delay-limited control bound lies in minimizing or compensating for pH measurement lags.

B. PRELIMINARY DESIGN OF A CENTRAL EFFLUENT PLANT

In this example, we illustrate the use of preliminary design tools in exploring design options efficiently. The design problem is represented schematically in Fig. 5.

The aim of the analysis was to determine the likely benefit from local containment of the main pulse flow disturbance in terms of requirements for the treatment system. The treatment system was being required to move from 95% compliance with discharge consents of 6–9.5 pH to 100% compliance. Two CSTRs with 10 minutes residence time at normal flows were already available.

Reagent selection was not an issue as NaOH/HCl were already being used. The reaction was assumed to be virtually instantaneous, eliminating steady-state considerations. A complete design analysis was not considered justified until the decision on local containment had been made. The preliminary control analysis (Section III.B.2) was therefore used as the sole design tool. As two CSTRs with 10 minutes residence time at normal flow already existed, the analysis was carried out based on n CSTRs of this size. Application of Hoyle's correlation (Eq.

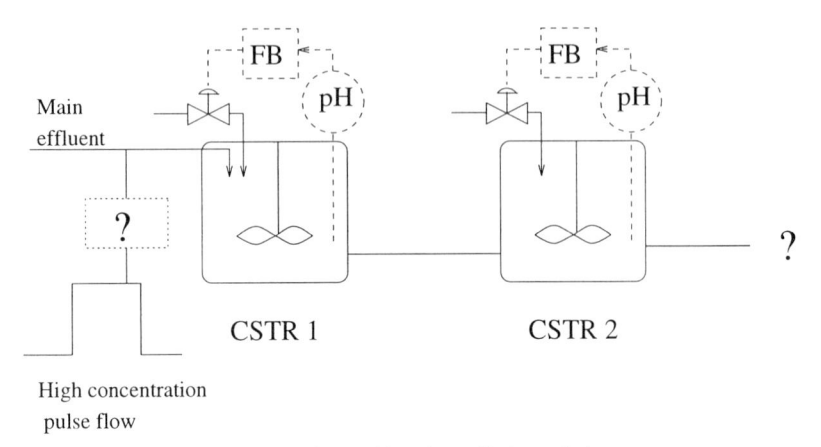

FIG. 5. Design problem for preliminary design.

49) to the existing tank design indicated a mixing delay of 10 seconds. Assuming typical industrial probes situated within the CSTRs, the measurement could contribute a maximum lag of about 30 seconds. Each CSTR therefore has a natural period, t_u, of about 2 minutes and a heuristic effective delay ($t_u/4$) of 30 seconds (see Section II.B.2).

Inlet effluent concentration was normally about ±.01 N, but there was a regular pulse flow discharge of 30 minutes duration, at comparable flowrate to the main effluent streams and between −1 and −3.5 N concentration. There was no available means of measuring the effect of this disturbance on the inlet concentration, so feedforward control was not considered as an option. This discharge gives the worst inlet concentration variation and the maximum flow and was therefore taken as the worst case. This disturbance gives a residence time per CSTR of about 5 minutes.

The first scenario explored was the plant requirement to handle all the streams with no local containment of the flow pulse. Extensive titration data were provided which indicated a minimum concentration change across the consent range of .0005 N. This compares to the strong-acid/strong-base concentration change of about .00003 N, showing that significant buffering is present even in the worst case. The slope of the titration curve ($\partial \text{pH}/\partial c_{net}$) was observed to vary by more than 10:1 between experimental samples. The inlet concentration change, Δ_{cnet}, in the worst-case disturbance is 1.75 N. The rising edge of the pulse is much more severe because of the higher associated flow; thus the setpoint can be biased toward the upper consent level, giving an allowable concentration deviation, δ_{cnet}, of approximately .0005 N. The required disturbance attenuation, δ_c, is therefore about .0003 (.0005/1.75). The attenuation available from n tanks whose design is the same as those already available can be examined by using the bounds on the performance computed using the pure delay and the effective delay (Eq. 30) and by using the heuristic estimate of attenuation with PI control (Eq. 52). The results of this analysis are tabulated below:

n	1	2	3	4
Pure delay bound	.03	.0005	6×10^{-6}	1.2×10^{-7}
Effective delay bound	.1	.004	.00015	3×10^{-6}
PI estimate	.15	.015	.0015	.00015

The rigorous bound implies that at least three tanks would be required, and the heuristic estimate indicates that a four-tank system would be required when using PI control. Even with a four-tank system, it can be expected that meeting the performance requirements will be difficult, considering the variation in titration curve slope—and hence process gain—noted.

If local containment is applied to the worst-case disturbance, the new worst-case disturbance can be assumed to be bounded by startup to maximum flow, giving $\Delta_{cnet} = .01\ N$. This disturbance may be in either the acid or the alkali direction, so δ_{cnet} is reduced to .00025 N. This corresponds to a required disturbance attenuation of .025. The residence time has been doubled and the achievable attenuation decreases by a factor of 2^n compared to the table above. The rigorous bound suggests that one tank might be just adequate (.017 versus .025), while the heuristic estimate indicates that two tanks would be required. The two-tank system has an estimated attenuation of .004 compared to a requirement of .025 and should be able to deliver the required performance despite the titration curve variability.

The rules presented in Section II.B.4 indicate that on–off control should be considered (maximum disturbance comparable to maximum load and titration curve slope varying by more than 5:1). The corresponding inlet concentration change, Δ_{cnet}, is at least .02 N (to allow effluent concentration variations to be accommodated). The residence time is 10 minutes, and the natural period is about 2 minutes. Applying Eq. (45) gives a predicted output concentration variation of $2.5 \times 10^{-5}\ N$, which is much less than the minimum concentration variation spanning the consent range ($\delta_{cnet} \approx .0005\ N$), confirming that on–off control could be a viable option.

In summary, the analysis indicates that, to achieve 100% compliance with the discharge pH consents, either (1) two CSTRs with PI or on–off control and local containment of the worst-case disturbance or (2) four CSTRs with careful controller design are required. The preliminary design analysis gives a useful picture of the tradeoff between the higher-level design decision as to whether to contain the main disturbance locally and the "end-of-pipe" treatment cost without requiring substantial design effort.

This example follows the procedures in Section III.B.2 quite closely, with some judgment being exercized in the selection of the appropriate worst-case scenarios. This application is fairly typical of preliminary designs in that good approximations to the worst-case disturbance can be identified *a priori,* allowing computation of the required disturbance attenuation for comparison to the predicted performance of series CSTR systems. The main experimental requirement was obtaining sufficient titration curves to provide a good estimate of the minimum buffering. Once the key high-level design decision has been made, these curves would allow a good uncertainty description for the pH characteristics to be developed for the worst-case design.

C. pH CONTROL OF SEVERAL STRONG ACID STREAMS WITH NaOH/HCl

1. Problem Definition

The process requirements and constraints are summarized in Fig. 6. The distinctive aspects of this problem definition are the large delay in measurement response due to the sampling arrangement employed, the tight constraint on total volume of the treatment system, and the presence of tanks upstream and downstream which assist in meeting the specification.

The pH relationship was taken as strong-acid/strong-base, as this was found to be a good approximation in the target pH range.

Mixing in treatment system CSTRs was modeled by a mixing delay, $t_{d_{mix}}$, of 7 seconds in series with a mixing lag $t_{c_{mix}}$. The 7-second delay was based on avoiding both air entrainment and flow short-circuiting with small residence times. Careful engineering is required to achieve this mixing delay. The first buffer tank had a mixing delay of 30 seconds. The second buffer tank was treated as well-mixed.

All delays were represented by 20 identical first-order lags in series. This representation is more precise than the default 10-lag model as the time delays

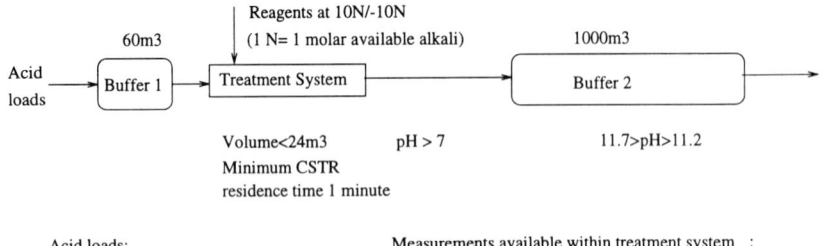

FIG. 6. Problem definition.

are unusually large compared to the mixing lags, which are less than 2 minutes at maximum flow.

Measurement biases were assumed constant for each simulation of the system. It should be noted that the measurement performance assumed corresponds to a "clean," well-maintained system in which the pH probes are placed in an in-line sampling system rather than in the CSTR.

The slowly varying flow was treated as constant but uncertain.

As in the previous example, reagent selection had already been made and this part of the design procedure was omitted. Reaction was effectively instantaneous.

2. Design

a. Preliminary Design. Due to the virtually instantaneous neutralization reaction, any nontrivial screening test for this problem must be based on dynamics (Section III.B.2). The fraction of the worst-case disturbance controllable with ideal delay-limited feedback was calculated to provide a controllability measure (Section II.B.2). Disturbance concentrations were contracted toward zero by the "disturbance fraction." The disturbances have separate effects due to flow and concentration changes. A change in flow at the inlet to the first buffer tank is assumed to propagate immediately to the treatment system, changing the net concentration following a reagent addition point and creating a disturbance propagating from this point. A change in concentration at the inlet of the buffer tank must propagate through the buffer tank mixing delay before beginning to affect the treatment system. The flow and concentration effects are therefore separated in time due to the mixing delay in the buffer tank and may be analyzed separately for this system, even though they are not independent.

The first control structure considered was pH measurement and reagent addition at the inlet to the treatment system. This gives a minimum delay of 20 seconds between the disturbance reaching the exit of the treatment system and the control response reaching the exit. All feedback control structures have at least this delay, owing to the delay in measurement response. Delay between reagent addition and the pH sampling point was assumed to be negligible, the required micromixing to attenuate noise being provided by turbulent diffusion in the sampling line. Optimization of a dynamic model was used to identify the worst-case disturbances and measurement bias, calculate the disturbance fraction, and select the controller setpoint. Solution of each problem took only a few minutes on a SPARC 2. The worst-case disturbances were found to involve both acidic flow pulses starting simultaneously for the concentration effect, and the lower-concentration, higher-flow pulse starting with the other pulse disturbance already active for the flow effect. The worst-case "slowly-varying" flow varied with the number of tanks in the disturbance path, being at its minimum value with no tanks in the path, and at its maximum value with one or more tanks in the path. The optimized controller setpoint was 11.6 to satisfy the steady-state pH constraint while giving maximum margin from neutral. The initial pH during the disturbances was 11.5 due to the measurement bias. If no backmixing is provided in the treatment system, only 1% of the flow effect or 8% of the concentration effect can be tolerated. Putting the maximum volume into a single tank allows 30% of the flow effect or 93.7% of the concentration effect to be tolerated. Splitting the volume evenly, giving two equal-sized tanks, allows 81.9% of the flow effect or 388% of the concentration effect to be handled. Uneven division of volume between two tanks or the use of more than two tanks is prevented by the requirement of at least one minute of residence time per CSTR.

The dominance of the flow effect suggests that flow information should be used directly to cancel this effect. This can be achieved by adding a flow measurement and controlling the *ratio* of the effluent and reagent flows based on the pH measurement (Section III.B.4). Despite the measurement lag and minor biases in flow measurement, this makes the concentration effect dominant. As only the two-tank configuration allows the concentration effect to be tolerated, we have two 12-m^3 tanks with ratio control of reagent flow at the inlet to the treatment system as the starting point for the search for an effective plant.

The results of the disturbance fraction analysis are summarized in Fig. 7. The worst-case disturbance (for the concentration effect) is indicated to be the maximum acid pulses occurring simultaneously, so this becomes the starting point for the worst-case design analysis.

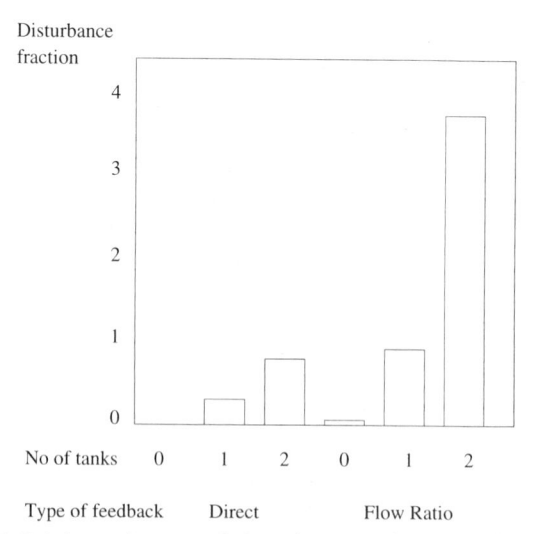

FIG. 7. Relative performance of alternative process/control configurations.

Identifying the worst-case condition was less clear-cut in this example than in the example presented in Section V.B. The worst-case disturbance combination varied with the process design and disturbance type considered. Using the optimization formulation from Section II.B.2 to identify the worst-case was therefore very useful. The preliminary control analysis is sufficient to eliminate most design options, indicating the need to use two tanks despite the fact that this implies unusually low minimum residence times (1 minute compared to the norm of at least 3 minutes) and indicating the need for ratio control of reagent addition.

b. Complete Design. PI controllers were used to implement feedback control. All controllers in the analysis below attempt cancelation of the strong-acid/strong-base characteristic, by passing the measured pH through the inverse of the nominal strong-acid/strong-base characteristic. As the screening analysis coupled with equipment constraints has defined the process parameters for the example considered, our optimization objective was to minimize control system cost plus the cost of excess reagent compared to the ideal delay-limited control case. The design parameters were the controller tuning and setpoints. The uncertain parameters were the disturbance characteristics and the varying measurement characteristics identified in Fig. 6. The disturbances were characterized by steps with switching times between 1 second and 10 minutes and variable step levels between zero and the maximum flow before and after the switch. The steady flow was allowed to vary between its minimum and maximum value. This gave a total of 11 uncertain parameters. In addition to the constraints specified in Fig. 6, it was required that the concentration recover to within $\pm 3 \times 10^{-4}$ N of the setpoint within 30 minutes to ensure well damped control.

The basic control scheme used for the perfect control analysis, a single in-line feedback loop, gives 10 times the allowable concentration variation at the exit of the second tank for the predicted worst case. Including feedforward reagent addition would be insufficient by itself to give the tenfold improvement required, due to a $\pm 20\%$ error in the estimated load (inferred from pH), so an additional in-line ratio feedback controller was added between the two tanks. As an additional actuator was therefore available at no extra cost, lead-lag feedforward from the load error at the first controller to the second controller actuator was added. To reduce feedforward dynamic mismatch, the lag was set to approximately the residence time of the first tank. The lead constant was added to the design parameters.

This scheme (shown in Fig. 8) was successful in meeting the performance requirements with the tuning parameters given below:

setpoint1	gain1	iat1	setpoint2	gain2	iat2	FF gain	lead constant
1.8	.061	23.3	11.31	.0082	2.	1.12	20.

378   STEVE WALSH AND JOHN PERKINS

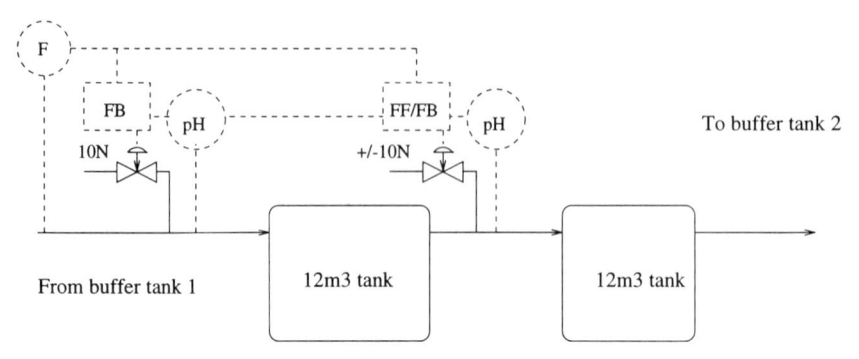

FIG. 8. Final process and control scheme.

Feedforward (FF) gain is given as a multiple of the gain based on the nominal steady-state model. All times are in seconds. Feedback controller gains are given in $m^3 \, m^{-3} \, N^{-1}$.

The outer-approximation algorithm (Section II.A) took six iterations to identify this solution, with a projection factor, ϵ_p, of .05 on the disturbance amplitude. Both vertex and nonvertex constraint maximizers were identified, confirming the need to consider nonvertex maximizers. The variables that contributed nonvertex maximizers were the step switching times (several times) and the measurement lags (once). Robustness was verified with respect to all vertex combinations of uncertain values and a random selection of interior points (*ivert* = 1, $nrand_{max}$ = 1000).

The ideal delay-limited control optimum gives a reagent cost of about £3,500,000 p.a. based on a unit reagent cost of £50/m^3 and a typical operating condition of 400 m^3/h acid flow at −.2 N. The scheme in Fig. 8 with the parameters above gave an additional reagent cost of £950 p.a. compared to the ideal control case. The robust design obtained seems likely to be close to the economic optimum.

Increasing the projection factor to .1 reduced the number of iterations to generate this solution to three, with a 20% increase in the excess reagent cost. This shows the effectiveness of an increased projection factor in generating an approximate solution more rapidly. The progress of the algorithm on this problem is discussed in more detail in Appendix B, which illustrates the operation of the algorithm more clearly.

3. Conclusions and Review

A robust design was developed in the face of strong requirements on disturbance rejection, tight constraints on the equipment, and unusually large measurement delay. The preliminary control analysis was effective in narrowing down the possible designs for consideration. The worst-case design algorithm allowed the development of a feasible design which was shown to be close to optimal. This example shows that the worst-case design algorithm developed is capable of solving realistic problems, confirms the existence of nonvertex constraint maximizers and indicates the use of a projection factor to trade off accuracy of solution and computation time to be effective.

D. Ca(OH)$_2$ Neutralization with Highly Variable Titration Characteristic (Central Effluent Plant)

1. Problem Definition

The objective of this case-study was to *evaluate* a proposed design for a treatment system for a central effluent plant dealing with waste streams from many sources.

The proposed design is a two-CSTR system with minimum residence time per tank of 24 minutes. Each tank is supplied with 5% w/w (1.45 N) hydrated lime (Ca(OH)$_2$) reagent controlled based on pH measured at the exit of the tank. Detailed tank design was such that a mixing delay of less than 10 seconds was expected. The neutralization tanks are followed by a flocculation tank with a minimum residence time of 10 minutes. There is an agitated level-controlled buffer tank upstream of the neutralization tanks, giving about 12 hours of concentration and flow equalization at maximum flow. The scheme to be evaluated is shown in Fig. 9.

The aim is to be able to control with setpoints between 7.5 and 8.5 pH ± .5 pH. The ±.5 pH bounds are applied *at the flocculation tank*. This has the effect of reducing variability in the concentration (as compared to the second tank) and taking account of the discrepancy between average pH in the second controlled tank and in the flocculation tank due to carryover of reactive lime.

The presence of the buffer tank upstream means that process disturbances are greatly attenuated. Nevertheless it was desired to be able to make step flow changes of at least 50% of design flow, introducing significant disturbances.

The pH probe response is described by a lag of 5–30 seconds. Biases of up to ±.25 pH are assumed to be possible on pH measurement.

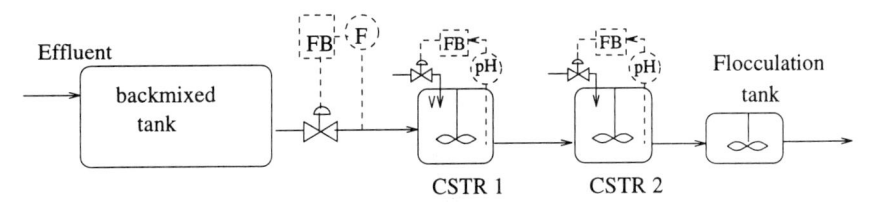

Fig. 9. Treatment system to be evaluated.

Twenty-four-hour flow-averaged titration curves collected over one month were taken as representative of the titration curve variability to be expected at the exit of the buffer tank. The curves are shown in Fig. 10 with pH plotted against net concentration in normals, pH 5 being taken as zero net concentration.

The key characteristics of titration data for design purposes are the slope of the titration curve ($\partial pH/\partial cnet$) and the variability of this slope. The titration data are shown in Fig. 11 in the form of ($\partial pH/\partial cnet$) as a function of pH. The titration curve slopes show a variability between 4:1 to 10:1 above pH 3; e.g., the slope between 7 and 9 varies 5:1 with values between 1000 pH/N and 200 pH/N. The maximum slope of 2000 pH/N was 1000 times less than the maximum slope for the strong-acid/strong-base titration characteristic, showing substantial buffering even in the worst case. The peak sensitivity in the target region (7–9 pH) was 1000 pH/N. Inlet concentration ranged from .08 M to .035 M acid ($-.08$ to $-.035$ N). For the analysis carried out, it is assumed that any titration

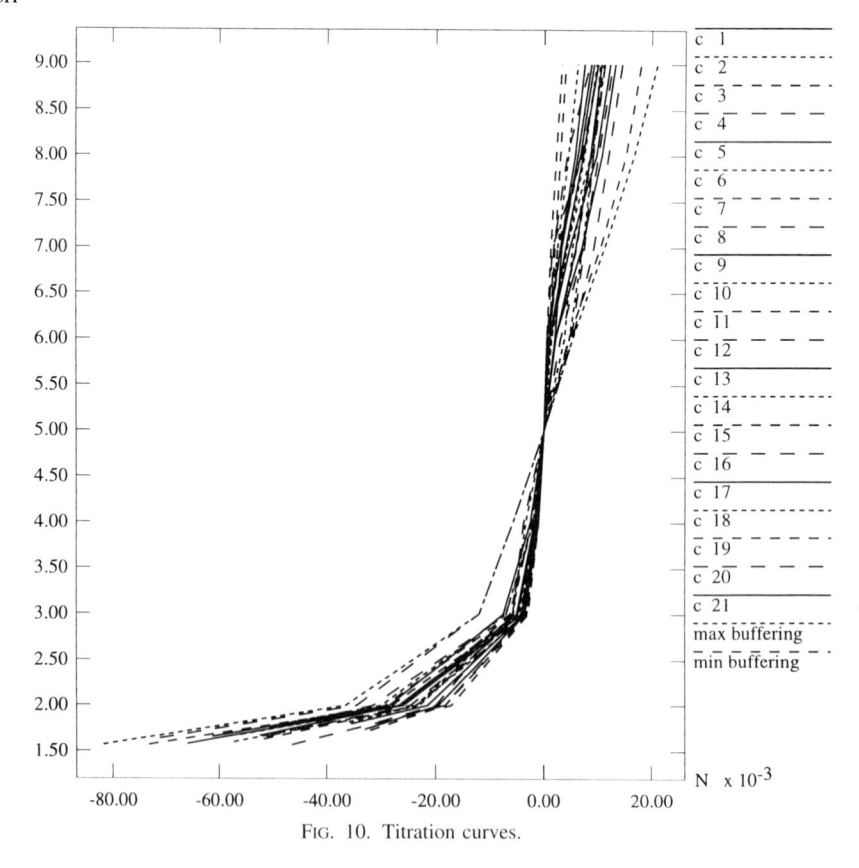

FIG. 10. Titration curves.

pH/M x 10^3

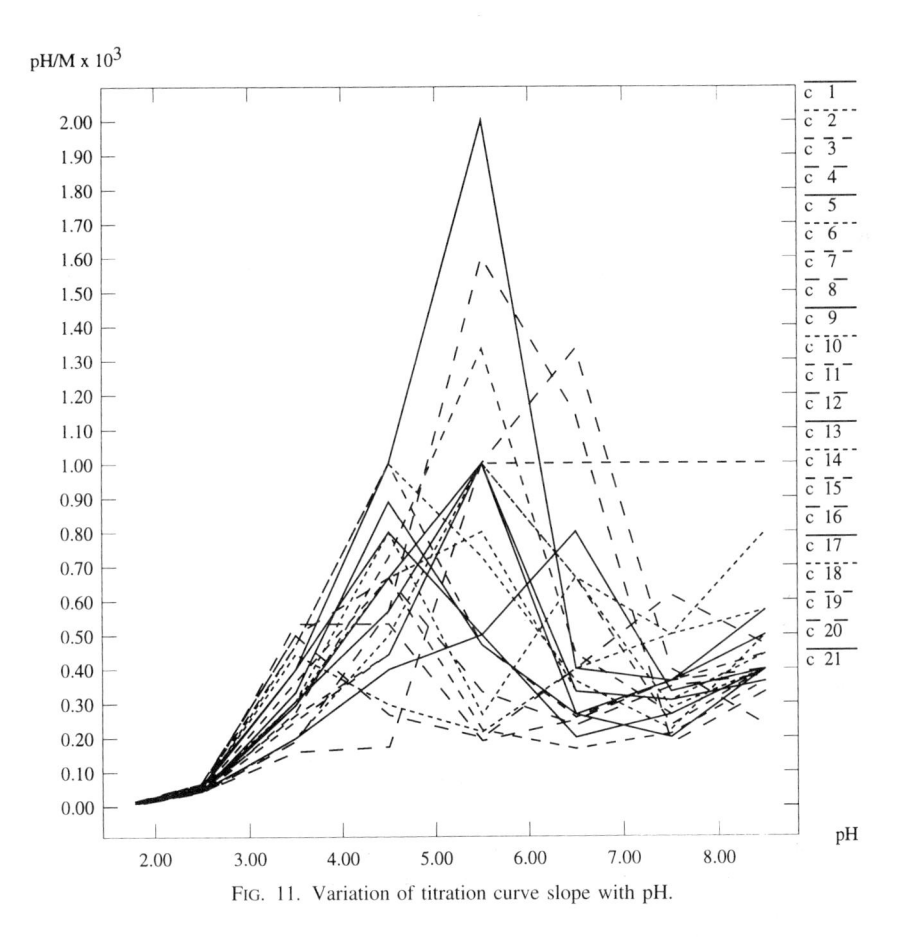

FIG. 11. Variation of titration curve slope with pH.

curve lying within the range of sensitivities observed may occur; i.e., there is no correlation between buffering at different pH levels. The "max buffering" and "min buffering" curves in Fig. 10 represent the extremes under this assumption and do not depart substantially from the observed curves.

The titration curves are given below in the form of ranges of concentration change spanning a given pH range. The reagent concentration required to reach pH 7 ($cnet = 0$) varies from about .035 N to .08 N, due to the inlet concentration variation noted above.

PH range	2–3	3–4	4–5	5–6
Concentration (N)	.015–.03	.0015–.007	.001–.005	.0005–.005

pH range	6–7	7–9
Concentration (N)	.0007–.005	.002–.01

Reagent dynamics were modeled using the standard model (Eq. 47):

$$\frac{\partial r_p}{\partial t} = -k_{mix}.\max\left(1, \frac{10}{r.k_{mix}}\right)(k_1 + k_2[H^+] + k_3 cweak) \ \mu/s$$

As the acid concentrations are moderate, k_2 and k_3 may be set to zero without introducing significant errors. k_1 was taken as .08–.12 to allow for modeling error and k_{mix} was taken as 1 as there was no reason to expect atypical mixing.

The particle size distribution was assumed to be given by

$$\phi(\ln(r_p)) = \frac{1}{1.15\sqrt{2\pi}} \exp(-((\ln(r_p) - 1.39)/1.15)^2/2), \qquad r_p < 250 \ \mu \qquad (57)$$

2. Analysis of Design

a. Preliminary Control Analysis. The maximum sensitivity of pH to concentration in the target pH range is 1000 pH/N. Allowing for the measurement bias of ±.25 pH the acceptable transient deviation is reduced from .5 to .25 pH corresponding to $\delta_{cnet} = .00025$ N. The worst-case disturbance will be a step from 50 to 100% of flow corresponding to a maximum inlet concentration change, Δ_{cnet}, of up to .04 N. This disturbance is worse than a step from 0 to 50% of flow because although it has $\Delta_{cnet} = .08$ N, this step has half the associated flowrate, which with 3 tanks in the disturbance propagation path implies 8 times the disturbance attenuation available in the worst case. The required disturbance attenuation at maximum flow is about .006 ($\delta_{cnet}/\Delta_{cnet}$).

The mixing delay is 10 seconds. The maximum measurement lag is 30 seconds. The reagent dynamics are complex, but the effective lag associated with the initial response is no more than 10 seconds (based on examination of time for 50% conversion of reagent using the model given above). The effective delay, calculated as $t_u/4$ (Section II.B.2), is therefore about 41 seconds. The estimated disturbance attenuation with PI control is 8×10^{-5} (Eq. 52). Transient control performance is not expected to be a problem for this system and a basic control scheme should be adequate.

b. Steady-State Analysis. To meet the performance requirements at steady-state, the reaction of $Ca(OH)_2$ downstream of the controlled tanks must not raise the pH by more than .5 pH in the flocculation tank. This assumes that the setpoint in the second tank is chosen .25 pH below the target value so that maximum margin for carryover is provided, given the measurement errors. This is equivalent to at least .625% of the reagent reacting downstream. Also, the reagent carrying over from the first tank to the second tank must not cause the

OPERABILITY AND CONTROL IN PROCESS SYNTHESIS AND DESIGN 383

pH in the second tank to rise above its setpoint, preventing effective control. For a standard control scheme with fixed setpoints, it is necessary to establish a choice of setpoints for the first and second tanks that meets these requirements.

Shortcut analysis can be used to indicate whether a problem is likely. The time from 97 to 99% reagent conversion is about 1 minute based on the dynamic titration curves shown in Fig. 12 and the associated titration curve. Using this value of 1 minute as the equivalent reaction lag, τ_r, the carryover of reagent may be estimated by using Eq. (46).

The following table indicates carryover of reagent added to one tank from the tank exits downstream.

	tank 1	tank 2	tank 3
tank 1	.04	.0016	.00015
tank 2		.04	.0036

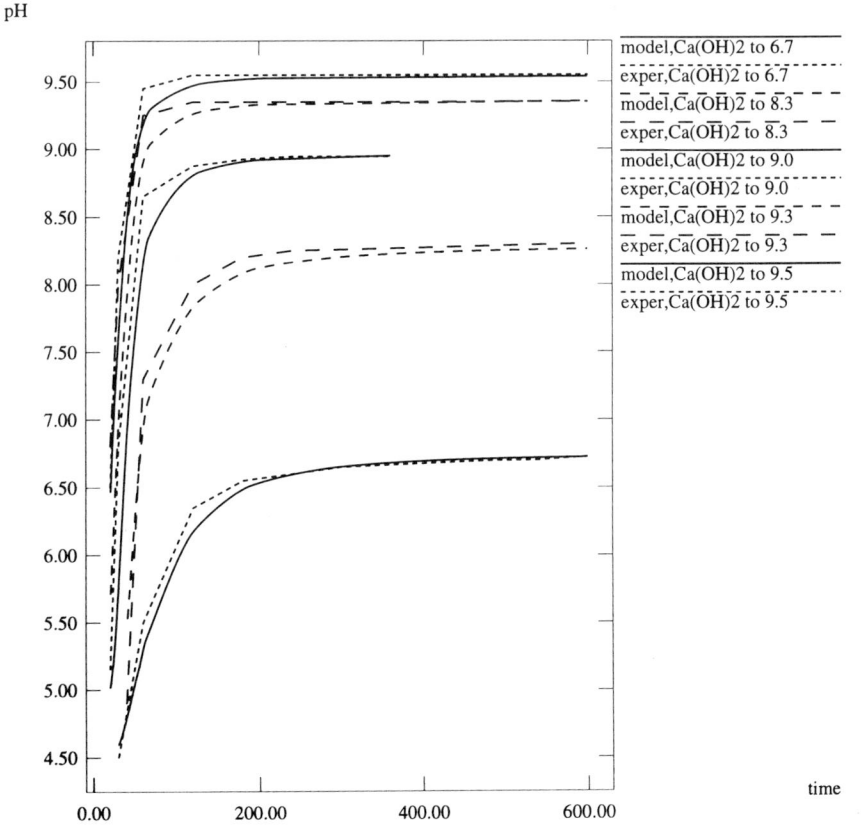

FIG. 12. Dynamic titration experiments with $Ca(OH)_2$ (1).

The shortcut calculation results indicate that there should be no problem with excessive carryover provided less than 11.6% [.0016(1 − x) + .04x ≤ .00625] of the reagent is added to the second tank. The second tank should not be swamped by reagent from the first tank, provided at least 4% of the total reagent is required to make the shift between the first and second tank pH setpoints. This margin between setpoints allows all the reagent to be added to the first tank without the 4% carryover to the second tank causing the pH in the second tank to increase above the second tank setpoint. Choosing setpoints to keep the reagent addition to the second tank between 4 and 11.6% of the total reagent addition is not possible as the titration curve slope shows variability of at least 5:1 when most of the reagent has been added. As the shortcut calculation should be pessimistic (at least for carryovers of the order of 1%), it is likely that the system will work. Analysis of the full reagent model is required to confirm this.

Using the full model gives the carryover table below as $k1$ varies from .12 to .08.

	tank 1	tank 2	tank 3
tank 1	.0084–.0124	.00027–.00057	.000035–.0001
tank 2		.0084–.0124	.0006–.0012

It should be noted that, as k_2 and k_3 are assumed to give no contribution to the rate, the calculated carryovers are independent of the particular pH setpoints in the tanks. If the pH setpoint in the first CSTR is below about 3, carryover of reagent added to this tank should be slightly lower owing to acid acceleration of the reaction.

The analysis indicates that up to about 50% of the reagent can be added to the second tank without excessive downstream pH drift in the worst case. Provided at least 1.2% of the reagent is required to make the shift between the first and the second tank pH setpoints, the second tank should always be adding some reagent. Despite the titration curve variability of up to 10:1, there should be no difficulty in accommodating these requirements. Steady-state conversion is not likely to be problematic.

The more precise analysis shows the shortcut analysis to have been pessimistic—as expected—although the results tally within an order of magnitude. If the effective time constant had been based on the initial part of the batch titration responses (τ_r less than 10 seconds), the shortcut analysis would have been optimistic compared to the full analysis.

3. Conclusions and Review

The proposed design was shown to be capable of meeting the performance requirements quite comfortably without having to carry out a complete design.

The preliminary analysis (Section III.B.2) indicated sufficient performance margin to allow confidence that the design was adequate.

The plant described has now been commissioned and is working satisfactorily as predicted. pH differences between the second tank and the flocculation tank have been observed to be small (less than .3 pH) and may be dominated by measurement errors. These observations are consistent with the analysis above.

E. Ca(OH)$_2$ Neutralization of a Single Effluent Stream at High Intensity

1. Problem Definition

This section discusses work carried out on the design of an effluent treatment system for neutralization of a highly concentrated (-11.8 N) acid stream, predominantly composed of HCl.

The design problem is to produce a neutralization system that can maintain a pH of between 6.5 and 7.5 after all reagent has reacted. Effluent load is high and low suspended solids are required. A good-quality high-calcium hydrated lime was used as it was the cheapest reagent meeting these requirements and reacting reasonably rapidly.

The reagent was to be delivered as a 20% w/w slurry (standard practice), giving a concentration of about 5.9 N. The concentration of acid in the neutralized stream is therefore about -4 N due to dilution effects (reagent flow is twice the effluent flow near neutral).

It is desired to vary the flow of effluent readily. This requirement was captured by requiring that the system be able to start up with a step to 50% flow, followed by a step to 100% of flow one hour later and a step to 25% of flow after a further hour.

Cost analysis indicated that the total treatment system residence time should be kept below about 30 minutes (qualitative reduction in civil costs) while minimizing the number of CSTRs (low marginal cost with volume and substantial cost per unit). Additional cost reductions could be obtained by reducing the total residence time further.

Imperfect CSTR mixing is estimated to introduce 10 seconds dead time between reagent addition and measurement response.

pH measurement properties are characterized by a ± 0.25 pH bias and a first-order lag between 5 and 30 seconds. To avoid degraded measurement performance, it is required that the pH controller setpoints lie between 2 and 12 pH.

The titration data are tabulated in terms of the alkali concentration change required for a given pH change.

PH range	1–2	2–2.5	2.5–3	3–5	5–6	6–8
Concentration change (N)	.01	.006	.003	.001	.001	.008–.02

The only significant variability in the buffering was between 6 and 8 pH. The effluent to be treated has a slowly varying acid concentration around $-11.8\ N$. Zero net acid/alkali concentration was taken as giving a pH of 7.

The dynamic titration data for this problem (Fig. 13) were consistent with the model discussed earlier, so the following model is taken to be appropriate (Eq. 47):

$$\frac{\partial r_p}{\partial t} = -k_{mix}.\max\left(1, \frac{10}{r.k_{mix}}\right)(k_1 + k_2[H^+] + k_3 cweak)\ \mu/s$$

There is uncertainty in the kinetics due to model mismatch, errors in fitting the model, and variations in mixing conditions, temperature, and particle size distribution. For this application, an overall uncertainty in the rate, $\partial r_p/\partial t$, of $\pm 40\%$ was included to allow for these factors, with $k_1 = 0.1$, $k_2 = 7.5$, and $k_3 = 1.5$.

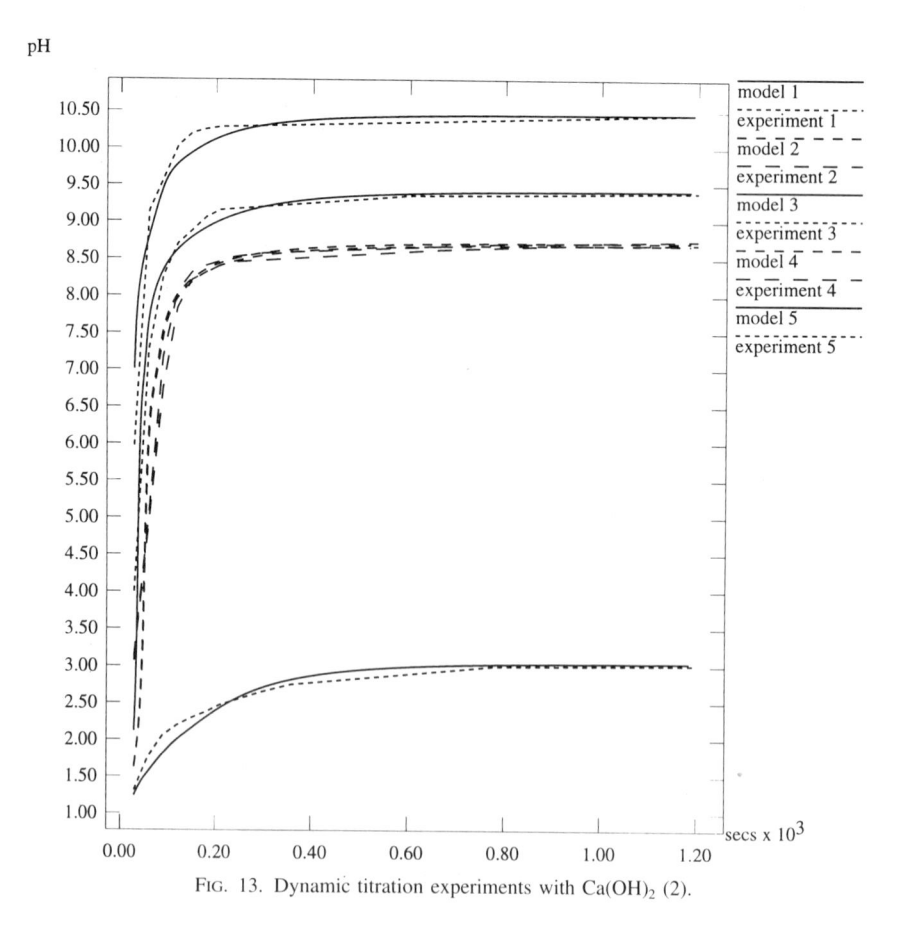

FIG. 13. Dynamic titration experiments with Ca(OH)$_2$ (2).

For CSTRs, k_{mix} was assumed to be 1, while for in-line mixers k_{mix} was assumed to be between 0.5 and 0.9. A surface reaction rate of 1.5 μ/s was assumed to allow for a possible change in reaction mechanism at high acid concentrations not ruled out by the experimental data. The particle size distribution was assumed to be given by Eq. (57):

$$\phi(\ln(r_p)) = \frac{1}{1.15\sqrt{2\pi}} \exp(-((\ln(r_p) - 1.39)/1.15)^2/2), \qquad r_p < 250\ \mu$$

(same reagent source as previous example).

2. Design

a. Preliminary Control Analysis. The maximum sensitivity of pH to concentration is 250 pH/N. Allowing for measurement bias, the acceptable transient deviation is about .25 pH, giving $\delta_{cnet} = .001\ N$. The worst-case disturbance will be the step from 50 to 100% of flow corresponding to a concentration change at the treatment system inlet, Δ_{cnet}, of about 2 N. The required disturbance attenuation, δ_c, at maximum flow is therefore about .0005.

Ideal delay-limited feedback analysis with n equal-sized tanks (Eq. 30) may be used to give a rigorous bound on achievable performance. This indicates that with simple control from CSTR pH to reagent addition, the minimum residence time required for one CSTR is 5 hours while the total residence time required for two CSTRs in series is 10 minutes. Combined feedforward/feedback control was also considered. The feedforward reagent addition was estimated to have a relative error made up of ±2% varying linearly with flow and ±3.5% varying slowly with time in the flow range of interest. This implies a potential benefit from feedforward of about a factor of 10 in disturbance rejection, requiring the feedback system to achieve a disturbance attenuation of 0.005. The minimum residence times for the combined control scheme are then 35 minutes for a single CSTR and 3 minutes for a two-CSTR system. A single CSTR system using combined feedforward/feedback control as above can therefore be eliminated on control grounds without the need for controller design and simulation.

Using Eq. (52) and an effective delay of about 41 seconds (as in Section V.D) to estimate the likely PI performance, indicates a total residence time of about 25 minutes for a two-CSTR system with combined feedforward and feedback and a total residence time of 75 minutes without feedforward. As the cost of the feedforward controller is small compared to the cost of an extra tank or the extra civil work to accommodate a residence time above 30 minutes, the scheme with feedforward control and two CSTRs seems likely to be the preferred option to achieve good control.

b. Steady-State Analysis. A second fundamental limit on system design is the steady-state reagent conversion. The pH downstream of the treatment system must lie between 6.5 and 7.5 for all values of the relevant uncertain parameters—measurement biases, rate of reaction, and the relationship between pH and concentration. If sufficient unreacted reagent carries over from the final stage, it will cause the pH to drift upward excessively downstream. Considering two CSTRs in series with the second tank using caustic reagent (NaOH) due to potential metering problems at low flows, the minimum total residence time is found to be 85 minutes. This value was generated using a steady-state worst-case optimization in which the downstream pH following complete conversion of reagent was required to be between 6.5 and 7.5 and the second tank was required to be adding reagent continuously at steady-state. The pH setpoint for the first tank was set to 2, the allowable value most favorable to reagent conversion. The second tank setpoint was set to 6.75 as this is the lowest value consistent with meeting the specification, given measurement errors. Flow was set to 100% of design rate. The CSTR residence times were optimized. The measurement biases, the uncertainties in the rate expression, and the uncertainty in the titration curve buffering were the relevant uncertain parameters. Ten CSTRs in series were used to model the plug-flow reactor.

The 85-minute residence time required for reagent conversion in a two-CSTR system greatly exceeds the desired maximum residence time of 30 minutes. It is therefore necessary to modify the treatment scheme so as to improve reagent conversion while retaining adequate control response. Previous design practice indicates adding a third CSTR, but this would result in an undesirable increase in cost. The approach adopted was to use a simple low-cost plug-flow reactor (PFR) before the first CSTR to achieve improved reagent conversion. A PFR has very poor feedback control characteristics as it increases the delay in the controller response considerably. It was therefore decided to use feedforward control to add most of the reagent to the PFR, while allowing feedback control from the first CSTR pH to add reagent directly to the CSTR. An additional feedback loop (feedforward trim) was included to monitor the ratio of reagents added to the PFR and CSTR and adjust the feedforward ratio so that the actual ratio was at its desired value. This minimizes the effect of feedforward calculation errors, allowing maximum benefit to be obtained. The maximum PFR residence time was estimated to be 2 minutes. The idea of using a PFR to assist in meeting steady-state requirements, which emerged in this case-study, has been incorporated into the design procedure (Section III.B.4). This scheme is shown in Fig. 14.

Repeating the steady-state worst-case design with the new scheme gave a minimum total residence time of 20.3 minutes, which is satisfactory and comparable to the requirement predicted from the control analysis. The PFR residence time went to its maximum value of 2 minutes. The fraction of total

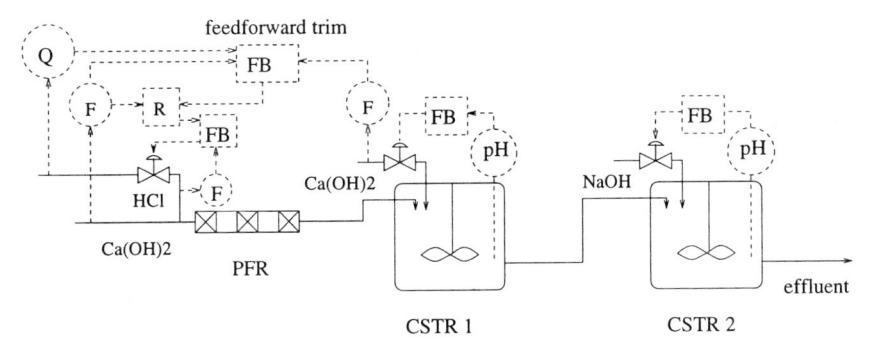

FIG. 14. Final process and control scheme.

Ca(OH)$_2$ added to the PFR was optimized along with the reactor volumes and found to have an optimum value of about 94.5%. Requiring the two tanks to be of equal size did not significantly alter the minimum residence time. Equal-sized tanks were therefore assumed for further analysis, as this has inherent benefits in terms of cost and should be optimal for control.

At this point in the design process, a structure that meets the steady-state requirements and is likely to meet the control requirements has been obtained.

Each steady-state worst-case optimization took about 15 minutes to run (on a SPARC 2). The worst-case combination was minimum rate of reaction with the measurement biases pushing the effective setpoints upward by .25 pH and the minimum buffering. This was a relatively simple worst-case optimization and no projection factor was used. This vertex worst case was in fact correctly chosen *a priori* based on physical considerations, and was promptly reidentified in the local phase of the constraint maximization. Most of the optimization time was spent verifying this choice of the worst case by vertex enumeration and random search.

c. Complete Design. At this point it becomes appropriate to carry out a full dynamic worst-case design, as a design structure has been identified that is likely to satisfy the design requirements and be close to optimal. The uncertainty in measurement dynamics and in the feedforward calculation was added to the sources of uncertainty considered when looking at the steady-state behavior. PI feedback controllers were assumed initially. The feedforward controller was started with a fixed ratio of 92% for the first increase in flow (feedback adjustment must be disabled when there is no flow), after which the feedback adjustment of the ratio was enabled to bring the fraction of Ca(OH)$_2$ added to the PFR to 95%. The first tank setpoint was set to 2 pH as before. The feedforward trim adjustment is predominantly integral, so the gain was set to a small value (.01 with a process gain of about 1) and only the integral action time was adjusted. The tuning of the three feedback controllers and the setpoint for the second tank were optimized along with the CSTR volume and the PFR volume.

The input conditioning function tabulated below was used on the first CSTR pH controller, as the steep titration curve slope between 3 and 6 pH poses a risk of limit-cycling (see Section III.B.4):

pH in	1.	2.5	3.	6.	8.
"pH" out	1.	2.5	2.75	3.	5.

The main effect of this conditioning function is to prevent the rapid pH shift between 3 and 6 pH, giving rise to a sharp kick in the control response which could otherwise cause oscillations. The input conditioning is assumed to be implemented as a piecewise linear function, as this is the method generally available in control software.

The optimization attempted to minimize total reactor volume, subject to maintaining the downstream pH between 6.5 and 7.5 and to the second CSTR pH recovering to within .005 pH of the setpoint 30 minutes after a load change. The latter requirement was imposed to ensure reasonable damping.

The optimization generated a minimum residence time of 27.4 minutes with 2 minutes residence time in the PFR and 12.7 minutes in each CSTR. The controller tuning obtained was

	K_c	iat (s)	setpoint
CSTR 1	.13	1000	2 pH*
CSTR 2	.08	300	6.88 pH
FF trim	.01*	8.4	95%*

The variables indicated by an asterisk (*) were assigned *a priori*. The feedforward (FF) trim controller gain is dimensionless as both the input and the output are ratios of flows. The CSTR controller gains are in m^3/h reagent per m^3/h acid at maximum flow per pH. Some of the tuning parameters were on bounds, but the associated Lagrange multipliers did not indicate a significant incentive to rerun the optimization.

This design is likely to be close to the optimum design for this problem as control scheme modifications have limited potential to reduce the equipment size and there is no apparent means to eliminate any of the reactors without violating the constraint on residence time.

The worst-case dynamic optimization took 4 days on a SPARC 2, highlighting the importance of using efficient screening tools rather than plunging into the full design analysis for each alternative design considered. There were 10 uncertain parameters altogether; feedforward error was characterized by a fixed bias (±2% of value) and biases at minimum and maximum flow (±3.5% of value) with linear interpolation at intermediate flow, two pH measurement lags, two pH measurement biases, the titration curve slope between 6 and 8 pH, the overall uncertainty in rate of reaction, and the uncertainty in the rate of reaction in the PFR. Vertex enumeration and up to 200 random points were allowed for the global phase of the search and $\rho_{min}^{max} = 1$ was used for the local search. As the model was slow to run (about 2 minutes for a function evaluation and 10 minutes for a gradient evaluation), careful *a priori* selection of worst-case combinations was used. Three worst cases were set *a priori* and an additional worst case was identified by vertex enumeration, although this did not greatly alter the solution. All the worst cases were vertex points. A projection factor of .05 on the maximum flow was used and only two iterations were needed. The Lagrange multipliers indicated two of the *a priori* worst cases to be dominant. In both of these the measurement lags were at their maximum value of 30 seconds. In one worst case the setpoints were biased high with low reaction rates (the steady-state worst case) and the feedforward error as a function of flow was set so that the effective feedforward ratio would increase with flow, tending to push the transient toward the high pH limit. In the other case the setpoints were biased low with high reaction rates, and the feedforward error as a function of flow was set so that the effective feedforward ratio would decrease with flow, tending to push the transient toward the low pH limit. The pH responses associated with these two worst cases are shown in Fig. 15, illustrating the diversity of response the treatment system is designed to accommodate.

The maximum transient deviation is actually somewhat less than expected from the earlier analysis (0.12 versus 0.25 pH). This largely reflects the fact that the feedforward trim allows a consistent benefit from feedforward of almost a factor of 20 compared to the factor of 10 assumed in the earlier analysis.

This example is at the limit of what could be handled on a SPARC 2, despite all the measures discussed in Section II.A to improve efficiency.

VI. Conclusions

The proposed approach to integrated design has been applied successfully to chemical wastewater neutralization systems. This approach has been shown to be effective in allowing the efficient development of robust low-cost designs. The results achieved show that an integrated attack on design problems using computer-aided process engineering tools is feasible.

A number of techniques have been presented that can be used in many other contexts. The delay-limited control analysis and the worst-case design method may be directly applied to a wide range of problems. Development of a suitable design procedure, models, and additional tools for any new problem area would require considerable effort. We consider that appropriate design procedures for

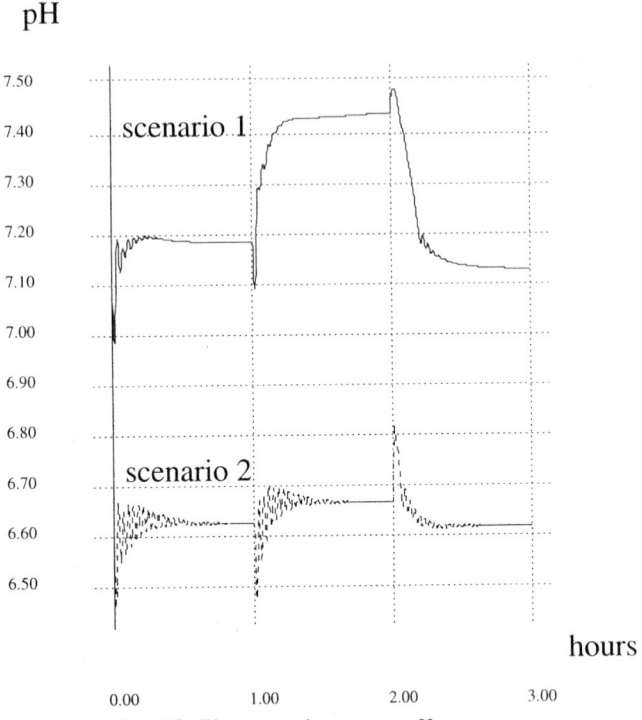

FIG. 15. Worst-case downstream pH responses.

many problems would retain the basic structure of preliminary control analysis, steady-state analysis, and full design with dynamic models and uncertainty. There are many opportunities for extension of the work described.

Dynamic optimization of realistic problems remains time-consuming and model reliability can be a major obstacle on large problems. Many industrial models require considerable hand-holding to get them to work at all. Improvements in computing power, solution techniques, and model-building tools are clearly needed for the widespread utilization of this technology. It is likely that no single approach to dynamic optimization will prove universally appropriate; thus, the software packages that intelligently select appropriate strategies are needed.

The algorithm presented for design with uncertainty seems quite effective, but a number of useful extensions can be identified.

1. The use of a single polyhedron for the set of uncertain parameters is restrictive and should be generalized to allow more complex restrictions on the allowable parameter set. While this is straightforward conceptually, its implementation needs careful attention to maintain the efficiency of the algorithm.

2. Logical constraints could be introduced to the vertex search to identify and eliminate redundant vertices and make the search more efficient.

The use of the design techniques in other problem areas should be explored. We are currently exploring the design of batch distillation columns and the design of distributed chemical plant.

Notation

Symbols

atol	absolute tolerance on integrator error control
c	vector of constraints
c_i	the ith constraint
c_x	concentration of acid/base x in normals
c_{sig}	the smallest significant constraint violation
cnet	the net acid/base concentration in normals (alkali = positive); taken as zero at the reference pH (pH$_{ref}$)
cweak	the concentration of weak acids that dissociate below the reference pH (pH$_{ref}$)
$\tilde{c}$	"majorized" constraints
d	disturbances
evtol	tolerance on location of time of discontinuity
$f(\dot{x}, x, z, \theta, t)$	residual function of DAE equations
$f(x)$	a general function of variable x
F	flexibility index
F_T	total flow
Fagit	recirculating flow generated by agitator
Fr	reagent flow
Fr_h	reagent flow corresponding to upper pH limit
Fr_l	reagent flow corresponding to lower pH limit
$g(x, z, \theta, t)$	residual function of algebraic equations
$g(\theta)$	gradient vector of objective, J, w.r.t. θ
H_t	height of tank
iat	controller integral action time
ivert	flag for whether vertex enumeration is used in constraint maximization
I_j^c	the set of i corresponding to inputs, u_i, connected by a controller to the jth output, y_j
I_j^s	the set of i corresponding to inputs, u_i, having a strong effect on the jth output, y_j
J	objective function
J_d	the set of j corresponding to outputs, y_j, from which the effect of the disturbance, d, can be compensated
k_1, k_2, k_3, k_{mix}	constants in solid alkali reaction equation
$L(x(\phi), z(\phi), \theta, \phi)$	equality interior point constraints
m	number of elements of constraint vector c
meq	number of equality constraint elements of constraint vector c
$M(x(\phi), z(\phi), \theta, \phi)$	inequality interior point constraints

neq	number of equations in the model
nloc1	number of local searches during local constraint maximization
nloc2	number of local searches during global constraint maximization
nmods	number of uncertain parameter vectors in V_0
nrand	number of random searches
nset	number of elements of V_0 that have been preset
np	number of parameters for which gradients are required
N_i	number of parameter sets in V_i
o	vector of n_o operating variables
optacc	specified optimization accuracy
O	set of operating variables
p	vector of n_p design variables
P	set of design variables
pH	the potential indicated by a pH probe
pH_{ref}	the reference pH used in defining *cnet* or *cweak*
$q(x, z, \theta, t)$	dynamic path constraints
r_p	particle radius
rtol	relative tolerance on integrator error control
simacc	specified simulation accuracy
sstol	tolerance on residual norm for initialization
t	time (seconds, unless otherwise specified)
t_d	delay between disturbance occurring and controller response reaching output
t'_d	delay between disturbance reaching output and controller response reaching output
$t_{dd,y}$	delay between disturbance and output
$t_{du,y}$	delay between input and output
$t_{c_{mix}}$	the lag time constant associated with mixing
$t_{d_{min}}$	the minimum deadtime associated with mixing
$t_{d_{mix}}$	the deadtime associated with mixing
t_f	final time of integration
t_u	the ultimate period of a control loop
v	vector of n_v varying or uncertain variables
$\tilde{v}$	the normalized deviations in v
v_0	the nominal value of v
v_i	the ith element of the vector v
v^i	vector of uncertain parameters lying in the subset V^i
v^v	a vertex element of V
V	set of varying or uncertain variables
V_0	outer approximation to V at first iteration of worst-case design
V_i	outer approximation to V at ith iteration of worst-case design
V^i	ith subset of the uncertain parameter space
V_x	the volume of solution x
V_t	the tank volume
W_u	weighting matrix on u
W_y	weighting matrix on y
x	state variables
x^s	state variables evaluated while evaluating sensitivities
u	process inputs/manipulated variables/controller outputs

y	process outputs/measured variables
y_s	target values or setpoints for measured variables
z	algebraic variables
z^s	algebraic variables evaluated while evaluating sensitivities
α	scaling factor on full optimization step, δ_k
δ	flexibility index for a single value of v
δ_a	estimated achievable disturbance attenuation
δ_c	required disturbance attenuation ($\delta_{cnet}/\Delta_{cnet}$)
δ_{cnet}	acceptable variation in $cnet$ at treatment system exit
δ_f	the fraction of a disturbance that can be rejected ($\delta_f = \partial_c/\partial_a$)
δ_l	permissible fractional carryover of reagent
δ_k	full optimization step at kth iteration
$\delta(y)$	change in variable value when sensitivities are evaluated
Δ_{cnet}	variation of $cnet$ at treatment system inlet
ϵ	a small number (defined more precisely where required)
ϵ_p	a projection factor to accelerate design with uncertainty
$\epsilon(y)$	precision error in variable y
ρ^{max}	threshold for testing for potential change in active set
θ	optimization variables
Θ	set of optimization variables
λ	Lagrange multiplier
Λ	Relative gain array matrix
μ	penalty factor in SQP
τ	time constants (various)
τ_{PFR}	time in plug-flow reactor
τ_r	first-order reaction time constant
ϕ	times in Φ
Φ	discrete subset of time

Operators

E	expected value of
$\| \ \|_n$	the nth norm
Δ	change in value
$[x]$	molar (g-moles/liter) concentration of component x

Superscripts

*	optimum value

Subscripts

max	maximum value
min	minimum value

Acronyms

CSTR	continuous (flow) stirred-tank reactor
PFR	plug-flow reactor
PID	proportional-integral-derivative control
psd	particle-size distribution (fraction by mass)

Units

M	concentration in moles per liter
N	concentration in normals [available OH$^-$ (M); acids have negative normality]
w/w	concentration measured as weight of one component over total weight
μ	length in microns [micrometers (μm)]

APPENDIX A: WORST-CASE DESIGN ALGORITHM

1: Initialization
Define the sets P and V.

$\rho^{max} = 1$, $nloc1 = 1$, $nloc2 = 1$, $nrand = 1$
Set ρ^{max}_{min}, $nloc1_{max}$, $nloc2_{max}$, $ivert$, $nrand_{max}$
to control the maximum and minimum effort in the constraint maximization algorithm.

For each v_i define $pvert_i$

Set ϵ_p

Set the initial number of models, $nmods$, the number of models defined by the user, $nset$, and the values of v_j, $j = 1 ... nset$

If $nmods > nset$
   Maximize each constraint separately in a local vertex search
      Generate a random initial point
      Move to vertex maximizing the constraint based on a local linearization
      Repeat until search fails to increase constraint
Eliminate duplicate vertex maximizers
If the projected model gives an increased violation replace the original model with the projected model.
Select the best $nadd$ vertex maximizers.
($nadd$ = max($nmods - nset$, no of distinct vertex maximizers))
Reset $nmods$ to $nset + nadd$.

Proceed to multimodel design (2:)

2: Multimodel design problem

A: Identify the predicted active set of models to be used in the design.
All models which have shown an active constraint in any of the three previous design iterations are predicted as active.
Initial models are held for at least 3 iterations.
Models which have been incorrectly dropped from the predicted active set are not dropped again.

B. Carry out design optimization subject to constraints from active models
If the optimization fails to find a feasible solution STOP
(Worst-case design may be infeasible)

C: Check violations for models not in the predicted active set.
If necessary update the predicted active set and return to B:

3: Constraint maximization

LOCAL SEARCH:
Local vertex search
lv1: Choose a random initial value of v
lv2: Update v using

$$v_i = v_i^l \text{ if } \frac{\partial c_{max}}{\partial v_i} \leq 0$$

$$v_i = v_i^u \text{ if } \frac{\partial c_{max}}{\partial v_i} > 0$$

until either c_{max} fails to increase as predicted or the search predicts a vertex already examined
lv3: For each $c_k < c_{max}$ compute the maximum increase Δc_k in the constraint, based on the constraint gradients, and identify the corresponding vertex.
Compute the projected fractional change $\rho = \Delta c_k/(c_{max} - c_k)$
If the maximum value of ρ is less than ρ^{max}
or no new vertices are identified than local vertex search is complete
Else select the new vertex value of v giving the maximum value of ρ and go back to lv2:
If no constraint violations have been found set ρ^{max} to ρ^{max}_{min}
to increase the depth of search and go to back to lv3:

Carry out local searches into the interior of V from up to $nloc1$ vertices identified during the local vertex search as having ascent directions into V which did not lead to an increased value of c_{max} on the corresponding vertex.
If no new constraint violation is obtained then increment $nloc1$ until either $nloc1 = nloc1_{max}$ or no more suitable vertices are left

If a constraint violation has not been found or
a global search has been used previously then

GLOBAL SEARCH:
Vertex search
If $ivert = 1$ test all the vertices not already examined
Carry out a local search from the largest new maximizer, with an ascent direction into V

Multistart random search
Carry out random trials until the number of trials exceeds *nrand*
  Generate trial points using

$$v_i = v_i^l + (v_i^u - v_i^l) \min\left(1, \max\left(0, \frac{r_i}{1 - pvert_i} + .5\right)\right)$$

  where r_i is a uniform random variable $\in [-0.5, 0.5]$.
  and v_i^u and v_i^l are the upper and lower limits on the *i*th parameter in v
  If a random point is found increasing the maximum constraint violation then
    If no new constraint violation has been found and *nloc2* local searches have been used in the global search increment *nloc2* up to a maximum of $nloc2_{max}$
    Carry out local search unless the number of local searches in the global search phase exceeds *nloc2*
  If no new constraint violation has been found and *nrand* random trials have been used increment *nrand* up to a maximum of $nrand_{max}$

OUTER APPROXIMATION UPDATE:
If a constraint violation has been found then
  Update models for design, V_i
    If projected model gives increased constraint violation then add projected model else add the unprojected model to active set
    Proceed to multimodel design (2:)
else
  Optimum of worst-case design has been found, STOP

APPENDIX B: EXAMPLE OF PROGRESS OF ALGORITHM

The discussion below details the operation of the worst-case design algorithm for the example discussed in Section V.C.2.

The tuning parameters for the optimization were set as follows: $\rho_{min}^{max} = 1$, $nloc1^{max} = 10$, $nloc2_{max} = 10$, $ivert = 1$, $nrand_{max} = 1000$. *pvert* was set to .99 on the initial and final levels of the two step disturbances and to .8 on all other variables. $\epsilon_p = 0.1$ is used in the example below, corresponding to increasing the concentration of the disturbance streams by 10%.

The nominal case was taken as a base flow of 400 m³/h with no disturbances and no measurement bias. As the disturbances were infrequent, the economic objective for a given control structure was taken to be minimize the excess reagent used at steady state compared to the reagent required with ideal delay-limited control. The ideal delay-limited controller had been able to operate with an exit pH setpoint of 11.3 (actual pH can drop to the limit of 11.2 due to measurement bias). Increasing this setpoint to accommodate disturbances or overneutralizing in the first tank results in increased reagent cost.

As the preliminary analysis suggested that the ratio control should make the concentration disturbance dominant over the flow disturbance, the initial scenarios were all based on the disturbances occurring simultaneously to give the maximum feed concentration variation. It was clearly necessary to consider both pulses starting at once and both pulses ending at once. Normally, the maximum measurement lag will be the worst case, so the measurement lags for the initial scenarios were set to 5 for the pH measurement on both tanks. The key pH measurement was that on the first tank, which was used for both feedback and feedforward control, so a bias of both .1 and −.1 pH units was considered. Finally, a bias of .1 was assumed on the last pH measurement as this reduces the effective pH at the treatment system exit, pushing the final setpoint up and increasing requirements for alkali reagent. This gave five scenarios in all (nominal case plus combinations of pulses starting/finishing bias positive/negative). This approach of choosing the initial scenarios based on a mixture of preliminary design analysis and qualitative reasoning is quite typical.

The progress of the design optimizations is shown below. The objective function increases monotonically as the optimization progresses and the outer approximation is refined by adding additional scenarios. All the additional scenarios added use the projection factor of 0.1.

iteration	setpoint1	gain1	iat1	setpoint2	gain2	iat2
0	11.32	.05	10	11.34	.05	10
1	1.89	.051	11.8	11.308	.0083	2
2	1.75	.073	41	11.31	.01	2
3	1.7	.07	47.4	11.312	.082	2

iteration	FF gain	lead constant	excess reagent cost
0	.8	30.	not feasible
1	1.16	19.4	£628
2	1.19	18.4	£833
3	1.18	18	£1100

The third design passed the test for feasibility and was accepted as the optimum.

The first scenario added involved the low-concentration, 80 m^3/h, disturbance being applied for the first 10 minutes and then removed. The high-concentration, 40 m^3/h, disturbance was initially absent, but was applied after 344 seconds. The first pH probe was biased so that the actual pH was .1 higher than the measured pH. The second probe was biased in the opposite way. The first probe was at maximum lag, while the second probe had a lag of 1.6 seconds.

This scenario was identified as follows. The local vertex search failed to identify a constraint violation. A local search from a vertex identified in the local vertex search as not being a local maximum converged to a solution that differed from one of the initial scenarios only in that the second measurement lag was reduced to 1 second. This gave a constraint violation of 8×10^{-6}, which was just less than the optimization tolerance of 10^{-5}. No other suitable vertices from which to carry out a local search had been found. The global phase of the maximization was therefore activated. Vertex enumeration identified a constraint violation of 40×10^{-5}. This corresponded to the low-concentration disturbance being removed and the high-concentration disturbance being applied simultaneously, with the true pH being biased high on the first stage and low on the second stage and the first probe having maximum lag and the second probe having minimum lag. This vertex was not a local maximizer, so a local search was initiated using it as a starting point. This local search found the scenario described above, giving a constraint violation of 96×10^{-5}. As the number of random trials in the global phase, *nrand*, is initialized to 1 and a significant constraint violation had been found, the search stopped after just one random trial which failed to give an increased constraint violation. Setting ϵ_p to 0.1 for the scenario identified increased the integrated squared constraint violation so the projected scenario was added to the scenarios for the next design iteration.

The second constraint maximization did not find a violation until the vertex enumeration phase and gave a violation of 92×10^{-5}. The maximizing vertex occurred with minimum base flow, the low-concentration disturbance being applied after 10 minutes and the high-concentration disturbance being applied after 1 second, with both pH values biased low and probe lags at their maximum. Again this vertex was not a local maximizer. A local search increased the constraint violation to 97×10^{-5} by delaying the onset of the high-concentration disturbance to 51 seconds into the simulation. Again, only one unsuccessful random trial was made. Increasing ϵ_p to 0.1 again increased the violation, so the projected constraint was added to the scenarios for design.

On the third constraint maximization the vertex enumeration and subsequent local search failed to identify a significant violation. The random search therefore was allowed to run until either a violation was found or $nrand_{max}$ trials were made. The allotted 1000 random trials failed to identify an increased constraint violation (from which a further local search would have been initiated), so the optimization terminated with the solution noted above.

In relation to the noise-control algorithm, it is of interest to note that the models corresponding to the scenarios for design were integrated with values of *simacc* ranging from 10^{-3} to 4×10^{-7}, illustrating the difficulty of optimizing *simacc* without an adaptation mechanism.

References

Asai, S., Konishi, Y., and Sasaki, Y. "Mass Transfer Between Fine Particles and Liquids in Agitated Vessels," *J. Chem. Eng. Jpn.* **21**, 107–112 (1988).

Biegler, L. T. "Strategies for Simultaneous Solution and Optimization of Differential-Algebraic Systems," *in* "Proceedings of the Third International Conference on Foundations of Computer-Aided Process Design, Snowmass, CO, 1989," p. 155–179 (1990).

Brucato, A., Brucato, V., Rizzuti, L., and Sanfilippo, M. "Particle Dissolution Kinetics with Batch Stirred Vessels, *Int. Chem. Eng. Symp. Ser.* **121**, 327–341 (1990).

Chan, W. K., and Prince, R. G. H. "Heuristic Evolutionary Synthesis with Non-Sharp Separators," *in* "Third International Symposium on Process Systems Engineering, Sydney, 1988," p. 300–312 (1988).

Chemineer. "Kenics Static Mixers: KTEK Series," Tech. Rep. K-TEK-4. Chemineer, 1988.

Chen, C. L. "A Class of Successive Quadratic Programming Methods for Flowsheet Optimisation," Ph.D. Thesis, University of London (1988).

Cushnie, G. C. "Removal of Metals from Wastewater: Neutralisation and Precipitation." Noyes Publications, Park Ridge, NJ, 1984.

Dunker, A. M. "The Decoupled Direct Method for Calculating Sensitivity Coefficients in Chemical Kinetics," *J. Chem. Phys.* **81**, 2385–2393 (1984).

Gill, P. E., Murray, W., and Wright, M. H. "Practical Optimization." Academic Press, New York, 1981.

Grossmann, I. E., and Floudas, C. A. "Active Constraint Strategy for Flexibility Analysis in Chemical Processes," *Comput. Chem. Eng.* **6**, 675–693 (1987).

Grossmann, I. E., Haleman, K. P., and Swaney, R. E. "Optimization Strategies for Flexible Chemical Processes," *Comput. Chem. Eng.* **7**, 439–462 (1983).

Gustafsson, T. K. "Calculation of the pH Value of a Mixture of Solutions," *Chem. Eng. Sci.* **37**, 1419–1421 (1982).

Gustafsson, T. K., and Waller, K. V. "Dynamic Modelling and Reaction Invariant Control of pH," *Chem. Eng. Sci.* **38**, 389–398 (1983).

Haslam, R. T., Adams, F. W., and Kean, R. H. "The Rate of Solution and Availability of Commerical Limes," *Ind. Eng. Chem.* **18**, 19–23 (1926).

Hershkovitch, H. Z., McAvoy, T. J., and Liapis, A. L. "Dynamic Modelling of pH Electrodes," *Can. J. Chem. Eng.* **56**, 346–353 (1978).

Holt, B. R., and Morari, M. "Design of Resilient Processing Plants. V. The Effect of Deadtime on Dynamic Resilience," *Chem. Eng. Sci.* **40**, 1229–1237 (1985).

Hoyle, D. "Designing for pH Control," *Chem. Eng.* **83**, 121–126 (1976).

Johansson, G., and Norberg, K. "Dynamic Response of the Glass Electrode," *J. Electronal. Chem.* **18**, 239–250 (1968).

Leis, J. R., and Kramer, M. A. "Sensitivity Analysis of Systems of Differential and Algebraic Equations," *Comput. Chem. Eng.* **9**, 93–96 (1985).

Levenspiel, O. "Chemical Reaction Engineering." Wiley, New York, 1972.

Levins, D. M., and Glastonbury, J. R. "Particle-Liquid Hydrodynamics and Mass Transfer in a Stirred Vessel. II. Mass Transfer," *Trans. Inst. Chem. Eng.* **50**, 132–146 (1972).

Luyben, W. L. "Process Modelling, Simulation and Control for Chemical Engineers." McGraw-Hill, New York, 1990.

Mayne, D. Q., Michalska, H., and Polak, E. "An Efficient Outer Approximations Algorithm for Solving Infinite Sets of Inequalities," *in* "Proceedings of the 29th Conference of Decision and Control, 1990," p. 960–965 (1990).

McAvoy, T. J. "Dynamic Modelling of pH in Aqueous Systems," *in* "Proceedings of AIChE Workshop in Industrial Process Control, 1979," p. 35–39 (1979).
McMillan, G. K. "Tuning and Control Loop Performance." Instrument Society of American Publications, 1983.
McMillan, G. K. "pH Control." Instrument Society of America Publications, 1984.
Mizsey, P., and Fonyó, Z. "Toward a More Realistic Overall Process Synthesis—The Combined Approach," *Comput. Chem. Eng.* **14**, 1213–1236 (1990).
Moore, R. L. "Neutralisation of Waste Water by PH Control." Instrument Society of America Publications, 1978.
Morari, M. "Effect of Design on the Controllability of Chemical Plants," *in* "Interactions between Process Design and Process Control (J. D. Perkins, ed.), p. 3–16. Pergamon, Oxford, 1992.
Perkins, J. D., and Wong, M. P. F. "Assessing Controllability of Chemical Plants," *Chem. Eng. Res. Des.* **63**, 358–362 (1985).
Polak, E. "An Implementable Algorithm for the Optimal Design Centering, Tolerancing, and Tuning Problem," *J. Optim. Theory Appl.* **37**, 45–67 (1982).
Polak, E., and Stimler, D. M. "Majorisation: A Computational Complexity Reduction Technique in Control System Design," *IEEE Trans. Autom. Control* **AC-31**, 1010–1021 (1988).
Prett, D. M., and Garcia, C. E. "Fundamental Process Control." Butterworth, London, 1988.
Prett, D. M., Garcia, C. E., and Ramaker, B. L. "The Second Shell Process Control Workshop." Butterworth, London, 1990.
Sargent, R. W. H., and Sullivan, G. R. "Development of Feed Changeover Policies for Refinery Distillation Columns," *Ind. Eng. Process Dis. Dev.* **18**, 113 (1979).
Shinskey, F. G. "pH and pIon Control," Wiley Environ. Sci. Technol. Ser. Wiley, New York, 1973.
Straub, D. A., and Grossmann, I. E. "Integrated Stochastic Metric of Flexibility for Systems with Discrete State and Continuous Parameter Uncertainties," *Comput. Chem. Eng.* **14**, 967–985 (1990).
Swaney, R. E., and Grossmann, I. E. "An Index for Operational Flexibility in Chemical Process Design. Part I. Formulation and Theory," *AIChE J.* **31**, 621–630 (1985a).
Swaney, R. E., and Grossmann, I. E. "An Index for Operational Flexibility in Chemical Process Design. Part II. Computatonal Algorithms," *AIChE J.* **31**, 631–641 (1985b).
Tits, A. L. "On the Optimal Design Centering, Tolerancing, and Tuning Problem," *J. Optim. Theory Appl.* **45**, 487–494 (1985).
Vasantharajan, S., and Biegler, L. T. "Simultaneous Strategies for Optimisation of Differential-Algebraic Systems with Enforcement of Error Criteria," *Comput. Chem. Eng.* **14**, 1083–1100 (1990).
Velguth, F. W., and Anderson, R. C. "Determination of Minimum Capacities for Control Applications," *ISA J.* **1**, 33–38 (1954).
Walsh, S. P. "Integrated Design of Chemical Waste Water Treatment Systems," PhD Thesis, Imperial College, London (1993).
Yagi, H., Motouchi, T., and Hikita, H., "Mass Transfer from Fine Particles in a Stirred Vessel," *Ind. Eng. Chem. Process Des. Dev.* **23**, 145–150 (1984).

INDEX

A

Acetate, production, 20, 23
Acetone
   acetone/chloroform/benzene separation, 108–121
   pentane/acetone/methanol/water separation, 122–131
Adaptive initial design synthesizer (AIDES), 17–19, 173
Adjoint network synthesis, 249
ADVENT, 175
Aggregated models, 184, 187, 217–218
AIDES process synthesis hierarchy, 17–19, 173
Alcohols
   distillation-based separation
      n-butanol/water mixture, 91–93
      ethanol dehydration, 41–50
      five alcohols, 80–81, 83, 84–89
      isopropanol/water mixture, 157–161
   ethanol mixtures, finding azeotropes, 136–137
Algorithmic process synthesis
   history, 172–178, 237–238
   mathematical programming, 171–174, 178–180, 237
   MINLP algorithms, 197–213
   MINLP modeling, 175, 177, 187–197, 224–236
   MINLP synthesis problems, 172, 178
      algebraic form, 197
      large size, 215–219
      nonconvexities, 174, 219–224
      zero flows, 213–215
   optimization, 180–187
Algorithmic systematic generation approaches, 58, 59
Algorithms
   AP/OA/ER algorithm, 226
   distillation, 176
   finding azeotropes, 134–135
   global optimization algorithms, 222–223
   infeasibility algorithm, 305
   local optimization algorithms, 297
   MINLP algorithms, 197–213, 343
   OA/ER algorithm, 227
   outer-approximation (OA) algorithm, 209–211, 307–308
   worst-case algorithm, 307–308, 310, 315–321, 396–401
Ammonia, production, process synthesis, 186
Analysis-driven synthesis, 94–97
AP/OA/ER algorithm, 226
Artificial intelligence, AIDES hierarchy, 18
Attainable regions, 250–253, 295, 297
Augmented penalty/equality relaxation (APER), 205–206
Azeotropes
   binary azeotropes, 41, 69–71
   finding, 132–137
   homogeneous, 41–42
   hydrogen bonding, 71–72
   infinite-dilution K-values, 69–71, 75, 98
   residue curve maps, 39–41, 99–100
      diethoxymethane production, 55
      ethanol dehydration, 42, 51–52, 53
      S-shaped curves, 149–150
      ternary systems, 101–105
   separation systems, 39–41, 177
      acetone/chloroform/benzene mixture, 108–121
      ethanol dehydration, 41–50
      isopropanol/water mixtures, 157–161
      pentane/acetone/methanol/water mixture, 122–131
      water/n-butanol, 91–93

INDEX

Azeotropes (*Continued*)
  ternary azeotropes, 42, 44, 101–105, 135–137
Azeotropic entrainer, 44–45

B

BALTAZAR, 17, 173
Benzene, acetone/chloroform/benzene separation, 108–121
Binary azeotropes, 41, 69–71
Binary saddle, 103
Bottoms products, 82, 154–155
Bounding, 343–344
Branch-and-bound method (BB), 198, 199–200, 202, 223–224
n-Butanol/water mixture, separating, 91–93

C

Calcium hydroxide, 355–357
CFRs, *see* Cross-flow reactors
Chen approximation, 196
Chloroform, acetone/chloroform/benzene separation, 108–121
Column reflux ratio, 163–165
Combined reaction–separation model, 284–290
Complex columns, 36–39, 176
Composite curves, 29
Computer software, *see* Software
Conceptual process design, 10, 13–23
Connector, 252
Continuous stirred tank reactors (CSTRs), 249, 250
  attainable regions, 252–253
  effluent plant, 387–388
  energy integration, 279
  in series, 332–333
  waste water neutralization, 361, 367, 370, 387–388
Cross-flow reactors (CFRs), 266, 296
  discretized representation, 269, 276

D

Dangerous species heuristic, 82
Decant aqueous layers, 45–46, 48
Delay-lag model, 358

Delays, 325–333, 366
Design, *see* Process design
DICOPT++, 174, 206, 213, 225, 235–236
Difference point, 157–162
Differential sidestream reactors (DSRs), attainable regions, 252–253
Dioxymethane, production, 54–58
Direct sequence heuristic, 87, 89
Discrete design decisions, 342–344
Distillation-based separation systems, 64–68, 176
  algorithms, 176
  analysis-driven synthesis, 94–97
  azeotropic behavior, 69–72
  case studies
    acetone/chloroform/benzene separation, 108–121
    n-butanol/water mixture, 91–93
    ethanol dehydration, 41–50
    five alcohol mixture, 80–81, 83, 84–89
    heat-integrated distillation synthesis, 28–39
    isopropanol/water separation, 157–161
    pentane/acetone/methanol/water separation, 122–131
  extractive distillation, 29, 64, 157–166
  heuristics, 89–90
  liquid/liquid behavior, 73–75
  multiple-effect, 36–39
  nearly ideal systems, 75–90
  nonideal mixtures, 90–93, 107–131
  post-analysis methods, 166
  pre-analysis methods, 98
    distillation columns, 105–107, 140–157
    equilibrium-phase behavior, 98–105
    extractive distillation, 64, 157–166
    species behavior, 131–140
Distillation boundaries, 40, 54, 103
  crossing, 151–154
  ethanol dehydration, 42–45
Distillation columns, 36–39, 176
  for binary separations, 36–39
  design calculations, 166
  limiting behavior, 105–107, 140–157
  side stripper, 67–68, 176
  total reflux column, 100
  types, 66–68
Distillation curves, 99, 100–101
  infinite reflux constraint, 141
  reachable regions, 141, 145–146
  S-shaped curves, 149–150

INDEX 405

Distillation regions, 40, 103
Disturbance, *see* Uncertainty
Disturbance attenuation, 348
Disturbance fraction, 331–332, 375, 376
DSRs, *see* Differential sidestream reactors
Dynamic optimization, 333–342, 392

E

Effluent plant, design, 371–391
Energy integration, 247–248
  heat integration, 28–29
    distillation, 33, 35, 65, 176–177
    optimization, 232–233
    reacting segment, 276
    software, 30
    superstructure, 232
    transshipment model, 187
  reactor networks, 274–283
Entrainers, 42–45
Equalities, MILP master problem, 204–205
Equilibrium-phase behavior, *see* Phase equilibrium
Error tolerance, 336–338
Ethanol, dehydration, 41–50
Ethanol/water mixture, azeotropes in, 136
Ethanol/water/toluene mixture, azeotropes in, 136–137
Evolutionary modification, 13, 23–26, 176
Extended cutting-plane (ECP) method, 198, 202
Extractive distillation, 29, 64, 157–166

F

Feasibility tests, 329–331
Final product heuristic, 82
Finite reflux, 143–148
Flexibility index, 311–312, 313
Flowsheets, 4, 10, 172–175
  means–ends analysis, 13–15, 17
Fractional tubularity model, 358

G

GAMS, 174, 206
GAMS/MINOS, 235
Generalized benders decomposition (GBD), 198, 201–202, 210–212

Generate-and-test design strategy, 8
Generate-evolve-optimize-critique synthesis strategy, 50
Gibbs free energy, liquid mixture, 73–74, 138, 139
Global optimization, 221–222, 297, 314, 321
Global optimization algorithms, 222–223
GLOBESEP, 234, 235
GLOPEQ, 140
Grand composite curves, 29

H

HDA process, 225–226, 229–231
Heat balance, heat exchanger network, 193–194
Heat cascade, 184, 185
Heat-exchanger network synthesis (HENS)
  distillation, 28–39, 175–176
  energy integration, 274
  MINLP model, 192–197, 221–222, 225, 232
  software, 175
  superstructure, 183, 184
Heat integration, 28–29
  distillation, 33, 35, 65, 176–177
    case study, 28–39
  optimization, 232–233
  reacting segment, 276
  software, 30
  superstructure, 232
  transshipment model, 187
Heuristics, 131, 173, 186, 237
  dangerous species heuristic, 82
  direct sequence heuristic, 87, 89
  distillation heuristics, 89–90
  fifty-fifty split heuristic, 90
  final product heuristic, 82
  major species heuristic, 90
  sequencing heuristics, 176
Hierarchical decomposition, 173
Hierarchical process synthesis systems, 16–19, 26–28
High-boiling node, 41
Homogeneous azeotropes, 41–42
Hydrogen bonding, azeotropes, 71–72

I

Infeasibility algorithm, 305
Infinite-dilution activity coefficients, 73–75

Infinite-dilution K-values, 69–71, 75, 98
Infinite reflux, 105, 141–143
Infinite reflux constraint, 141
Innovation process, 2, 3–10
Integrator error tolerances, 338–339
Isopropanol/water mixture, separation, 157–161
Isothermal systems
   α-pinene, 264–265
   reactor network synthesis, 254–265
      optimization, 258–262
      segregated-flow approximation, 254–258
      Trambouze reaction, 263–264
      van de Vusse reaction, 263, 265, 292–294
Isovolatility curves, 134

L

Linear programming, AIDES hierarchy, 18
Liquid/liquid behavior, infinite-dilution activity coefficients, 73–75
Liquid/liquid extraction, 123, 125, 130
Liquid/liquid(/liquid) behavior, 137–140
Local approximation, dynamic optimization, 340–341
Local optimization algorithms, 297
Logic-based methods, 207–209
Low-boiling node, 40

M

MAGNETS, 175, 215
Majorization, 314
Major species heuristic, 90
M-APER, 205–206
Marginal cost separation design, 81–84
Marginal vapor flow, 83–84
Margules equation, 74
Mass-exchanger networks, 177
Material balance constrains, 141–142
Mathematical programming, 171–174, 178–180, 237
McCabe–Thiele diagram, 166
Means-ends analysis, 13–15, 17
Methanol, pentane/acetone/methanol/water separation, 122–131
Methyl acetate, process design, 20–28
M-GBD, 205
MILP, 174, 213, 313
MILP master problem, 199, 200–201, 203

Minimum delay, 327–329
Minimum reflux, 105
Minimum vapor flow, 85
MINLP algorithms, 197–213, 343
MINLP modeling, 175, 177, 187–197, 224–236
MIQP, 203
Mixed-integer linear program (MILP), 174
Mixed-integer nonlinear programming (MINLP), 171, 172, 174
   discrete design decisions, 343
   math programming, 171–174, 178–180
   MINLP algorithms, 197–213, 343
   MINLP modeling, 187–197, 224–236
      simultaneous, 175, 177
   MINLP synthesis problems, 172, 178
      algebraic form, 197
      large size, 217–219
      nonconvexities, 174, 219–224
      zero flows, 213–215
   nonconvexities, 174
   optimization, 180–187
   reactor network synthesis, 249–250, 290–292
Mixing, delay-lag model, 358–360
Mixtures, *see also* Azeotropes
   azeotropic behavior, 69–72
   finding azeotropes, 132–137
   liquid/liquid behavior, 73–75
   liquid/liquid(/liquid) behavior, 137–140
M-MIP, 199, 200–201, 203
M-MIQP, 203, 205
Modeling/decomposition (M/D) strategy, 175, 213–215, 226
Modeling systems
   energy integration, 274–279
   GAMS, 174
   MINLP modeling, 175, 177, 187–197, 224–236
   operability and control, 304
   waste water neutralization, 353–360
Mole-fraction-averaged relative volatility, 76
Multiobjective programming, AIDES hierarchy, 18

N

Nearly ideal systems
   separating, 75
   pinch, 76–77
   process system synthesis, 80–90
   Underwood's method, 77–80

NLP, 174, 175
  energy integration, 278
  reactor network synthesis, 249, 296
  worst-case design, 313
NLP subproblems, 198–202, 205–206, 221, 313
Nodes, 53, 102, 103
Noise control, dynamic optimization, 335, 365
Nonconvexities
  MILP master problem, 205–206
  MINLP, 174, 219–224
Nonideal mixtures
  separating, 41, 90–91
    process synthesis, 107–131
    water/n-butanol, 91–93
Nonisothermal systems, reactor network synthesis, 265–275, 296
Nonlinear programming, *see* NLP

modeling/decomposition (M/D) strategy, 175
NLP, 174, 175
process integration, 290–292
reactor network synthesis, 248, 249–250
  isothermal systems, 258–262
superstructure optimization, 13, 59
  algorithmic methods, 173
  complex columns, 176
  distillation processes, 130–131, 176
  reactor network synthesis, 249–250
  representation of alternatives, 180–184
  time delays, 330–331
Outer-approximation (OA) algorithm, 209–211, 307–308
Outer-approximation (OA) method, 198, 200–201, 210–212, 315
Overall disturbance attenuation, 348

O

OA algorithm, 209–211, 307–308
OA/ER algorithm, 227
Operability and control, 301–306, 391–393
  controllability, 322, 331–332
  dynamic optimization, 333–342, 392
  screening tools for disturbance, 321–333
  uncertainty, 303, 304, 306–321
  wastewater neutralization, 342
    design examples, 360–391
    modeling, 353–360
  worst-case design
    algorithm, 315–321, 396–401
    Grossman, 307–308
    majorization, 314
    outer-approximation algorithm, 307–308, 310
    Polak, 313, 314
Optimization, 58–60
  aggregated models, 184, 187, 217–218
  dynamic optimization, 333–342, 392
  energy integration, 281
  global optimization, 221–222, 297, 314, 321
  heat integration, 232–233
  heuristics, 82, 87–90, 131, 176, 186, 237
  local approximation, 297
  logic-based methods, 207–209
  mathematical programming, 171–174, 178–180, 237
  mixed-integer optimization techniques (MINLP), 180–187

P

Path constraint representation, 339–340
Pentane, pentane/acetone/methanol/water separation, 122–131
PFRs, *see* Plug flow reactors
Phase equilibrium, 98–99
  distillation curves, 100
  residue curve maps, 39–41, 99–100
Pinch, 76–78
Pinch point, 75, 77, 78, 145, 153, 177
Pinch point curves, 64, 146
Pinch technology, 17, 29, 175, 290–292
α-Pinene, 264–265
PIP, 17, 173
Plug flow reactors (PFRs), 249, 250
  attainable regions, 252–253
  segregated flow networks, 256–258
Ponchon–Savarit diagram, 166
Post-analysis separation, 166
Predictor-based bounding, 343–344
Pre-synthesis analysis, 94
Probe lag, 370
Process design, 5, 7, 8, 367
  controllability, 322, 331–332
  discrete design decisions, 342–344
  dynamic optimization, 333–342, 392
  generate-and-test design strategy, 8
  generate-evolve-optimize-critique synthesis strategy, 50
  innovation, 2, 3–10
  operability and control, 301–401

Process design (*Continued*)
  propose-critique-modify design strategy, 8
  screening tools for disturbance, 321–333
  uncertainty, 303, 304, 306–321
  wastewater neutralization, 342
    design examples, 360–391
    modeling, 353–360
  worst-case design
    algorithm, 315–321, 396–401
    Grossman, 307–308, 310
    majorization, 314
    outer-approximation algorithm, 307–308, 310
    Polak, 313, 314
Process integration, 175, 283–294, 296, 298
  operability and control, 302–304
Process synthesis, 1–3
  adjoint network synthesis, 249
  algorithmic process synthesis, 171–239
  analysis-driven synthesis, 94–97
  case studies
    acetone/chloroform/benzene separation, 108–121
    n-butanol/water separation, 91–93
    dioxymethane production, 54–58
    ethanol dehydration, 41–50
    heat-integrated distillation synthesis, 28–39
    methyl acetate production, 20–28
    pentane/acetone/methanol/water separation, 122–131
    wastewater neutralization, 342, 353–391
  conceptual process engineering, 10, 13–19, 20–23
  controllability, 322, 331–332
  definition, 10
  distillation-based separation systems, 64–68, 176–177
    ethanol dehydration, 41–50
  dynamic optimization, 333–342, 392
  flowsheets, 4, 10, 172–175
    means–ends analysis, 13–15, 17
  four or more species, 97
  heat-exchanger network synthesis
    distillation, 28–29, 175–176
    energy integration, 274
    MINLP model, 192–197, 221–222, 225, 232
    software, 175
    superstructure, 183, 184
  heterogeneous models, 186–187
  hierarchical approaches, 16–19, 26–28
  hierarchical decomposition, 173
  homogeneous models, 181–184
  industrial applications, 1–61
  innovation process, 2, 3–10
  mass-exchanger networks, 177
  mathematical programming, 171–174, 178–180, 237
  means-ends analysis, 13–15, 17
  nearly ideal systems, 75–90
  nonideal mixtures, 90–93, 107–131
  operability and control, 301–401
  reactor network synthesis, 177–178, 247–298
  residue curve maps, 39–41, 99–100
    diethoxymethane production, 55
    ethanol dehydration, 42, 51–52, 53
    S-shaped curves, 149–150
    ternary systems, 101–105
  screening tools for disturbance, 321–333
  separations systems synthesis, 41–54
  uncertainty, 303, 304, 306–321
  worst-case design
    algorithm, 315–321, 396–401
    Grossman, 307–308, 310
    majorization, 314
    outer-approximation algorithm, 307–308, 310
    Polak, 313, 314
Property-changing operators, 14–15
Property hierarchy, 16, 19
Propose-critique-modify design strategy, 8
PROSYN, 175, 215, 227

R

Ramp disturbances, 333
Raoult's law, 72, 76
Reachable regions, 105–107, 140–157
Reactor network synthesis, 177–178, 248–250, 295–298
  attainable regions, 250–253, 295, 297
  definition, 248
  energy integration, 274–283
  isothermal systems, 254–265
  nonisothermal systems, 265–275
  simultaneous reaction, separation, and energy system synthesis, 175, 283–294
Recursive strategy, 15
Recycle reactors (RRs), 270–271
  segregated flow networks, 256–258

Recycles, 17
Reflux
  column reflux ratio, 163–165
  finite reflux, 143–148
  infinite reflux, 105, 141–143
  infinite reflux constraint, 141
  minimum reflux, 105
  total reflux, 105–107
Reflux ratio, 162–164
Relative volatility, 76
Residence time distribution (RTD), 286–290, 358–359
Residue curve maps (RCMs), 39–41, 99–100
  diethoxymethane production, 55
  ethanol dehydration, 42, 51–52, 53
  S-shaped curves, 149–150
  ternary systems, 101–105
RTD function, 286–290, 358–359

S

Saddles, 41, 52–53, 102, 103
Sargent–Gaminbandara superstructure, 184, 185
Screening tools, for disturbance and uncertainty, 321–333
Segregated-flow model, 254–258, 296
Separability factors, 123–124
Separation systems, distillation-based, *see* Distillation-based separation systems
Separation systems synthesis method, 41–54
Separatrices, 40
Side stripper, 67–68, 176
Simultaneous reaction–separation model, 284–290
Single-species node, 103
Software
  ADVENT, 175
  AIDES process design procedure, 18–19, 173
  BALTAZAR, 17, 173
  DICOPT++, 174, 206, 213, 225, 235–236
  GAMS, 174, 206
  GAMS/MINOS, 235
  GLOBESEP, 234, 235
  GLOPEQ, 140
  heat integration, 30
  HYCON, 65
  MAGNETS, 175, 215
  MAYFLOWER, 54

MILP, 174
MINLP, 171, 172, 174
NLP, 174, 175
PIP, 17, 173
PROSYN, 175, 215, 227
SPLIT, 54, 177
SUPERTARGET, 175
SYNHEAT, 175, 225, 228
Solvent ratio, 162–166
Species, 94–97, 131–140
  four or more species, 94, 155–157
  major species heuristic, 90
SPLIT, 54, 177
Sulfur dioxide, catalytic oxidation, 272–274
Superstructure optimization, 13, 59
  algorithmic methods, 173
  complex columns, 176
  distillation processes, 130–131
  reactor network synthesis, 249–250
  representation of alternatives, 180–184
Superstructures, 237–238
  ammonia process, 186
  HDA process, 225, 229
  heat integration, 232
  heterogeneous systems, 186–187
  homogeneous systems, 181–184
SUPERTARGET, 175
SYNHEAT, 175, 225, 228
Systematic generation, 13–17, 61

T

Targeting, 5, 176, 178
  combined reaction-separation, 284, 287
  energy integration, 175, 279
  isothermal systems, 254–265
  nonisothermal systems, 265–275
  reactor network synthesis, 248, 249–250, 297
  segregated flow, 255
Targeting algorithm, AIDES hierarchy, 18
Targets, 237
Task identification, 17, 19
  heat-integrated distillation, 30, 32, 35
  methyl acetate production, 28
Task integration, 17, 19, 28
  heat-integrated distillation, 33, 35
Ternary azeotropes, 42, 44
  finding, 135–137
  residue curve maps, 101–105

Thermodynamics
  distillation sequencing, 176
  residue curve maps, 39–41
Time delays, 325–333, 366
Titration curves, process design, 354–355
Total reflux, 105–107
Trambouze reaction, 263–264
Transshipment model, 184–186, 187
Two-species node, 103
Two-species saddle, 103

U

Uncertainty
  process design, 303, 304, 306–321
  titration curves, 355
Underwood's method, 78–80, 84

V

van de Vusse reaction, 263, 265, 292–294

W

Waste water neutralization, 342
  process design examples, 360–391
  modeling, 353–360
Worst-case design
  algorithm, 307–308, 310, 315–321, 396–401
  Grossman, 307–308, 310
  majorization, 314
  outer-approximation algorithm, 307–308, 310
  Polak, 313, 314

Contents of Volumes in This Serial

Volume 1

J. W. Westwater, *Boiling of Liquids*
A. B. Metzner, *Non-Newtonian Technology: Fluid Mechanics, Mixing, and Heat Transfer*
R. Byron Bird, *Theory of Diffusion*
J. B. Opfell and B. H. Sage, *Turbulence in Thermal and Material Transport*
Robert E. Treybal, *Mechanically Aided Liquid Extraction*
Robert W. Schrage, *The Automatic Computer in the Control and Planning of Manufacturing Operations*
Ernest J. Henley and Nathaniel F. Barr, *Ionizing Radiation Applied to Chemical Processes and to Food and Drug Processing*

Volume 2

J. W. Westwater, *Boiling of Liquids*
Ernest F. Johnson, *Automatic Process Control*
Bernard Manowitz, *Treatment and Disposal of Wastes in Nuclear Chemical Technology*
George A. Sofer and Harold C. Weingartner, *High Vacuum Technology*
Theodore Vermeulen, *Separation by Adsorption Methods*
Sherman S. Weidenbaum, *Mixing of Solids*

Volume 3

C. S. Grove, Jr., Robert V. Jelinek, and Herbert M. Schoen, *Crystallization from Solution*
F. Alan Ferguson and Russell C. Phillips, *High Temperature Technology*
Daniel Hyman, *Mixing and Agitation*
John Beek, *Design of Packed Catalytic Reactors*
Douglass J. Wilde, *Optimization Methods*

Volume 4

J. T. Davies, *Mass-Transfer and Interfacial Phenomena*
R. C. Kintner, *Drop Phenomena Affecting Liquid Extraction*
Octave Levenspiel and Kenneth B. Bischoff, *Patterns of Flow in Chemical Process Vessels*
Donald S. Scott, *Properties of Concurrent Gas-Liquid Flow*
D. N. Hanson and G. F. Somerville, *A General Program for Computing Multistage Vapor-Liquid Processes*

Volume 5

J. F. Wehner, *Flame Processes—Theoretical and Experimental*
J. H. Sinfelt, *Bifunctional Catalysts*
S. G. Bankoff, *Heat Conduction of Diffusion with Change of Phase*
George D. Fulford, *The Flow of Liquids in Thin Films*
K. Rietema, *Segregation in Liquid–Liquid Dispersions and Its Effect on Chemical Reactions*

Volume 6

S. G. Bankoff, *Diffusion-Controlled Bubble Growth*
John C. Berg, Andreas Acrivos, and Michel Boudart, *Evaporation Convection*
H. M. Tsuchiya, A. G. Fredrickson, and R. Aris, *Dynamics of Microbial Cell Populations*
Samuel Sideman, *Direct Contact Heat Transfer between Immiscible Liquids*
Howard Brenner, *Hydrodynamic Resistance of Particles at Small Reynolds Numbers*

Volume 7

Robert S. Brown, Ralph Anderson, and Larry J. Shannon, *Ignition and Combustion of Solid Rocket Propellants*
Knud Østergaard, *Gas–Liquid–Particle Operations in Chemical Reaction Engineering*
J. M. Prausnitz, *Thermodynamics of Fluid–Phase Equilibria at High Pressures*
Robert V. Macbeth, *The Burn-Out Phenomenon in Forced-Convection Boiling*
William Resnick and Benjamin Gal-Or, *Gas–Liquid Dispersions*

Volume 8

C. E. Lapple, *Electrostatic Phenomena with Particulates*
J. R. Kittrell, *Mathematical Modeling of Chemical Reactions*
W. P. Ledet and D. M. Himmelblau, *Decomposition Procedures for the Solving of Large Scale Systems*
R. Kumar and N. R. Kuloor, *The Formation of Bubbles and Drops*

Volume 9

Renato G. Bautista, *Hydrometallurgy*
Kishan B. Mathur and Norman Epstein, *Dynamics of Spouted Beds*
W. C. Reynolds, *Recent Advances in the Computations of Turbulent Flows*
R. E. Peck and D. T. Wasan, *Drying of Solid Particles and Sheets*

Volume 10

G. E. O'Connor and T. W. F. Russell, *Heat Transfer in Tubular Fluid–Fluid Systems*
P. C. Kapur, *Balling and Granulation*
Richard S. H. Mah and Mordechai Shacham, *Pipeline Network Design and Synthesis*
J. Robert Selman and Charles W. Tobias, *Mass-Transfer Measurements by the Limiting-Current Technique*

Volume 11

Jean-Claude Charpentier, *Mass-Transfer Rates in Gas–Liquid Absorbers and Reactors*
Dee H. Barker and C. R. Mitra, *The Indian Chemical Industry–Its Development and Needs*
Lawrence L. Tavlarides and Michael Stamatoudis, *The Analysis of Interphase Reactions and Mass Transfer in Liquid–Liquid Dispersions*
Terukatsu Miyauchi, Shintaro Furusaki, Shigeharu Morooka, and Yoneichi Ikeda, *Transport Phenomena and Reaction in Fluidized Catalyst Beds*

Volume 12

C. D. Prater, J. Wei, V. W. Weekman, Jr., and B. Gross, *A Reaction Engineering Case History: Coke Burning in Thermofor Catalytic Cracking Regenerators*
Costel D. Denson, *Stripping Operations in Polymer Processing*
Robert C. Reid, *Rapid Phase Transitions from Liquid to Vapor*
John H. Seinfeld, *Atmospheric Diffusion Theory*

Volume 13

Edward G. Jefferson, *Future Oportunities in Chemical Engineering*
Eli Ruckenstein, *Analysis of Transport Phenomena Using Scaling and Physical Models*
Rohit Khanna and John H. Seinfeld, *Mathematical Modeling of Packed Bed Reactors: Numerical Solutions and Control Model Development*
Michael P. Ramage, Kenneth R. Graziani, Paul H. Schipper, Frederick J. Krambeck, and Byung C. Choi, *KINPTR (Mobil's Kinetic Reforming Model): A Review of Mobil's Industrial Process Modeling Philsophy*

Volume 14

Richard D. Colberg and Manfred Morari, *Analysis and Synthesis of Resilient Heat Exchanger Networks*
Richard J. Quann, Robert A. Ware, Chi-Wen Hung, and James Wei, *Catalytic Hydrometallation of Petroleum*
Kent Davis, *The Safety Matrix: People Applying Technology to Yield Safe Chemical Plants and Products*

Volume 15

Pierre M. Adler, Ali Nadim, and Howard Brenner, *Rheological Models of Suspensions*
Stanley M. Englund, *Opportunities in the Design of Inherently Safer Chemical Plants*
H. J. Ploehn and W. B. Russel, *Interactions between Colloidal Particles and Soluble Polymers*

Volume 16

Perspectives in Chemical Engineering: Research and Education

Clark K. Colton, *Editor*

Historical Perspective and Overview

L. E. Scriven, *On the Emergence and Evolution of Chemical Engineering*
Ralph Landau, *Academic–Industrial Interaction in the Early Development of Chemical Engineering*
James Wei, *Future Directions of Chemical Engineering*

Fluid Mechanics and Transport

L. G. Leal, *Challenges and Opportunities in Fluid Mechanics and Transport Phenomena*
William B. Russel, *Fluid Mechanics and Transport Research in Chemical Engineering*
J. R. A. Pearson, *Fluid Mechanics and Transport Phenomena*

Thermodynamics

Keith E. Gubbins, *Thermodynamics*
J. M. Prausnitz, *Chemical Engineering Thermodynamics: Continuity and Expanding Frontiers*
H. Ted Davis, *Future Opportunities in Thermodynamics*

Kinetics, Catalysis, and Reactor Engineering

Alexis T. Bell, *Reflections on the Current Status and Future Directions of Chemical Reaction Engineering*
James R. Katzer and S. S. Wong, *Frontiers in Chemical Reaction Engineering*
L. Louis Hegedus, *Catalyst Design*

Environmental Protection and Energy

John. H. Seinfeld, *Environmental Chemical Engineering*

T. W. F. Russell, *Energy and Environmental Concerns*
Janos M. Beer, Jack B. Howard, John P. Longwell, and Adel F. Sarofim, *The Role of Chemical Engineering in Fuel Manufacture and Use of Fuels*

Polymers

Matthew Tirrell, *Polymer Science in Chemical Engineering*
Richard A. Register and Stuart L. Cooper, *Chemical Engineers in Polymer Science: The Need for an Interdisciplinary Approach*

Microelectronic and Optical Materials

Larry F. Thompson, *Chemical Engineering Research Opportunities in Electronic and Optical Materials Research*
Klavs F. Jensen, *Chemical Engineering in the Processing of Electronic and Optical Materials: A Discussion*

Bioengineering

James E. Bailey, *Bioprocess Engineering*
Arthur E. Humphrey, *Some Unsolved Problems of Biotechnology*
Channing Robertson, *Chemical Engineering: Its Role in the Medical and Health Sciences*

Process Engineering

Arthur W. Westerberg, *Process Engineering*
Manfred Morari, *Process Control Theory: Reflections on the Past Decade and Goals for the Next*
James M. Douglas, *The Paradigm After Next*
George Stephanopoulous, *Symbolic Computing and Artificial Intelligence in Chemical Engineering: A New Challenge*

The Identity of Our Profession

Morton M. Denn, *The Identity of Our Profession*

Volume 17

Y. T. Shah, *Design Parameters for Mechanically Agitated Reactors*
Mooson Kwauk, *Particulate Fluidization: An Overview*

Volume 18

E. James Davis, *Microchemical Engineering: The Physics and Chemistry of the Microparticle*
Selim M. Senkan, *Detailed Chemical Kinetic Modeling: Chemical Reaction Engineering of the Future*
Lorenz. T. Biegler, *Optimization Strategies for Complex Process Models*

Volume 19

Robert Langer, *Polymer Systems for Controlled Release of Macromolecules, Immobilized Enzyme Medical Bioreactors, and Tissue Engineering*
J. J. Linderman, P. A. Mahama, K. E. Forsten, and D. A. Lauffenburger, *Diffusion and Probability in Receptor Binding and Signaling*
Rekesh K. Jain, *Transport Phenomena in Tumors*
R. Krishna, *A Systems Approach to Multiphase Reactor Selection**
David T. Allen, *Pollution Prevention: Engineering Design at Macro-, Meso-, and Microscales*
John H. Seinfeld, Jean M. Andino, Frank M. Bowman, Hali J. L. Forstner, and Spyros Pandis, *Tropospheric Chemistry*

Volume 20

Arthur M. Squires, *Origins of the Fast Fluid Bed*
Yu Zhiqing, *Application Collocation*
Youchu Li, *Hydrodynamics*
Li Jinghai, *Modeling*
Yu Zhiquing and Jin Yong, *Heat and Mass Transfer*
Mooson Kwauk, *Powder Assessment*
Li Hongzhong, *Hardware Development*
Youchu Li and Xuyi Zhang, *Circulating Fluidized Bed Combustion*
Chen Junwu, Cao Hanchang, and Liu Taiji, *Catalyst Regeneration in Fluid Catalytic Cracking*

Volume 21

Christopher J. Nagel, Chonghun Han, and George Stephanopoulos, *Modeling Languages: Declarative and Imperative Descriptions of Chemical Reactions and Processing Systems*
Conghun Han, George Stephanopoulos, and James M. Douglas, *Automation in Design: The Conceputal Synthesis of Chemical Processing Schemes*
Michael L. Mavrovouniotis, *Symbolic and Quantitative Reasoning: Design of Reaction Pathways through Recursive Satisfaction of Constraints*
Christopher Nagel and George Stephanopoulos, *Inductive and Deductive Reasoning: The Case of Identifying Potential Hazards in Chemical Processes*
Kevin G. Joback and George Stephanopoulos, *Searching Spaces of Discrete Solutions: The Design of Molecules Possessing Desired Physical Properties*

Volume 22

Chonghun Han, Ramachandran Lakshmanan, Bhavik Bakshi, and George Stephanopoulos, *Nonmonotonic Reasoning: The Synthesis of Operating Procedures in Chemical Plants*
Pedro M. Saraiva, *Inductive and Analogical Learning: Data-Driven Improvement of Process Operations*
Alexandros Koulouris, Bhavik R. Bakshi, and George Stephanopoulos, *Empirical Learning through Neural Networks: The Wave-Net Solution*
Bhavik R. Bakshi and George Stephanopoulos, *Reasoning in Time: Modeling, Analysis, and Pattern Recognition of Temporal Process Trends*
Matthew J. Realff, *Intelligence in Numerical Computing: Improving Batch Scheduling Algorithms through Explanation-Based Learning*

Volume 23

Jeffrey J. Siirola, *Industrial Applications of Chemical Process Synthesis*
Arthur Westerberg and Oliver M. Wahnschafft, *The Synthesis of Distillation-Based Separation Systems*
Ignacio E. Grossman, *Mixed-Integer Optimization Techniques for Algorithmic Process Synthesis*
Subash Balakrishna and Lorenz T. Biegler, *Chemical Reactor Network Targeting and Integration: An Optimization Approach*
Steve Walsh and John Perkins, *Operability and Control in Process Synthesis and Design*

ISBN 0-12-008523-2